OPTIMIZATION TECHNIQUES

OPTIMIZATION TECHNIQUES

A.K. Malik

S.K. Yadav

S.R. Yadav

Dept. of Mathematics

B K Birla Institute of Engg. & Tech., Pilani (Raj.)

TECHSAR™

NEW DELHI

Published and Printed by
TechSar Pvt. Ltd.
4435-36/7, Ansari Road, Daryaganj
New Delhi-110 002 (India)
E-mail: tech_info@techsarworld.com
Website: www.techsarworld.com

ISBN 978-93-81141-81-6

Reprint 2025

PREFACE

The **Optimization Techniques** is the outcome of my long experience in teaching the engineering and science students at various levels. The text can be used for the courses offered by various universities in the different branches of engineering, management, economics and science as well. The basic philosophy in writing this text has been mainly to avoid to writing a purely theoretical book. Further, a pure algorithmic text would become a rigid compartmentalization. Thus, presenting the basic concepts with requisite derivations, the main emphasis has been to illustrate the concepts through examples. A lot many problems have been given in exercises for practice. Doing these problems will enhance the understanding of the basic concepts presented in the text.

Suggestions from students, teachers or users are welcome for further improvement in the material presented.

ACKNOWLEDGEMENTS

We are greatly thankful to Dr. P. S. Bhatnagar, Director, B. K. Birla Institute of Engineering & Technology, Pilani (Rajasthan) for providing motivation, encouragement and all over support in completing this book. We have freely consulted the books written by many authors on the subject. We gratefully acknowledge these writers. Last but not the least we are very much thankful to our family members for their moral support without which it was not possible for us to complete this project in time.

CONTENTS

1 Chapter Introduction to Optimization Techniques

1.1 INTRODUCTION

In the present chapter the historical developments of Operations Research, important definitions of Operations Research, applications of optimization techniques, optimization problems and classification of optimization problems have been discussed.

1.2 INTRODUCTION TO OPERATIONS RESEARCH

Operations Research is a branch of mathematics concerned with the application of scientific methods and techniques to decision-making problems with the best or optimal solution. In the years 1914-1915 Thomas Edison made an effort to use a tactical game board to find a way to minimize shipping losses from enemy submarines instead of risking ships in actual war conditions. He used a particular model and techniques of Operations Research for this purpose.

Operations research has become the most important instrument in the organization and management in various institutions. The formal initiation of operational research was given in the Second World War and arose from a study of the effectiveness of military operations, primarily in the royal air force. The term of operations research was first used by Mc Clospy and Trefthen in the year 1940. They coined the term 'Operations Research' as a result of research on military operations during the second World War. The application of mathematical models for solving such business problems first attracted the attention of the Russian mathematician, L. V. Kantorovich, who published a monograph "Mathematical Methods in the Organization and Planning of Production", in 1939. In this monologue he pointed out that there are many classes of production problems that can be defined mathematically and that can therefore, be solved numerically.

In the year 1948 an Operations Research club was founded in England. Later it came to be known as the Operational Research Society of U.K. It started publishing a quarterly journal and its first issue appeared in 1950. A big step in the field of operational research was the establishment of Operations Research Society of America (ORSA) in 1952 and the publication of its first journal *Operations Research* in 1953. Operations research in India came into existence in the year 1949 with the establishment of an operations research unit at Regional Research Laboratory, Hyderabad for the purpose of planning and organizing research. Of equal importance was the setting up of an operations research team at Defence Science Laboratory by Prof. R. S. Verma for the specific purpose of solving the problems of store, planning and purchase.

In recent years service organizations such as hospitals, airlines, banks, railways, and libraries have started recognizing the usefulness of operations research for improving efficiency. Operations research techniques provide effective base for management decisions. Operations research received a further boost with the setting up of operations research team in the Indian Statistical Institute, Kolkata by Prof. P. C. Mahalanobis in 1953 for the purpose of solving the problems related to national planning and survey. Later the Operations Research Society of India (ORSI) was established in 1957. This society started publishing its journal *OPSEARCH* from 1964.

1.3 DEFINITIONS OF OPERATIONS RESEARCH

It is very difficult to give a precise definition of operation research but it has a wide scope of applications. However, we are giving below a few definitions by some great scholars which are generally used:

H. M. Wagner: "Operations Research is a scientific approach to problems solving for executive management."

Auther Clark: "Operations Research is the art of winning wars without actually fighting them."

Churchman, Ackoff and Arnoff: "Operations Research is the application of scientific methods techniques and tools to problems involving the operation of a system so as to provide those in control of the system with optimum solution to the problem."

T. L. Saaty: "Operations Research is the art of giving bad answers to problems which otherwise have worse answers."

F. S. Hiller and G. J. Lieberman: "Operations Research may be described as a scientific approach to decision-making that involves the operations of organizational system."

D. W. Miller and M. K. Starr: "Operations Research is applied decision theory. It uses many scientific, mathematical or logical means to attempt to scope with the problems that confront the executive when he tries to achieve a through-going rationality in dealing with his decision problems."

H. A. Taha: "Operations Research is a scientific knowledge through interdisciplinary team effort for the purpose of determining the best utilization of limited resources."

P. M. Morse and G. E. Kimball: "Operations Research is a scientific method of providing executive departments with a quantitative basic for decisions under their control."

1.4 OPTIMIZATION TECHNIQUES

Let us discuss the importance and use of optimization techniques. Optimization techniques are very important and they are used in diverse branches of knowledge as electrical engineering, civil engineering, electronics engineering, mechanical engineering, telecommunication engineering, automotive engineering, chemical and biochemical engineering, aerospace engineering, computer engineering, medical science, education, biotechnology, management, banking, manufacturing industries and information technology industries etc. Optimization means getting the best output which may be maximum or minimum value of the criterion.

Optimization techniques have been used from the ancient past for solving various mathematical, chemical, biological and physical problems etc. They were used by great scientists like Newton, Cauchy and Lagrange. Newton and Leibniz established some important optimization methods in differential calculus. Cauchy was the first mathematician who gives the first application of unconstrained minimization problem solved by steepest descent method. Lagrange developed a method of optimization for constrained problems, which involves the addition of unknown multiplies, known by the name of Lagrange's method of undetermined multipliers.

Optimization techniques are generally studied as a part of operations research, also known as mathematical programming techniques. Mathematical programming techniques are very important and useful to finding the maximum or minimum of a function of several variables under the given set of constraints. The various mathematical programming techniques along with other well-defined area of operations research are given below:

1. Mathematical programming techniques
   - (i) Linear Programming
   - (ii) Non-linear Programming
   - (iii) Dynamic Programming
   - (iv) Integer Programming
   - (v) Geometric Programming
   - (vi) Multi-objective Programming
   - (vii) Quadratic Programming
   - (viii) Goal Programming
   - (ix) Stochastic Programming
   - (x) Separable Programming
   - (xi) Information theory
   - (xii) Sequencing theory
   - (xiii) Game theory
   - (xiv) Assignment Methods
   - (xv) Transportation Methods
   - (xvi) Inventory Control Methods
   - (xvii) Network Scheduling Methods
   - (xviii) Differential Calculus Methods
   - (xix) Neural Networks
   - (xx) Genetic Algorithms
   - (xxi) Simulated Annealing
   - (xxii) Calculus of Variations

2. Stochastic process techniques
   (i) Queuing Theory
   (ii) Reliability Theory
   (iii) Statistical Decision Theory
   (iv) Renewal Theory
   (v) Simulation Methods
   (vi) Markov Process
3. Statistical methods
   (i) Correlation Analysis
   (ii) Regression Analysis
   (iii) Cluster Analysis
   (iv) Factor Analysis
   (v) Design of Experiments

1.5 APPLICATIONS OF OPTIMIZATION TECHNIQUES

The importance of applications of optimization techniques in solving various problems in the engineering field is universally accepted now. They are used in solving the problem in such fields as:

(i) Optimal inventory cost.
(ii) Optimal designing of control system.
(iii) Optimal planning, scheduling and controlling.
(iv) Optimal designing of chemical processing equipment and plant.
(v) Optimal selection of a new site for an industry.
(vi) Optimal designing of pipeline networks for process industries.
(vii) Optimal designing of computer structure for minimum cost.
(viii) Optimal designing of plastic structures.
(ix) Optimal planning to get maximum profit in the presence of one or more competitor.
(x) Optimal designing of electrical networks.
(xi) Optimal designing of aircraft and aerospace structures for minimum weight.
(xii) Optimal designing of earthquake structures for minimum weight.
(xiii) Optimal controlling the waiting and idle times and queuing in production lines to reduce the costs.
(xiv) Optimal designing of civil engineering structures such as bridge, tower, dam, frames, chimneys etc., for minimum cost.
(xv) Optimal designing of material handling equipment such as conveyors, trucks and cranes for minimum cost.
(xvi) Optimal designing of water resources system for maximum benefit.
(xvii) Optimal designing of pumps heat transfer and turbines equipment for maximum efficiency.
(xviii) Optimal designing of electrical machinery such as generators, transformers and motors.
(xix) Optimal designing of mechanical components such as gears, cams and machine tools etc.

1.6 OPTIMIZATION PROBLEMS

An optimization problem is of the form

$$\text{Max or Min } Z = f(X) \qquad \ldots(1)$$

Subject to constraints

$$g_j(X) \geq \; = \; \leq 0, j = 1, 2, 3, \ldots\ldots, m. \qquad \ldots(2)$$

and $X \geq 0$

where $X = (x_1, x_2, x_3, \ldots\ldots, x_n)^T$ ...(3)

In general such problems are called mathematical programming problems. The function $f(X)$ in (1) which is to be maximized or minimized is called the objective function. Conditions given in (2) are called constraints. Variables $x_1, x_2, x_3, \ldots\ldots, x_n$ are called decision (design) variables and conditions in (3) are called non-negativity restrictions. Thus, any mathematical problem has three main parts:

(i) Objective function
(ii) Constraints
(iii) Non-negative restrictions.

1.7 CLASSIFICATION OF OPTIMIZATION PROBLEMS

How optimization problems can be classified in different ways are discussed below:

1.7.1 Classification Based on the Number of Objective Functions

(i) Single Objective Programming Problem

An optimization problem involving only single objective function is known as single objective programming problem.

(ii) Multi-objective Programming Problem

An optimization problem involving two or more objective functions is known as multi-objective programming problem.

1.7.2 Classification Based on the Existence of Constraints

(i) Constrained Optimization Problem

An optimization problem involving constraints is known as constrained optimization problem. For example,

$$\text{Max } Z = f(X)$$

Subject to the constraints $g_j(X) \geq \; = \; \leq 0, j = 1, 2, 3, \ldots\ldots, m.$

and $X \geq 0,$ where $X = (x_1, x_2, x_3, \ldots\ldots, x_n)^T$

(ii) Unconstrained Optimization Problem

An optimization problem without constraints is known as unconstrained optimization problem. For example,

$$\text{Max } Z = f(X) \text{ or Min } Z = f(X)$$

1.7.3 Classification Based on the Nature of the Constraints Involved

(i) Linear Programming Problem

An optimization problem involving objective function $f(X)$ and all the constraints $g_j(X)$ are linear is known as linear programming problem.

(ii) Non-Linear Programming Problem

An optimization problem involving either objective function $f(X)$ or at least one of the constraints $g_j(X)$ are non-linear is known as non-linear programming problem.

(iii) Quadratic Programming Problem

An optimization problem involving objective function $f(X)$ is quadratic and the constraints $g_j(X)$ are linear is known as quadratic programming problem.

(iv) Geometric Programming Problem

An optimization problem involving objective function $f(X)$ and constraints $g_j(X)$ are expressed as polynomial is known as non-linear programming problem.

1.7.4 Classification Based on the Permissible Values of the Design

(i) Integer Programming Problem

A linear programming problem involving some or all the design variables $x_1, x_2, x_3, \ldots\ldots, x_n$ are restricted to take on only integer values is known as integer programming problem.

(ii) Real-Valued Programming Problem

An optimization problem involving all the design variables $x_1, x_2, x_3, \ldots\ldots, x_n$ are permitted to take any real value is known as non-linear programming problem.

1.7.5 Classification Based on the Physical Structure of the Problem

(i) Optimal Control Problem

An optimization problem involving a number of stages in which each stage evolves from the preceding stage in a specific manner is called optimal control problem.

(ii) Non-Optimal Control Problem

An optimization problem which is not a optimal control problem is called non-optimal control problem.

1.7.6 Classification Based on the Nature of the Design Variables

(i) Static (Parameter) Optimization Problem

If in the optimization problem we find the value of a set of design parameters, in which some prescribed function of these parameters is made that minimizes subject to certain constraint is known as static optimization problem.

(ii) Dynamic Optimization Problem

An optimization problem involving each variable is a function of one or more parameters is known as dynamic optimization problem.

1.7.7 Classification Based on the Deterministic Nature of the Variables

(i) Deterministic Programming Problem

An optimization problem involving deterministic design variables is known as deterministic programming problem.

(ii) Stochastic Programming Problem

An optimization problem involving some or all the design variables are probabilistic is known as stochastic programming problem.

1.7.8 Classification Based on the Separability of the Functions

(i) Separable Programming Problem

An optimization problem involving objective function $f(X)$ and the constraints $g_j(X)$ are separable is known as separable programming problem.

(ii) Non-Separable Programming Problem

An optimization problem involving objective function $f(X)$ and the constraints $g_j(X)$ are non-separable is known as non-separable programming problem.

EXERCISE 1.1

1. Write a short note on the historical development of operations research?
2. What is optimization? Explain applications of optimization in engineering.
3. Write a short note on optimization techniques.
4. Given ten engineering applications of optimization techniques.
5. Write a short note on the classification of optimization problems.

Chapter 2
Classical Optimization Techniques

2.1 INTRODUCTION

In the present chapter, we shall discuss the classical optimization techniques with necessary and sufficient conditions for obtaining the optimum solution of unconstrained single and multivariable optimization problems. Constrained multivariable problems with equality and inequality constraints have also been discussed in detail with examples. The classical optimization techniques are very useful to obtain the optimal solution of problems involving continuous and differentiable functions. Such types of techniques are analytical in nature to obtain maximum and minimum points for unconstrained and constrained continuous objective functions. For equality constrained problems we use Lagrange's multiplier method and for inequality constrained, the Kuhn-Tucker conditions, for getting optimum solutions.

2.2 UNCONSTRAINED OPTIMIZATION PROBLEMS

2.2.1 Single Variable Optimization Problems

Let $f(x)$ be a continuous function of single variable x defined in interval $[a, b]$.

Local Maxima

A function $f(x)$ with single variable is said to have a local (relative) maxima at $x = x_o$ if

$$f(x_0) \geq f(x_0 + h)$$

for all sufficiently small positive and negative values of h.

Local Minima

A function $f(x)$ with single variable is said to have a local (relative) minima at $x = x_o$ if

$$f(x_0) \leq f(x_0 + h)$$

for all sufficiently small positive and negative values of h.

Global Maxima

A function $f(x)$ with single variable is said to have a global (absolute) maxima at $x = x_o$ in $[a, b]$ if

$$f(x_0) \geq f(x)$$

for all x defined in the interval $[a, b]$.

Global Minima

A function $f(x)$ with a single variable is said to have a global (absolute) minima at $x = x_o$ in $[a, b]$ if

$$f(x_0) \leq f(x)$$

for all x defined in the interval $[a, b]$.

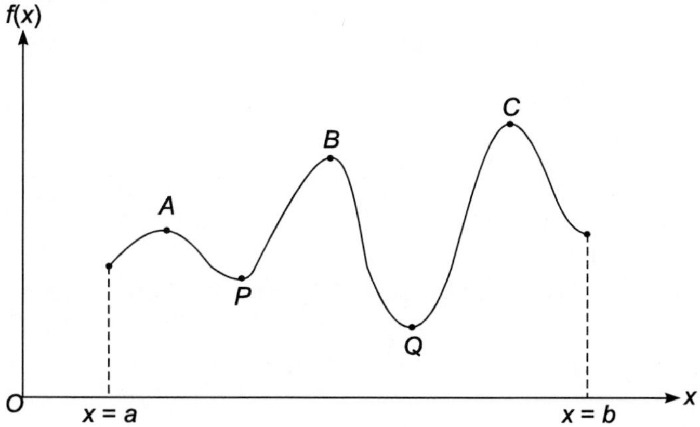

Here points A, B and C are local maxima; P and Q are local minima; C global maxima and Q global minima.

2.2.2 Condition for Local Maxima or Minima of Single Variable Function

Necessary Condition

If a function $f(x)$ is defined in the interval $[a, b]$ and has a local maxima or local minima at $x = x_0$, where $a < x_0 < b$ and $f'(x)$ exists as a finite number at $x = x_0$ then $f'(x_0) = 0$.

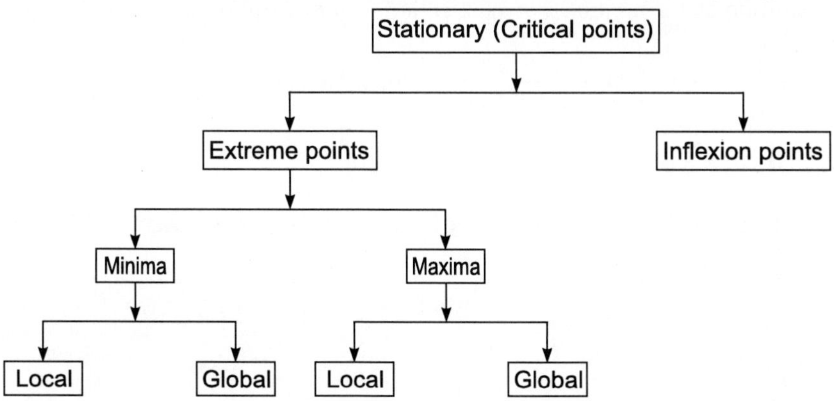

Sufficient Conditions

Let $f'(x_0) = f''(x_0) = f'''(x_0) = \ldots = f^{(n-1)}(x_0)$, but $f^n(x_0) \neq 0$ then $f(x_0)$ has

(i) A minimum value of $f(x)$ if $f^{(n)}(x_0) > 0$ and n is even.

(ii) A maximum value of $f(x)$ if $f^{(n)}(x_0) < 0$ and n is even.

(iii) Neither maxima nor minima, i.e., point of inflexion if $f^{(n)}(x_0) \neq 0$ and n is odd.

2.2.3 Working Rule to Finding the Extreme Points

Let us consider a single variable function $y = f(x)$ ...(1)

Step 1: Differentiate both sides of (1) with respect to x to get

$$\frac{dy}{dx} = f'(x) \qquad \ldots(2)$$

Step 2: For extreme point, we have $\frac{dy}{dx} = 0$

$$\Rightarrow \qquad f'(x) = 0 \qquad \ldots(3)$$

We get values of x which is satisfied $f'(x) = 0$

Solve (3) for x to get $x = x_0$ (say)

Step 3: Differentiate both sides of (2) again with respect to x both side, we get

$$\frac{d^2y}{dx^2} = f''(x)$$

Step 4: If $\left(\frac{d^2y}{dx^2}\right)_{x=x_0} > 0$ then $f(x)$ has minima at $x = x_0$.

Step 5: If $\left(\frac{d^2y}{dx^2}\right)_{x=x_0} < 0$ then $f(x)$ has maxima at $x = x_0$.

Step 6: If $\left(\frac{d^2y}{dx^2}\right)_{x=x_0} = 0$ then find $\frac{d^3y}{dx^3}$ and we have $\left(\frac{d^3y}{dx^3}\right)_{x=x_0} \neq 0$ then there is neither a maximum nor minima at $x = x_0$, i.e., $x = x_0$ is called point of inflexion.

Step 7: If $\left(\frac{d^3y}{dx^3}\right)_{x=x_0} = 0$ then find $\frac{d^4y}{dx^4}$ and repeat steps 4, 5 and 6.

2.3 MULTIVARIABLE OPTIMIZATION PROBLEMS

2.3.1 Working Rule to Find the Extreme Points of Functions of Two Variables

Let $u = f(x_1, x_2)$ be a function of x_1 and x_2.

Step 1: Differentiate $f(x_1, x_2)$ partially with respect to x_1 and x_2, we get

$$\frac{\partial f}{\partial x_1} \quad \text{and} \quad \frac{\partial f}{\partial x_2}$$

Step 2: For extreme points, we have

$$\frac{\partial f}{\partial x_1} = 0$$

and

$$\frac{\partial f}{\partial x_2} = 0$$

Solving these equations we get some points as (a_1, b_1), (a_2, b_2),...etc.

Step 3: Differentiate again partially to get

$$r = \frac{\partial^2 f}{\partial x_1^2}, \quad s = \frac{\partial^2 f}{\partial x_1 \partial x_2} \quad \text{and} \quad t = \frac{\partial^2 f}{\partial x_2^2}$$

Step 4: At (a_1, b_1), calculate r and $rt - s^2$

Case I: If $rt - s^2 > 0$ and $r < 0$ then $f(x_1, x_2)$ is maximum at (a_1, b_1).

Case II: If $rt - s^2 > 0$ and $r > 0$ then $f(x_1, x_2)$ is minimum at (a_1, b_1).

Case III: If $rt - s^2 < 0$ then $f(x_1, x_2)$ has neither maxima nor minima at (a_1, b_1).

Case IV: If $rt - s^2 = 0$ then $f(x_1, x_2)$ has point of inflexion at (a_1, b_1).

2.3.2 Working Rule to Find the Extreme Points of Functions of n-Variables

Let us consider $u = f(x_1, x_2, x_3, \ldots, x_n)$ as a function of $x_1, x_2, x_3, \ldots, x_n$.

Necessary Conditions

$$\frac{\partial f}{\partial x_1} = 0, \quad \frac{\partial f}{\partial x_2} = 0, \quad \ldots, \quad \frac{\partial f}{\partial x_n} = 0$$

Sufficient Conditions

The Hessian matrix at P for n variables will be

$$H = \begin{bmatrix} \frac{\partial^2 f}{\partial x_1^2} & \frac{\partial^2 f}{\partial x_1 \partial x_2} & \cdots & \frac{\partial^2 f}{\partial x_1 \partial x_n} \\ \frac{\partial^2 f}{\partial x_2 \partial x_1} & \frac{\partial^2 f}{\partial x_2^2} & \cdots & \frac{\partial^2 f}{\partial x_2 \partial x_n} \\ \cdots & \cdots & \cdots & \cdots \\ \cdots & \cdots & \cdots & \cdots \\ \frac{\partial^2 f}{\partial x_n \partial x_1} & \frac{\partial^2 f}{\partial x_n \partial x_2} & \cdots & \frac{\partial^2 f}{\partial x_n^2} \end{bmatrix}$$

Its leading minors are defined as

$$H_1 = \left| \frac{\partial^2 f}{\partial x_1^2} \right|,$$

$$H_2 = \begin{vmatrix} \dfrac{\partial^2 f}{\partial x_1^2} & \dfrac{\partial^2 f}{\partial x_1 \partial x_2} \\ \dfrac{\partial^2 f}{\partial x_2 \partial x_1} & \dfrac{\partial^2 f}{\partial x_2^2} \end{vmatrix},$$

$$H_3 = \begin{vmatrix} \dfrac{\partial^2 f}{\partial x_1^2} & \dfrac{\partial^2 f}{\partial x_1 \partial x_2} & \dfrac{\partial^2 f}{\partial x_1 \partial x_2} \\ \dfrac{\partial^2 f}{\partial x_2 \partial x_1} & \dfrac{\partial^2 f}{\partial x_2^2} & \dfrac{\partial^2 f}{\partial x_2 \partial x_3} \\ \dfrac{\partial^2 f}{\partial x_3 \partial x_1} & \dfrac{\partial^2 f}{\partial x_3 \partial x_2} & \dfrac{\partial^2 f}{\partial x_3^2} \end{vmatrix},$$

Hence, the following cases will arise:

Case I: If H_1, H_2, H_3,... are positive (i.e., H is positive definite) then $f(x_1, x_2, x_3, \ldots, x_n)$ has minimum at P.

Case II: If H_1, H_2, H_3,... are alternately negative, positive, negative (i.e., H is negative definite) then $f(x_1, x_2, x_3, \ldots, x_n)$ has maximum at P.

Case III: H_1 and H_3,... are not of same sign and $H_2 = 0$ (i.e., semidefinite or indefinite) then $f(x_1, x_2, x_3, \ldots x_n)$ has a saddle point at P.

SOLVED EXAMPLES

Example 1: Assume the following relationship for revenue and cost functions. Find out at what level of output x, where x is measured in tons per week, profit is maximum.

$$R(x) = 1000x - 2x^2 \text{ and } C(x) = x^3 - 59x^2 + 1315x + 5000.$$

Solution: The profit function is

$$\begin{aligned} P(x) &= R(x) - C(x) \\ &= 1000x - 2x^2 - x^3 + 59x^2 - 1315x - 5000 \\ &= -x^3 + 57x^2 - 315x - 5000 \qquad \ldots(1) \end{aligned}$$

Differentiating both sides of (1) with respect to x, we get

$$\frac{dP}{dx} = -3x^2 + 114x - 315 \qquad \ldots(2)$$

For maxima and minima, we have

$$\frac{dP}{dx} = 0$$

$\Rightarrow$ $$-3x^2 + 114x - 315 = 0$$

$\Rightarrow$ $$x = 3, 35.$$

Differentiating both sides of (2) again with respect to x, we get

$$\frac{d^2P}{dx^2} = -6x + 114$$

At $x = 3, \dfrac{d^2P}{dx^2} = 96 > 0$, i.e., P is minimum at $x = 3$.

At $x = 35, \dfrac{d^2P}{dx^2} = 96 < 0$, i.e., P is maximum at $x = 35$.

Hence, the profit is maximum at $x = 35$ tons per week.

Example 2: The profit P earned, by a company, on some item is function of its units produced say x and is given by

$$P = 800x - 2x^2$$

If the company's expenditure or interest, rent and salary of the staff be Rs. 1 lakh, show that the company will always be in loss.

Solution: Given that

$$P = 800x - 2x^2 \qquad \ldots(1)$$

Differentiate both sides of (1) with respect to x, we get

$$\frac{dP}{dx} = 800 - 4x \qquad \ldots(2)$$

For maxima and minima, we have

$$\frac{dP}{dx} = 0$$

$$800 - 4x = 0$$

$$x = 200$$

Differentiate both sides of (2) again with respect to x, we get

$$\frac{d^2P}{dx^2} = -4 \text{ (negative)}$$

So, profit (P) is maximum for ($x = 200$).

The net profit = P – expenditure

$$= 800(200) - 2(200)^2 - 1,00,000$$

$$= -20,000.$$

Hence, the company will always be in loss.

Example 3: Assuming that the petrol burnt (per hour) in driving a motor boat varies as the cube of its velocity, show that the most economical speed when going against a current of c kmph is $\dfrac{3c}{2}$ km per hour.

Solution: Let the velocity of the boat be v km per hour, the current's velocity is c km per hour and the relative velocity of the boat is $(v - c)$ km per hour.

If the total distance travelled 'a' km then the time taken $= \dfrac{a}{v-c}$ hour.

According to given the petrol burnt in one hour is proportional to v^3, i.e. $= \lambda v^3$ where λ is a suitable positive constant.

The petrol burnt for distance 'a' is given by

$$P = \lambda v^3 . \frac{a}{v-c} \qquad \ldots(1)$$

Differentiate both sides of (1) with respect to v, we get

$$\frac{dP}{dv} = \lambda\alpha\left[\frac{3v^2(v-c) - v^3}{(v-c)^2}\right] \qquad \ldots(2)$$

$$= \lambda\alpha\left[\frac{2v^3 - 3cv^2}{(v-c)^2}\right]$$

For maxima and minima, we have

$$\frac{dP}{dv} = 0$$

$$\Rightarrow \quad \lambda\alpha\left[\frac{2v^3 - 3cv^2}{(v-c)^2}\right] = 0$$

$$\Rightarrow \quad 2v^3 - 3cv^2 = 0$$

$$\Rightarrow \quad v = 0, \frac{3c}{2}$$

Differentiate both sides of (2) again with respect to v, we get

$$\frac{d^2P}{dv^2} = \lambda\alpha\left[\frac{(6v^2 - 6vc)(v-c)^2 - (2v^3 - 3cv^2).2(v-c)}{(v-c)^4}\right]$$

$$\frac{d^2P}{dv^2} = \lambda\alpha\left[\frac{6v(v-c)^2 - 2v^2(2v-3c)}{(v-c)^3}\right]$$

At $v = 0$, $\dfrac{d^2P}{dv^2} = 0$, i.e., $P = 0 \Rightarrow$ no petrol is burnt.

At $v = \dfrac{3c}{2}$, $\dfrac{d^2P}{dv^2}$ is positive, i.e., P is minimum at $v = \dfrac{3c}{2}$.

Hence, the most economical speed is $v = \dfrac{3c}{2}$ km per hour.

Example 4: A beam of length l is supported at one end. If w is the uniformly distributed load per unit length, the bending moment M at a distance x from the end is given by $M = \frac{1}{2} lx - \frac{1}{2} wx^2$.

Solution: Given the bending moment is

$$M = \frac{1}{2} lx - \frac{1}{2} wx^2 \qquad \ldots(1)$$

Differentiate both sides of (1) with respect to x, we get

$$\frac{dM}{dx} = \frac{1}{2} l - wx$$

For maxima, we have

$$\frac{dM}{dx} = 0$$

$$\Rightarrow \qquad \frac{1}{2} l - wx = 0$$

$$\Rightarrow \qquad x = \frac{l}{2w}$$

Differentiate both sides of (2) with respect to x, we get

$$\frac{d^2M}{dx^2} = -w \text{(negative)}$$

So the bending moment M is maximum at $x = \frac{l}{2w}$ from the end.

At $\qquad x = \frac{l}{2w}, \quad M = \frac{1}{2} l\left(\frac{l}{2w}\right) - \frac{1}{2} w\left(\frac{l}{2w}\right)^2 = \frac{l^2}{8w}.$

Example 5: Find the maximum and minimum value of $y = 3x^5 - 5x^3$.

Solution: Given that $y = 3x^5 - 5x^3 \qquad \ldots(1)$

Differentiate both sides of (1) with respect to x, we get

$$\frac{dy}{dx} = 15x^4 - 15x^2 \qquad \ldots(2)$$

For maxima and minima, we have

$$\frac{dy}{dx} = 0$$

$$\Rightarrow \qquad 15x^4 - 15x^2 = 0$$

$$\Rightarrow \qquad 15x^2(x - 1)(x + 1) = 0$$

$$x = 0, 1, -1.$$

Differentiate both sides of (2) again with respect to x, we get

$$\frac{d^2y}{dx^2} = 60x^3 - 30x$$

At $x = 0$, $\frac{d^2y}{dx^2} = 0$ i.e., x is a point of inflexion.

At $x = 1$, $\frac{d^2y}{dx^2} = 30 > 0$, i.e., y is minimum at $x = 1$.

At $x = -1$ $\frac{d^2y}{dx^2} = -30 < 0$, i.e., y is maximum at $x = -1$.

Example 6: In a submarine telegraph cable, the speed of signaling varies as $x^2 \log\left(\frac{1}{x}\right)$ where x is the ratio of the radius of the cube to that of the covering. Show that the greatest speed is attained when this ratio is $1:\sqrt{e}$.

Solution: Let u be the speed of signaling, then

$$u = x^2 \log\left(\frac{1}{x}\right), \lambda > 0, x \neq 0$$

$$u = -\lambda x^2 \log x \qquad \ldots(1)$$

Differentiate both sides of (1) with respect to x, we get

$$\frac{du}{dx} = -2\lambda x \log x - \lambda x^2(1/x)$$

$$= -\lambda[2x \log x + x] \qquad \ldots(2)$$

For maxima and minima, we have

$$\frac{du}{dx} = 0$$

$\Rightarrow$ $$-\lambda[2x \log x + x] = 0$$

$\Rightarrow$ $$\log x = -\frac{1}{2}$$

$\Rightarrow$ $$x = e^{-1/2} = 1/\sqrt{e}.$$

Differentiate both sides of (2) again with respect to x, we get

$$\frac{d^2u}{dx^2} = -\lambda[2x.(1/x) + 2 \log x]$$

$$= -\lambda[2 \log x + 3]$$

A $x = 1/\sqrt{e}$, $\frac{d^2u}{dx^2} = -2\lambda$ (negative), i.e., u is maximum.

Hence, u is maximum, ratio for $x = 1:\sqrt{e}$.

Example 7: A rectangular sheet of metal has four equal square portions removed at the corners and the sides are then turned up so as to form an open rectangular box. Show that when the volume contained in the box is maximum, the depth will be

$$\frac{1}{6}[(a + b) - (a^2 - ab + b^2)^{1/2}]$$

Where a and b are sides of the original dimensions of the rectangular.

Solution: Let x be the length of each side of the squares removed at the corners. Then the dimensions of the box will be $(a - 2x)$, $(b - 2x)$ and x. Let V be the volume of the box, then we have

$$V = (a - 2x)(b - 2x)x$$
$$= 4x^3 - 2x^2(a + b) + abx \quad \ldots(1)$$

Differentiate both sides of (1) with respect to x, we get

$$\frac{dv}{dx} = 12x^2 - 4x(a + b) + ab \quad \ldots(2)$$

For maxima and minima, we have

$$\frac{dv}{dx} = 0$$

$$\Rightarrow \quad 12x^2 - 4x(a + b) + ab = 0$$

$$\Rightarrow \quad x = \frac{4(a+b) \pm \sqrt{16(a+b)^2 - 48ab}}{24}$$

$$\Rightarrow \quad x = \frac{1}{6}[(a + b) \pm (a^2 - ab + b^2)^{1/2}]$$

Differentiate both sides of (2) again with respect to x, we get

$$\frac{d^2V}{dx^2} = 24x - 4(a + b)$$

At $x = \frac{1}{6}[(a + b) \pm (a^2 - ab + b^2)^{1/2}]$, $\frac{d^2V}{dx^2} = \pm\sqrt{(a-b)^2 + ab}$

If $x = \frac{1}{6}[(a + b) - (a^2 - ab + b^2)^{1/2}]$ then $\frac{d^2V}{dx^2}$ is negative, i.e., V is maximum.

Example 8: The efficiency of a screw jack is given by $\eta = \tan\alpha\cot(\alpha + \phi)$ where ϕ is constant. Prove that the efficiency is maximum at $\alpha = \frac{\pi}{4} - \frac{\phi}{2}$ and $\eta = \frac{1-\sin\phi}{1+\sin\phi}$.

Solution: The efficiency of a screw jack is

$$\eta = \tan\alpha\cot(\alpha + \phi) \quad \ldots(1)$$

Differentiate both sides of (1) with respect to α, we get

$$\frac{d\eta}{d\alpha} = -\tan\alpha\,\text{cosec}^2(\alpha + \phi) + \sec^2\alpha\cot(\alpha + \phi)$$
$$= \sec^2\alpha\cot(\alpha + \phi) - \tan\alpha\,\text{cosec}^2(\alpha + \phi)$$
$$= \frac{1}{\cos^2\alpha\sin^2(\alpha+\phi)}[\sin(\alpha + \phi)\cos(\alpha + \phi) - \sin\alpha\cos\alpha] \quad \ldots(2)$$

For maxima and minima, we have

$$\frac{d\eta}{d\alpha} = 0$$

$$\Rightarrow \quad \frac{1}{\cos^2 \alpha \sin^2(\alpha+\phi)}[\sin(\alpha+\phi)\cos(\alpha+\phi) - \sin\alpha\cos\alpha] = 0$$

$$\Rightarrow \quad \sin 2(\alpha+\phi) - \sin 2\alpha = 0$$

$$\Rightarrow \quad 2(\alpha+\phi) = \pi - 2\alpha$$

$$\Rightarrow \quad \alpha = \frac{\pi}{4} - \frac{\phi}{2}$$

By (2), we have

$$\cos^2\alpha \sin^2(\alpha+\phi)\frac{d\eta}{d\alpha} = [\sin(\alpha+\phi)\cos(\alpha+\phi) - \sin\alpha\cos\alpha]$$

$$= \frac{1}{2}[\sin 2(\alpha+\phi) - \sin 2\alpha]$$

$$= \cos(2\alpha+\phi)\cos\phi. \quad \ldots(3)$$

Differentiate both sides of (3) again with respect to α, we get

$$\cos^2\alpha \sin^2(\alpha+\phi)\frac{d^2\eta}{d\alpha^2} + \frac{d}{d\alpha}\{\cos^2\alpha\sin^2(\alpha+\phi)\}.\frac{d\eta}{d\alpha} = \frac{d}{d\alpha}[\cos(2\alpha+\phi)\cos\phi]$$

$$\cos^2\alpha \sin^2(\alpha+\phi)\frac{d^2\eta}{d\alpha^2} + \frac{d}{d\alpha}\{\cos^2\alpha\sin^2(\alpha+\phi)\}.\frac{d\eta}{d\alpha} = -[\sin(2\alpha+\phi)\cos\phi]$$

At $\quad \alpha = \frac{\pi}{4} - \frac{\phi}{2}, \frac{d\eta}{d\alpha} = 0$ then $\frac{d^2\eta}{d\alpha^2}$ is negative, i.e., η is maximum.

At $\quad \alpha = \frac{\pi}{4} - \frac{\phi}{2}$ the value of η by (1), we have

$$\eta = \frac{\sin(2\alpha+\phi-\sin\phi)}{\sin(2\alpha+\phi-\sin\phi)}$$

$$\eta = \frac{\sin(2\alpha+\phi-\sin\phi)}{\sin(2\alpha+\phi-\sin\phi)}$$

$$\eta = \frac{1-\sin\phi}{1+\sin\phi}.$$

Example 9: Show that the right circular cylinder of given surface (including its ends) and maximum volume is such that its height is equal to twice its radius.

Solution: We know that

$$V = \pi r^2 h \quad \ldots(1)$$

and $S = 2\pi rh + 2\pi r^2$ (let constant according to given)

$$\Rightarrow \quad 2\pi rh = 2k^2\pi - 2\pi r^2$$

$$h = \frac{k^2 - r^2}{r} \quad \ldots(2)$$

From (1) and (2), we have

$$V = \pi r\,(k^2 - r^2) \qquad \ldots(3)$$

Differentiate both sides of (3) with respect to r, we get

$$\frac{dV}{dr} = \pi(k^2 - 3r^2) \qquad \ldots(4)$$

For maxima and minima, we have

$$\frac{dV}{dr} = 0$$

$$\Rightarrow \quad \pi(k^2 - 3r^2) = 0$$

$$\Rightarrow \quad r = \frac{k}{\sqrt{3}} \qquad \ldots(5)$$

Differentiate both sides of (4) with respect to r, we get

$$\frac{d^2V}{dx^2} = -6\pi r\,(\text{negative}) \text{ i.e., } V \text{ is maximum.}$$

Using (2) and (5), we have

$$hr = k^2 - r^2$$

$$\Rightarrow \quad h = 2r.$$

Example 10: A person being in a boat 'a' miles from the nearest point of the beach, wishes to reach as quickly as possible, a point 'b' miles from that point along the sea shore. The ratio of his rate of walking to his rate of rowing is $\sec\alpha$. Prove that he should land at a distance $b - \alpha \cot\alpha$ from the place to be reached.

Solution: Let the person stand at A and reach on point B. Let C be the point where AC and BC meet.

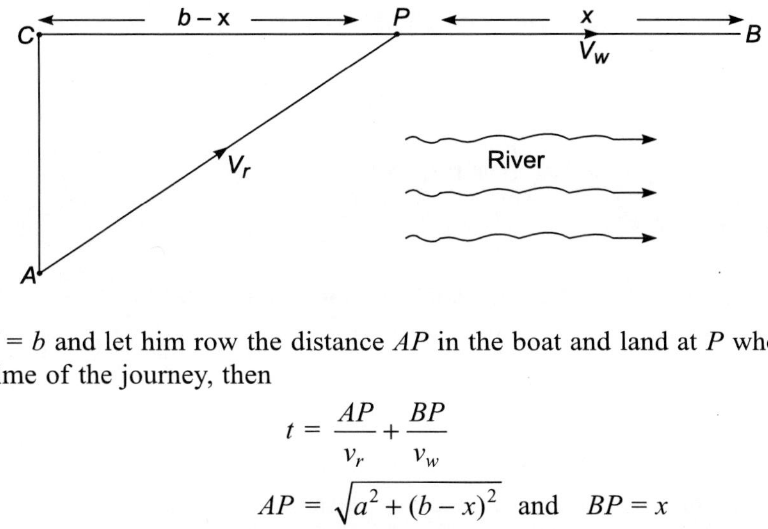

Let $BC = b$ and let him row the distance AP in the boat and land at P where $PB = x$.

If t is the time of the journey, then

$$t = \frac{AP}{v_r} + \frac{BP}{v_w}$$

where $\quad AP = \sqrt{a^2 + (b - x)^2}$ and $BP = x$

Given that $\quad \dfrac{v_r}{v_w} = \sec\alpha.$

so $$t = \frac{1}{v_r}\left[\sqrt{a^2+(b-x)^2} + \frac{x}{\sec\alpha}\right] \qquad \ldots(1)$$

Differentiate both sides of (1) with respect to x, we get

$$\frac{dt}{dx} = \frac{1}{v_r}\left[\frac{-(b-x)}{\sqrt{a^2+(b-x)^2}} + \frac{1}{\sec\alpha}\right] \qquad \ldots(2)$$

For maxima or minima, we have

$$\frac{dt}{dx} = 0$$

$$\Rightarrow \quad \frac{1}{v_r}\left[\frac{-(b-x)}{\sqrt{a^2+(b-x)^2}} + \frac{1}{\sec\alpha}\right] = 0$$

$$\Rightarrow \quad -(b-x)\sec\alpha + \sqrt{a^2+(b-x)^2} = 0$$

$$\Rightarrow \quad (b-x)\sec\alpha = \sqrt{a^2+(b-x)^2}$$

$$\Rightarrow \quad (b-x)^2\sec^2\alpha = a^2 + (b-x)^2$$

$$\Rightarrow \quad (b-x)^2\tan^2\alpha = a^2$$

$$\Rightarrow \quad b - x = a\cot\alpha$$

$$\Rightarrow \quad x = b - a\cot\alpha$$

By (2), we have

$$v_r\sec\alpha\sqrt{a^2+(b-x)^2}\,\frac{dt}{dx} = -(b-x)\sec\alpha + \sqrt{a^2+(b-x)^2} \qquad \ldots(3)$$

Differentiate both sides of (3) again with respect to x, we get

$$v_r\sec\alpha\sqrt{a^2+(b-x)^2}\,\frac{d^2t}{dx^2} + \frac{d}{dt}\left\{v_r\sec\alpha\sqrt{a^2+(b-x)^2}\right\} = \sec\alpha - \frac{(b-x)}{\sqrt{a^2+(b-x)^2}}$$

At $x = b - a\cot\alpha$, $\frac{dt}{dx} = 0$ and $\frac{d^2t}{dx^2}$ is positive.

i.e., t is minimum at $x = b - a\cot\alpha$.

Example 11: Prove that the minimum radius vector of the curve $\frac{a^2}{x^2} + \frac{b^2}{y^2} = 1$ is of length $(a + b)$.

Solution: It is given that

$$\frac{a^2}{x^2} + \frac{b^2}{y^2} = 1 \qquad \ldots(1)$$

Changing the variable co-ordinate Cartesian to polars by taking $x = r\cos\theta$, $y = r\sin\theta$ in (1), we get

$$\frac{a^2}{r^2\cos^2\theta} + \frac{b^2}{r^2\sin^2\theta} = 1 \qquad \ldots(2)$$

or $$r^2 = a^2 \sec^2\theta + b^2 \operatorname{cosec}^2\theta$$

Let $$R = a^2 \sec^2\theta + b^2 \operatorname{cosec}^2\theta \qquad \ldots(3)$$

Differentiate both sides of (3) with respect to θ, we get

$$\frac{dR}{d\theta} = 2a^2 \sec\theta.\ \sec\theta\tan\theta + 2b^2 \operatorname{cosec}\theta(-\operatorname{cosec}\theta\cot\theta)$$

or $$\frac{dR}{d\theta} = 2a^2 \sec^2\theta\tan\theta - 2b^2 \operatorname{cosec}^2\theta\cot\theta \qquad \ldots(4)$$

For maxima and minima, we have

$$\frac{dR}{d\theta} = 0$$

$$\Rightarrow \quad 2a^2 \sec^2\theta\tan\theta - 2b^2 \operatorname{cosec}^2\theta\cot\theta = 0$$

$$\Rightarrow \quad \tan^4\theta = \frac{b^2}{a^2}$$

$$\Rightarrow \quad \tan\theta = \sqrt{\frac{b}{a}}.$$

Differentiate both sides of (4) with respect to θ, we get

$$\frac{d^2R}{d\theta^2} = 2a^2[\sec^2\theta\sec^2\theta + 2\sec^2\theta\tan\theta.\ \tan\theta] - 2b^2[\operatorname{cosec}^2\theta(-\operatorname{cosec}^2\theta) - 2\operatorname{cosec}^2\theta\cot^2\theta]$$

$$= 2a^2[\sec^4\theta + 2\sec^2\theta\tan^2\theta] + 2b^2[\operatorname{cosec}^4\theta + 2\operatorname{cosec}^2\theta\cot^2\theta] > 0$$

i.e., R or r^2 is minimum at $\tan\theta = \sqrt{\frac{b}{a}}$

$$\Rightarrow \quad \sin\theta = \sqrt{\frac{b}{a+b}} \text{ and } \cos\theta = \sqrt{\frac{a}{a+b}}$$

By equation (2), we have

$$\frac{a^2}{r^2\left(\frac{a}{a+b}\right)} + \frac{b^2}{r^2\left(\frac{a}{a+b}\right)} = 1$$

$$\Rightarrow \quad r^2 = a(a+b) + b(a+b) = (a+b)^2$$

At $\tan\theta = \sqrt{\frac{b}{a}}$, the value of r is $(a + b)$.

Example 12: ADC generator has integral resistance R ohms and has an open circuit voltage of V volts. Find the lead resistance r for which the power delivered by the generator is maximum.

Solution: We know that the ohm's law

$$V = i\,(R + r)$$

$$\Rightarrow \quad i = \frac{V}{R+r}$$

The power generated $P = i^2 r = \dfrac{V^2 r}{(R+r)^2}$ ...(1)

Here V, R being constant.

Differentiate both sides of (1) with respect to r, we get

$$\frac{dP}{dr} = V^2\left[\frac{(R+r)^2.1 - r.2(R+r)}{(R+r)^4}\right]$$

$$= V^2\left[\frac{(R+r) - 2r}{(R+r)^3}\right]$$

$$= V^2\left[\frac{R-r}{(R+r)^3}\right] \quad \text{...(2)}$$

For maxima and minima, we have

$$\frac{dP}{dr} = 0$$

$$\Rightarrow \quad V^2\left[\frac{R-r}{(R+r)^3}\right] = 0$$

$$\Rightarrow \quad R = r$$

Differentiate both sides of (2) again with respect to r, we get

$$\frac{d^2P}{dr^2} = V^2\left[\frac{(R+r)^3.(-1) - (R-r).3(R+r)^2}{(R+r)^6}\right]$$

$$= V^2\left[\frac{-4R+2r}{(R+r)^4}\right]$$

At $R = r$, $\dfrac{d^2P}{dr^2} = \dfrac{-V^2}{8R^3}$ (negative) i.e., P is maximum for $r = R$.

Hence, $P_{max} = \dfrac{V^2}{4R}$

Example 13: A given quantity of metal is to be cast into a half cylinder, i.e., with rectangular base and semicircular ends. Show that in order to have minimum surface area, the ratio of the height of the cylinder to the diameter of semicircular ends is $\pi{:}\pi + 2$.

Solution: Suppose the volume of the half cylinder is

$$V = \frac{1}{2}\pi r^2 h \quad \text{...(1)}$$

where r and h are the radius and height of the half cylinder respectively

Surface area of rectangular base $= 2rh$

Curved surface $= \pi rh$

Two semicircular ends $= \pi r^2$

The total surface area

$$S = 2rh + \pi rh + \pi r^2$$
$$= rh\,(2 + \pi) + \pi r^2 \qquad \ldots(2)$$

From equation (1), we have

$$h = \frac{2V}{\pi r^2}$$

Then
$$S = \frac{r.2V}{\pi r^2}(2 + \pi) + \pi r^2$$
$$= \frac{2V}{\pi r}(2 + \pi) + \pi r^2 \qquad \ldots(3)$$

Differentiate both sides of (3) with respect to r, we get

$$\frac{dS}{dr} = 2\pi r - \frac{2V}{\pi r^2}(\pi + 2) \qquad \ldots(4)$$

For maxima and minima we have

$$\frac{dS}{dr} = 0$$

$$\Rightarrow \qquad 2\pi r - \frac{2V}{\pi r^2}(\pi + 2) = 0$$

Using equation (1), we have

$$2\pi r - h(\pi + 2) = 0$$

$$\Rightarrow \qquad \frac{h}{2r} = \frac{\pi}{\pi + 2} \qquad \ldots(5)$$

Differentiate both sides of (4) again with respect to r, we get

$$\frac{d^2S}{dr^2} = 2\pi - \frac{4V}{\pi r^3}(\pi + 2)$$
$$= 6\,\pi \text{ (positive)} \quad \text{i.e., } S \text{ is maximum.}$$

Hence, surface area S is minimum at $\frac{h}{2r} = \frac{\pi}{\pi + 2}$

Example 14: Find the extreme points of the function $u = x^2 + 4y^2 + 4z^2 + 4xy + 4xz + 16yz$.

Solution: Given that

$$u = x^2 + 4y^2 + 4z^2 + 4xy + 4xz + 16yz \qquad \ldots(1)$$

Differentiate partially both sides of (1) with respect to x, y and z respectively, we get

$$\frac{\partial u}{\partial x} = 2x + 2y + 2z$$

$$\frac{\partial u}{\partial y} = 8y + 4x + 16z$$

and
$$\frac{\partial u}{\partial z} = 8z + 4x + 16y$$

For extreme points, we have

$$\frac{\partial u}{\partial x} = 0, \ \frac{\partial u}{\partial y} = 0 \text{ and } \frac{\partial u}{\partial z} = 0$$

$$\Rightarrow \quad 2x + 2y + 2z = 0 \text{ or } 2(x + y + z) = 0$$

$$8y + 4x + 16z = 0 \text{ or } 4(x + 2y + 4z) = 0$$

and

$$8z + 4x + 16y = 0 \text{ or } 4(x + 4y + 2z) = 0$$

Solving above these expression, we get $x = 0, y = 0, z = 0$.
Let the point P be (0, 0, 0)
Differentiate partially (2) and we get

$$\frac{\partial^2 u}{\partial x^2} = 2, \ \frac{\partial^2 u}{\partial y^2} = 8, \ \frac{\partial^2 u}{\partial z^2} = 8, \ \frac{\partial^2 u}{\partial x \partial y} = 4, \ \frac{\partial^2 u}{\partial y \partial x} = 4, \ \frac{\partial^2 u}{\partial y \partial z} = 16, \ \frac{\partial^2 u}{\partial z \partial y} = 16, \ \frac{\partial^2 u}{\partial x \partial z} = 4,$$

$$\frac{\partial^2 u}{\partial z \partial x} = 4.$$

The Hessian matrix of $u(x, y, z)$ is

$$H = \begin{bmatrix} 2 & 4 & 4 \\ 4 & 8 & 16 \\ 4 & 16 & 8 \end{bmatrix}$$

The leading minors of H are

$$H_1 = |2| = 2, \ H_2 = \begin{vmatrix} 2 & 4 \\ 4 & 8 \end{vmatrix} = 0 \text{ and } H_3 = \begin{vmatrix} 2 & 4 & 4 \\ 4 & 8 & 16 \\ 4 & 16 & 8 \end{vmatrix} < 0.$$

Here H_1 and H_3 are not of same sign and H_2 = (i.e., semi-definite). Hence u has a saddle point at (0, 0, 0).

Example 15: Find the maximum and minimum value of u, where $u = \sin x \sin y \sin(x + y)$.

Solution: Given that

$$u = \sin x \sin y \sin(x + y) \qquad \ldots(1)$$

Differentiate partially (1) with respect to x and y both sides respectively, we get,

$$\frac{\partial u}{\partial x} = \sin y \, [\sin x \cos(x + y) + \cos x \sin(x + y)] \qquad \ldots(2)$$

and

$$\frac{\partial u}{\partial y} = \sin x \, [\sin y \cos(x + y) + \cos y \sin(x + y)] \qquad \ldots(3)$$

For maxima and minima, we have

$$\frac{\partial u}{\partial x} = 0, \text{ and } \frac{\partial u}{\partial y} = 0$$

$\sin y[\sin x \cos(x + y) + \cos x. \sin(x + y)] = 0$
and $\sin x[\sin y \cos(x + y) + \cos y, \sin(x + y)] = 0$

Solving above these expressions, we get

$$\tan(x + y) = -\tan x \qquad \ldots(4)$$

and

$$\tan(x + y) = -\tan y \qquad \ldots(5)$$

Now, we have

$$\tan 2x = -\tan x = \tan(\pi - x)$$
$$2x = \pi - x$$
$$x = \pi/3 = y$$

Also

$$\sin y = 0 \;\Rightarrow\; y = 0$$

and

$$\sin x = 0 \;\Rightarrow\; x = 0$$

Thus, the stationary points are (0, 0), ($\pi/3$, $\pi/3$).

Differentiate partially (2) and (3) again, we get

$$r = \frac{\partial^2 u}{\partial x^2} = 2\sin y \cos(2x + y)$$

$$s = \frac{\partial^2 u}{\partial x \partial y} = \sin 2(x + y)$$

and

$$t = \frac{\partial^2 u}{\partial y^2} = 2\sin x \cos(2y + x)$$

At (0, 0), we get $r = 0$, $s = 0$, $t = 0$.

$rt - s^2 = 0$, i.e., u has a saddle point at (0, 0).

Now at ($\pi/3$. $\pi/3$), we get $r = 2\sin(\pi/3)\cos\pi = -\sqrt{3}$.

$$s = \sin(4\pi/3) = -\sin(\pi/3) = -\frac{\sqrt{3}}{2}$$

and

$$t = 2\sin(\pi/3)\sin\pi = -\sqrt{3}$$

$\Rightarrow$

$$rt - s^2 = \frac{9}{4} \text{ is positive and } r < 0.$$

Hence, u is maximum at ($\pi/3$, $\pi/3$).

Example 16: Find the extreme points $f(x_1, x_2) = 20x_1 + 26x_2 + 4x_1x_2 - 4x_1^2 - 3x_2^2$.

Solution: Given that

$$f(x_1, x_2) = 20x_1 + 26x_2 + 4x_1x_2 - 4x_1^2 - 3x_2^2 \qquad \ldots(1)$$

Differentiate partially (1) with respect to x_1 and x_2 both sides respectively, we get

$$\frac{\partial f}{\partial x_1} = 20 + 4x_2 - 8x_1 \qquad \ldots(2)$$

and

$$\frac{\partial f}{\partial x_2} = 26 + 4x_1 - 6x_2 \qquad \ldots(3)$$

For extreme points, we have

$$\frac{\partial f}{\partial x_1} = 0 \quad \text{and} \quad \frac{\partial f}{\partial x_2} = 0$$

$$\Rightarrow \qquad 20 + 4x_2 - 8x_1 = 0$$

And

$$26 + 4x_1 - 6x_2 = 0$$

Solving these, we get $x_1 = 7$, $x_2 = 9$.

Differentiating again partially (2) and we get

$$r = \frac{\partial^2 f}{\partial x_1^2} = -8$$

$$s = \frac{\partial^2 f}{\partial x_1 \partial x_2} = 4$$

and

$$t = \frac{\partial^2 f}{\partial x_2^2} = -6$$

We have $rt - s^2 = (-8)(-6) - (4)^2 = 32$.

At $(7, 9)$, $rt - s^2 > 0$ and $r < 0$, i.e., f is maximum at $(7,9)$.

Example 17: Find the point (x_1, x_2, x_3) at which is functions $f(x_1, x_2, x_3) = -x_1^2 - x_2^2 - x_3^2 + x_1x_2 + x_1 + 2x_3$ has optimum values.

Solution: Given that

$$f(x_1, x_2, x_3) = -x_1^2 - x_2^2 - x_3^2 + x_1x_2 + x_1 + 2x_3 \qquad \ldots(1)$$

Differentiate partially (1) with respect to x_1, x_2 and x_3 both sides respectively, we get

$$\frac{\partial f}{\partial x_1} = -2x_1 + 1$$

$$\frac{\partial f}{\partial x_2} = -2x_2 + x_3$$

and

$$\frac{\partial f}{\partial x_3} = -2x_3 + x_2 + 2 \qquad \ldots(2)$$

For extreme points, we have

$$\frac{\partial f}{\partial x_1} = 0, \quad \frac{\partial f}{\partial x_2} = 0 \text{ and } \frac{\partial f}{\partial x_3} = 0$$

$$\Rightarrow \qquad -2x_1 + 1 = 0$$

$$-2x_2 + x_3 = 0$$

and

$$-2x_3 + x_2 + 2 = 0$$

Solving these we get $x_1 = ½$, $x_2 = 2/3$, $x_3 = 4/3$.

Let the point P be $= \left(\frac{1}{2}, \frac{2}{2}, \frac{4}{3}\right)$.

Differentiate partially (2) again and we get

$$\frac{\partial^2 f}{\partial x_1^2} = -2, \quad \frac{\partial^2 f}{\partial x_2^2} = -2, \quad \frac{\partial^2 f}{\partial x_3^2} = -2, \quad \frac{\partial^2 f}{\partial x_1 \partial x_2} = 0, \quad \frac{\partial^2 f}{\partial x_2 \partial x_1} = 0, \quad \frac{\partial^2 f}{\partial x_1 \partial x_3} = 0, \quad \frac{\partial^2 f}{\partial x_3 \partial x_1} = 0,$$

$$\frac{\partial^2 f}{\partial x_2 \partial x_3} = 1, \quad \frac{\partial^2 f}{\partial x_3 \partial x_2} = 1$$

The Hessian matrix of $f(x_1, x_2, x_3)$ is

$$H = \begin{bmatrix} -2 & 0 & 0 \\ 0 & -2 & 1 \\ 0 & 1 & -2 \end{bmatrix}$$

The leading minors of H are

$$H_1 = |-2| = 2,\ H_2 = \begin{vmatrix} -2 & 0 \\ 0 & -2 \end{vmatrix} = 4 \text{ and } H_3 = \begin{vmatrix} -2 & 0 & 0 \\ 0 & -2 & 1 \\ 0 & 1 & -2 \end{vmatrix} = -6.$$

Here H_1, H_2 and H_3 are not negative definite. So f has a saddle point on $P = \left(\frac{1}{2}, \frac{2}{2}, \frac{4}{3}\right)$.

Example 18: Show that the minimum value of $u = xy\left(\frac{a^3}{x}\right)+\left(\frac{a^3}{y}\right)$ is $3a^2$.

Solution: Given that

$$u = xy\left(\frac{a^3}{x}\right)+\left(\frac{a^3}{y}\right) \qquad \dots(1)$$

Differentiate partially (1) with respect to x and y both sides respectively, we have

$$\frac{\partial u}{\partial x} = y - \frac{a^3}{x^2} \qquad \dots(2)$$

and
$$\frac{\partial u}{\partial y} = x - \frac{a^3}{y^2} \qquad \dots(3)$$

For maxima and minima, we have

$$\frac{\partial u}{\partial x} = 0, \quad \text{and} \quad \frac{\partial u}{\partial y} = 0$$

$\Rightarrow$
$$y - \frac{a^3}{x^2} = 0$$

and
$$x - \frac{a^3}{x^2} = 0$$

Solving these, we get $x = a$, $y = a$.

Differentiate partially (2) again and we get

$$r = \frac{\partial^2 f}{\partial x_1^2} = \frac{2a^3}{x^3}$$

$$s = \frac{\partial^2 f}{\partial x_1 \partial x_2} = 1$$

and
$$t = \frac{\partial^2 f}{\partial x_2^2} = \frac{2a^3}{y^3}$$

At (a, a), we get $r = 2$, $s = 1$, $t = 2$.

Then $rt - s^2 = 3 > 0$

Since at (a, a), $rt - s^2 > 0$ and $r > 0$, then u is minimum at (a, a).

The minimum value of $u = a.\, a + \dfrac{a^3}{a} + \dfrac{a^3}{a} = 3a^2$.

Example 19: Find the extreme points of the function $f(x_1, x_2) = -x_1^3 + 2x_2^3 + 3x_1^2 + 12x_2^2 + 24$. And determine their nature also.

Solution: Given that

$$f(x_1, x_2) = x_1^3 + 2x_2^3 + 3x_1^2 + 12x_2^2 + 24 \qquad \ldots(1)$$

Differentiate partially (1) with respect to x_1 and x_2 both sides respectively, we get

$$\frac{\partial f}{\partial x_1} = 3x_1^2 + 6x_1 \qquad \ldots(2)$$

and

$$\frac{\partial f}{\partial x_2} = 6x_2^2 + 24x_2 \qquad \ldots(3)$$

For extreme points, we have

$$\frac{\partial f}{\partial x_1} = \text{ and } \frac{\partial f}{\partial x_2} = 0$$

$\Rightarrow$ $3x_1^2 + 6x_1 = 0$

And $6x_2^2 + 24x_2 = 0$

Solving these, we get $x_1 = 0, -2$, $x_2 = 0, -4$.

Differentiating again partially (2) and we get

$$r = \frac{\partial^2 f}{\partial x_1^2} = 6(x_1 + 1)$$

$$s = \frac{\partial^2 f}{\partial x_1 \partial x_2} = 0$$

and

$$t = \frac{\partial^2 f}{\partial x_2^2} = 12(x_2 + 2)$$

At $(0, 0)$ $rt - s^2 = 72(x_1 + 1)(x_2 + 2) = 72 > 0$ and $r > 0$ i.e., f is minimum at $(0, 0)$.

At $(0, -4)$ $rt - s^2 = 72(x_1 + 1)(x_2 + 2) = -144 < 0 \Rightarrow$ no extreme point, i.e., f has a saddle point at $(0, -4)$.

At $(-2, 0)$ $rt - s^2 = 72(x_1 + 1)(x_2 + 2) = -144 < 0 \Rightarrow$ no extreme point, i.e., f has a saddle point at $(-2, 0)$.

At $(-2, -4)$ $rt - s^2 = 72(x_1 + 1)(x_2 + 2) = 144 > 0$ and $r < 0$ i.e., f is maximum at $(-2, -4)$.

Example 20: Find the dimension of a box of the largest volume that can be inscribed in a sphere of radius 3 meters.

Solution: Let the volume of the box be

$$V = 2x.\, 2y.\, 2z \quad \Rightarrow \quad V = 8xyz \qquad \ldots(1)$$

Changing the variables Cartesian to spherical polar co-ordinates as $x = r \sin\theta \cos\phi$, $y = r \sin\theta \sin\phi$, $z = r \cos\theta$ and using $r = 3$ (given).

$$V = 216 \sin^2\theta \cos\theta \cos\phi \sin\phi$$
$$= 108 \sin^2\theta \cos\theta \sin 2\phi \quad \ldots(2)$$

Differentiate partially (2) with respect to θ and ϕ both sides respectively, we get

$$\frac{\partial V}{\partial \theta} = 108 \sin 2\phi\, [\sin^2\theta(-\sin\theta) + 2\sin\theta \cos^2\theta]$$
$$= 108 \sin 2\phi \sin\theta\{-\sin^2\theta + 2\cos^2\theta\} \quad \ldots(3)$$

and
$$\frac{\partial V}{\partial \phi} = 216 \cos 2\phi \sin^2\theta \cos\theta \quad \ldots(4)$$

For maxima and minima, we have

$$\frac{\partial V}{\partial \theta} = 0 \quad \text{and} \quad \frac{\partial V}{\partial \phi} = 0$$

$\Rightarrow$
$$108 \sin 2\phi \sin\theta\{-\sin^2\theta + 2\cos^2\theta\} = 0$$

i.e.,
$$\sin\theta = 0 \text{ or } \tan\theta = \sqrt{2} \text{ or } \sin\phi = 0$$

i.e.,
$$\theta = 0,\ \tan^{-1}\sqrt{2},\ \phi = 0.$$

and
$$216 \cos 2\phi \sin^2\theta \cos\theta = 0$$

i.e.,
$$\theta = 0,\ \pi/4.$$

Because $\theta = 0$ or $\phi = 0$ gives $V = 0$, so we take only $\theta = \tan^{-1}\sqrt{2}$ and $\phi = \pi/4$.

Differentiate partially (3) again and we get

$$\frac{\partial^2 V}{\partial \theta^2} = 108 \sin 2\phi[\sin\theta(-4\cos\theta \sin\theta - 2\sin\theta \cos\theta) + \cos\theta(2\cos^2\theta - \sin^2\theta)]$$

At $(\tan^{-1}\sqrt{2}, \pi/4)$,
$$\frac{\partial^2 V}{\partial \theta^2} = 108 \left[\frac{\sqrt{2}}{\sqrt{3}}\left(-6 \times \frac{1}{\sqrt{3}} \times \frac{\sqrt{2}}{\sqrt{3}}\right) + \frac{1}{\sqrt{3}}\left(1\left(\frac{1}{\sqrt{3}}\right)^2 - \left(\frac{\sqrt{2}}{\sqrt{3}}\right)^2\right)\right]$$
$$= -\frac{432}{\sqrt{3}}$$

$$\frac{\partial^2 V}{\partial \theta \partial \phi} = 216 \cos 2\phi[\sin^2\theta(-\sin\theta) + \cos^2\theta(2\sin\theta)]$$
$$= 216 \cos 2\phi \sin\theta[2\cos^2\theta - \sin^2\theta]$$

At $(\tan^{-1}\sqrt{2}, \pi/4)$,
$$\frac{\partial^2 V}{\partial \theta \partial \phi} = 0$$

$$\frac{\partial^2 V}{\partial \phi^2} = -432 \sin 2\phi[\sin^2\theta \cos\theta]$$

At $(\tan^{-1}\sqrt{2}, \pi/4)$,
$$\frac{\partial^2 V}{\partial \phi^2} = -432 \times \left(\frac{\sqrt{2}}{\sqrt{3}}\right)^2 \times \left(\frac{1}{\sqrt{3}}\right) = -\frac{864}{3\sqrt{3}}$$

Now at ($\tan^{-1}\sqrt{2}$, $\pi/4$), we have

$$\frac{\partial^2 V}{\partial \theta^2}\cdot\frac{\partial^2 V}{\partial \phi^2}-\left(\frac{\partial^2 V}{\partial\theta\partial\phi}\right)^2=\left(\frac{-432}{\sqrt{3}}\right)\left(\frac{-864}{3\sqrt{3}}\right)-0>0$$

i.e., V is maximum.

The dimensions of the maximum inscribed box are

$$\begin{aligned}2x &= 2\,r\sin\theta\cos\phi\\ &= 2(3).\,\frac{\sqrt{2}}{\sqrt{3}}\cdot\frac{1}{\sqrt{2}}\\ &= 2\sqrt{3}\\ 2y &= 2\,r\sin\theta\sin\theta\\ &= 2(3).\,\frac{\sqrt{2}}{\sqrt{3}}\cdot\frac{1}{\sqrt{2}}\\ &= 2\sqrt{3}\\ 2z &= 2\,r\cos\theta\\ &= 2(3).\,\frac{1}{\sqrt{3}}\\ &= 2\sqrt{3}\,.\end{aligned}$$

EXERCISE 2.1

1. Find the output x, which maximizes profit P given by the relationship $P = 5000 + 1200x - x^2$.
2. The price of a commodity is the function of its quantity q to be purchased which is given as $10.59\,e^{-0.01q}$. Find the quantity for which the total revenue is maximum.
3. The total revenue R of a firm is given by $R = 20x - 2x^2$, where x represents the quantity sold and $C = x^2 - 4x + 20$. Find the value of x for which the revenue will be maximum and also find the profit price and total revenue.
4. The average cost per unit of manufacturing for Bharat food products is given by the following relationship $C(x) = 4000 - 16x + 0.2x$ where x is the quantity produced. Find the value of x that will minimize average cost per unit.
5. Show that the function $y = x^3$ has a point of inflexion at (0, 0).
6. Determine the maximum and minimum values of the functions $f(x) = 12x^5 - 45x^4 - 40x^3 + 5$.
7. Find maxima or minima, if any of the function $f(x) = x^4(x-1)^{-1}(x-3)^{-1}$, $x \neq 1$ and $x \neq 3$.
8. Show that $f(x) = 4x^3 - 18x^2 + 27x - 7$ is never optimal in a given interval at its end points.
9. Prove that $\dfrac{x}{1+x\tan x}$ is a maximum when is given by the equation $x = \cos x$.
10. Find the local maxima or minima of the function

$$f(x) = \frac{x^5}{5} - 7\frac{x^4}{4} + 17\frac{x^3}{3} - 17\frac{x^2}{2} = + 6x + 10 \text{ for } x \in R.$$

11. Show that the semi-vertical angle of a cone of maximum volume and given slant height is $\tan^{-1}\sqrt{2}$.
12. Find the area of the largest rectangle, if its perimeter is to be constant.
13. Find the volume of the greatest right circular cone described by the revolution of triangle of hypotenuse C about one of its sides.
14. The lower corner of a leaf in a book is folded over so as just to reach the inner edge of the page. Show that the fraction of the width folded over when the area of the folded part is minimum is 2/3.
15. The cost of fuel for running a train is proportional to the square of the speed generated in km/hour. If it costs Rs. 480 per hour at 16 km/hour. What is the most economical speed, if the fixed charges are Rs. 3000 per hours?
16. Find the dimensions of a minimum cost cylindrical refrigeration tank of volume 50 m^3, if the sheet for the circular ends costs Rs. 100 per m^2 and that of the cylindrical well costs Rs. 60 per m^2 and the cost of refrigeration over the area comes of Rs. 800 per m^2.
17. An electric light is placed directly over the centre of a circular plot of lawn 100 meters in diameter. Assuming the intensity of light varies directly as the sine of the angle at which it strikes an illuminated surface and inversely as the square of the distance of the points from the source. At how much height should the light source be hung in order that the intensity at any point on the circumference of the plot may be maximum?
18. The horse power (HP) generated by a Pelton wheel is proportional to $kv(V - v)$, where the velocity of the runner is v m/sec and velocity of jet is V m/sec, which is constant. Prove that the max HP generated by the wheel is $V = 2v$.
19. In an open tank with a square base and a vertical side is to be constructed of sheet metal to hold a given quantity of water. Show that the cost of the material will be least when the depth is half the width.
20. Determine the dimension of an open box of maximum volume that can be constructed from a sheet of 210 mm × 297 mm by cutting four squares of size $x \times x$ from the corners and folding and gluing the edges.
21. Find the maxima and minima of the function $xy(a - x - y)$.
22. Find the maxima and minima of the function $u = x^3 + y^3 - 3x - 12y + 25$.
23. Find the extreme points $f(x_1, x_2) = x_1^3 + x_2^3 + 2x_1^2 + 4x_2^2 + 6$.
24. Find the maxima and minima of the function $u = 2(x - y)^2 - x^4 - y^4$.
25. Find the extremum points of the function $f(x_1, x_2, x_3) = \sin x_1 \sin x_2 \sin x_3$, where x_1, x_2 and x_3 are the vertex angle of a triangle.
26. A rectangular box open at the top is to have a given capacity. Find the dimensions of the box requiring least material for its construction.
27. Find the maxima and minima of the function $u = x^2 + y^2 + z^2 + x - 2z - xy$.
28. Find the stationary point of u, where $u = x^2 + y^2$.

ANSWERS

1. 3,65,000.
2. 100.
3. At $x = 5$ total revenue = 50, profit = 25, price = 5 per unit.

4. At $x = 400$.
6. At $x = 0$ give point of inflection; at $x = 1$ give maxima and value is 12; at $x = 2$ give minima and value is – 11.
7. At $x = 6/5$ $f(x)$ give maxima.
8. Diameter = 3.93 meters, length = 4.12 meters.
10. At $x = 1$ give point of inflexion, at $x = 2$ give local maxima and $x = 3$ give local minima.
12. $y = x = C/4$, i.e., square.
13. At $h = C/\sqrt{3}$ volume $= \dfrac{2\pi c^3}{9\sqrt{3}}$.
15. Speed 40 km/hour.
17. The intensity is maximum for $\tan\theta = \sqrt{2}$ height $= 25\sqrt{2}$ meters.
20. Height = 40.423 mm, length = 216.154 mm, width = 129.154 mm and maximum volume = 1128.5 cm^3.
21. At (0, 0), (0, a) and (a, 0) u is neither maxima nor minima. At $\left(\frac{a}{3}, \frac{a}{3}\right)$ = u is minimum if $a < 0$ and u is maximum if $a > 0$.
22. At (1, 2) u is minimum and at (–1, –2) u is maximum.
23. At (0, 0) f is minimum and $\left(-\frac{4}{3}, -\frac{8}{3}\right)$ f is maximum.
24. At $\left(\sqrt{2}, \sqrt{-2}\right)$ u is maximum.
25. At $\left(\frac{\pi}{3}, \frac{\pi}{3}, \frac{\pi}{3}\right)$ f is maximum.
26. At $x = y = z = (2\,v_0)^{1/3}$ give minima.
27. At $\left(-\frac{2}{3}, -\frac{1}{3}, 1\right)$ u is minimum.
28. At (0, 0) u give saddle point.

2.4 CONSTRAINED MULTIVARIABLE OPTIMIZATION PROBLEMS WITH EQUALITY CONSTRAINTS

The optimization problem of a continuous and differentiable function subject to equality constraints:

Optimize (Max or Min) $Z = f(X)$

Subject to constraints (s.t.) $g_j(X) = 0$; $j = 1, 2, 3, \ldots, mg$

where
$$X = \begin{pmatrix} x_1 \\ x_2 \\ \cdots \\ \cdots \\ x_n \end{pmatrix}$$

Here m is less than or equal to n. There are several methods for solving this type of problem. Here we discuss only two methods:

1. Direct substitution method
2. Lagrange multipliers method

2.4.1 Direct Substitution Method

In this method, the value of any variable from the constraint set is put into the objective function. The problem reduces to unconstrained optimization problem and can be solved by unconstrained optimization method.

2.4.2 Lagrange Multiplier Method

Consider a general problem with n variables and m equality constraints:

Optimize $Z = f(X)$

s.t. $g_j(X) = 0 \; ; j = 1, 2, 3, \ldots, m \; (m < n)$

$$X \geq 0$$

where
$$X = \begin{pmatrix} x_1 \\ x_2 \\ \cdots \\ \cdots \\ x_n \end{pmatrix}$$

Now we define a function

$$L(x_1, x_2, \ldots, x_n, \lambda_1, \lambda_2, \ldots, \lambda_m) = f(X) + \sum_{j=1}^{m} \lambda_j g_j(X) \qquad \ldots(1)$$

Here, $\lambda_1, \lambda_2, \ldots, \lambda_m$ are known as Lagrange's undetermined multipliers.

The necessary conditions for extreme of L are

$$\frac{\partial L}{\partial x_i} = \frac{\partial f}{\partial x_i} + \sum_{j=1}^{m} \lambda_j \frac{\partial g_j}{\partial x_i} = 0 \qquad \ldots(2)$$

and
$$\frac{\partial L}{\partial \lambda_j} = 0; \begin{pmatrix} i = 1, 2, \ldots n \\ j = 1, 2, \ldots m \end{pmatrix} \qquad \ldots(3)$$

Solving equation (4) and (5), we get

$$X = \begin{pmatrix} x_1^* \\ x_2^* \\ \cdots \\ \cdots \\ x_m^* \end{pmatrix} \text{ and } \lambda^* = \begin{pmatrix} x_1^* \\ x_2^* \\ \cdots \\ \cdots \\ x_m^* \end{pmatrix}$$

The sufficient condition for the function to have extreme at point X^*, is that the values of k obtained from equation

$$\begin{bmatrix} L_{11}-k & L_{12} & \cdots & L_{1n} & g_{11} & g_{21} & \cdots & g_{m1} \\ L_{21} & L_{22}-k & \cdots & L_{2n} & g_{12} & g_{22} & \cdots & g_{m2} \\ \cdots & \cdots & \cdots & \cdots & \cdots & \cdots & \cdots & \cdots \\ \cdots & \cdots & \cdots & \cdots & \cdots & \cdots & \cdots & \cdots \\ L_{n1} & L_{n2} & \cdots & L_{nn}-k & g_{1n} & g_{2n} & \cdots & g_{mn} \\ g_{11} & g_{12} & \cdots & g_{1n} & 0 & 0 & \cdots & 0 \\ \cdots & \cdots & \cdots & \cdots & \cdots & \cdots & \cdots & \cdots \\ \cdots & \cdots & \cdots & \cdots & \cdots & \cdots & \cdots & \cdots \\ g_{m1} & g_{m2} & \cdots & g_{mn} & 0 & 0 & \cdots & 0 \end{bmatrix} = 0$$

must be of the same sign. If all the eigen values of k are negative then it is a maxima and if all the eigen values k are positive then it is a minima. But if some eigen values are zero or of different sign then that is a saddle point. In above L_{ij} and g_{ij} denoted by $\dfrac{\partial^2 L}{\partial x_i \partial x_j}$ and $\dfrac{\partial g_j}{\partial x_i}$ respectively.

2.5. CONSTRAINED MULTIVARIABLE OPTIMIZATION PROBLEMS WITH INEQUALITY CONSTRAINTS

Let us consider a problem

Optimization (Max or Min) $Z = f(X)$ ...(1)

s.t. $g_j(X) \le 0$; $j = 1, 2, 3, \ldots, m$ ...(2)

where $X = \begin{pmatrix} x_1 \\ x_2 \\ \cdots \\ \cdots \\ x_n \end{pmatrix}$

The inequality constraints in equation (2) can be converted into equality constraints by adding slack variables as

$$g_j(X) + y_j^2 = 0; j = 1, 2, 3, \ldots, m$$

Now the problem becomes

Optimize (Max or Min) $Z = f(X)$

s.t. $G_j(X, Y) = g_j(X) + y_j^2 = 0; j = 1, 2, 3, \ldots, m$

where $X = \begin{pmatrix} x_1 \\ x_2 \\ \cdots \\ \cdots \\ x_n \end{pmatrix}$ and $Y = \begin{pmatrix} y_1 \\ y_2 \\ \cdots \\ \cdots \\ y_m \end{pmatrix}$

Now it can be solved by Lagrange's multipliers method.

2.6 CONVEX AND CONCAVE FUNCTIONS

A function $f(X)$ is said to be convex over a convex set S, if for any two points x_1 and x_2 in S

$$f[\lambda x_1 + (1 - \lambda)x_2] \le \lambda f(x_1) + (1 - \lambda) f(x_2); \quad 0 \le \lambda \le 1$$

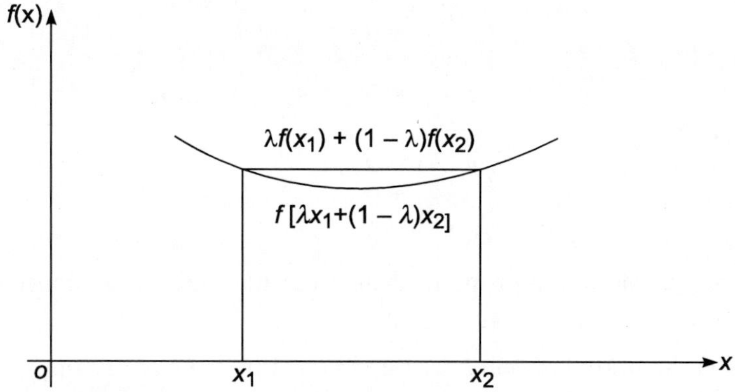

In other words, a function $f(X)$ is said to be convex if and only if the Hessian matrix $H(x)$ of the second partial derivatives is positive semi-definite.

A function $f(X)$ is said to be concave over a convex set S if for any two points x_1 and x_2 in S

$$f[\lambda x_1 + (1 - \lambda)x_2] \ge \lambda f(x_1) + (1 - \lambda) f(x_2); \quad 0 \le \lambda \le 1$$

In other words, a function $f(X)$ is said to be concave if and only if the Hessian matrix $H(x)$ of the second partial derivatives is negative semi-definite.

2.7 KUHN-TUCKER CONDITION

Consider the optimization problem

Optimization (Max or Min) $\quad Z = f(X) \quad$ ...(1)

s.t. $\quad g_j(X) \le 0; j = 1, 2, 3,\ldots, m \quad$ ...(2)

Converting the inequality constraints into equality constraints by adding slack variables y_j^2, we have

$$g_j(X) + y_j^2 = 0 \qquad (3)$$

Now we define a Lagrangian function

$$L(x_1, x_2,\ldots, x_n, \lambda_1, \lambda_2,\ldots, \lambda_m) = f(X) + \sum_{j=1}^{m} \lambda_j \{g_j(X) + y_j^2\} \qquad \ldots(4)$$

The Kuhn-Tucker necessary conditions for extreme points are

$$\frac{\partial L}{\partial x_i} = 0; \; i = 1, 2, 3,\ldots n$$

$$\frac{\partial f}{\partial x_i} + \sum_{j=1}^{m} \lambda_j \frac{\partial g_j}{\partial x_i} = 0 \qquad \ldots(5)$$

$$\frac{\partial L}{\partial \lambda_j} = 0$$

$\Rightarrow$ $$g_j(X) + y_j^2 = 0; j = 1, 2, 3,\ldots, m \qquad \ldots(6)$$

and $$\frac{\partial L}{\partial y_j} = 0$$

$\Rightarrow$ $$2\lambda_j y_j = 0; j = 1, 2, 3,\ldots, m \qquad \ldots(7)$$

From (6) and (7), we have

$$\lambda_j g_j(X) = 0$$

$\Rightarrow$ $$\lambda_j = 0$$

or $$g_j(X) = 0.$$

Case I: If $g_j(X) = 0$ at the optimum point then it is called the active constraint and we can find optimum solution.

Case II: If $\lambda_j = 0$ at the optimum point then it is called inactive constraints.

Note: If the given optimization problem is a minimization problem with constraints of the form $g_j(X) \geq 0$ then $\lambda_j \leq 0$ but if the problem is a maximization problem with constraints of the form $g_j(X) \leq 0$ then $\lambda_j \leq 0$. Let us consider some maximization or minimization problems given in the following terms.

(i) Maximize $Z = f(X)$

s.t. $g_j(X) \leq 0$; $j = 1, 2, 3,\ldots, m$

For the function $f(X)$ to have maxima, we have

$$\frac{\partial f}{\partial x_i} + \sum_{j=1}^{m} \lambda_j \frac{\partial g_j}{\partial x_i} = 0; i = 1, 2, 3,\ldots, m$$

$$\lambda_j g_j(X) = 0; j = 1, 2, 3,\ldots, m$$

and $$\lambda_j \leq 0.$$

(ii) Maximize $Z = f(X)$

s.t. $g_j(X) \geq 0; j = 1, 2, 3,\ldots, m$

For the function $f(X)$ to have maxima, we have

$$\frac{\partial f}{\partial x_i} + \sum_{j=1}^{m} \lambda_j \frac{\partial g_j}{\partial x_i} = 0; i = 1, 2, 3,\ldots, m$$

$$\lambda_j g_j(X) = 0; j = 1, 2, 3,\ldots, m$$

and $$\lambda_j \geq 0.$$

(iii) Minimize $Z = f(X)$

s.t. $g_j(X) \leq 0; j = 1, 2, 3,\ldots, m$

For the function $f(X)$ to have maxima, we have

$$\frac{\partial f}{\partial x_i} + \sum_{j=1}^{m} \lambda_j \frac{\partial g_j}{\partial x_i} = 0; i = 1, 2, 3, \ldots, m$$

$$\lambda_j g_j(X) = 0; j = 1, 2, 3, \ldots, m$$

and $$\lambda_j \leq 0.$$

(iv) Minimize $Z = f(X)$

s.t. $g_j(X) \geq 0; j = 1, 2, 3, \ldots, m$

For the function $f(X)$ to have maxima, we have

$$\frac{\partial f}{\partial x_i} + \sum_{j=1}^{m} \lambda_j \frac{\partial g_j}{\partial x_i} \; ; i = 1, 2, 3, \ldots, m$$

$$\lambda_j g_j(X) = 0; j = 1, 2, 3, \ldots, m$$

and $\lambda_j \leq 0.$

SOLVED EXAMPLES

Example 21: Find the optimum solution of the following constrained multivariable problem minimize $Z = x_1^2 + (x_2 + 1)^2 + (x_3 - 1)^2$ s.t. $x_1 + 5x_2 - 3x_3 = 6$.

Solution: Given that

$$Z = x_1^2 + (x_2 + 1)^2 + (x_3 - 1)^2 \quad \ldots(1)$$

and $$x_1 + 5x_2 - 3x_3 = 6$$

$\Rightarrow$ $$x_3 = \frac{x_1 + 5x_2 + 6}{3} \quad \ldots(2)$$

Using (1) and (2), we get

$$Z = x_1^2 + (x_2 + 1)^2 + \frac{1}{9}(x_1 + 5x_2 - 9)^2 \quad \ldots(3)$$

Differentiate partially both sides of (3) with respect to x_1 and x_2 respectively, we get

$$\frac{\partial Z}{\partial x_1} = 2x_1 + \frac{2}{9}(x_1 + 5x_2 - 9)$$

and $$\frac{\partial Z}{\partial x_2} = 2(x_2 + 1) + \frac{10}{9}(x_1 + 5x_2 - 9)$$

For maxima and minima, we have

$$\frac{\partial Z}{\partial x_1} = 0 \quad \text{and} \quad \frac{\partial Z}{\partial x_2} = 0$$

$\Rightarrow$ $$2x_1 + \frac{2}{9}(x_1 + 5x_2 - 9) = 0$$

and $$2(x_2 + 1) + \frac{10}{9}(x_1 + 5x_2 - 9) = 0$$

Solving these, we get

$$x_1 = \frac{2}{5} \text{ and } x_2 = 1$$

Differentiate again partially (4) and we get

$$r = \frac{\partial^2 Z}{\partial x_1^2} = 2 + \frac{2}{9} = \frac{20}{9}$$

$$t = \frac{\partial^2 Z}{\partial x_2^2} = 2 + \frac{50}{9} = \frac{68}{9}$$

and
$$\frac{\partial^2 Z}{\partial x_1 x_2} = \frac{10}{9}$$

At $\left(\frac{2}{5}, 1\right)$,
$$rt - s^2 = \left(\frac{20}{9}\right)\left(\frac{68}{9}\right) - \left(\frac{10}{9}\right)^2 = 1260 > 0 \text{ and } r > 0$$

i.e., Z is minimum at $\left(\frac{2}{5}, 1\right)$ the minimum value of Z is $\frac{28}{5}$.

Example 22: Find the dimensions of a box of large volume that can be inscribed in a sphere of radius a.

Solution: Suppose x, y and z are the dimensions of the box with respect to origin O and OX, OY, OZ are reference axes. The volume of box is

$$V = 8xyz \qquad \ldots(1)$$

Given that the box is to be inscribed in a sphere of radius 'a' i.e.,

$$x^2 + y^2 + z^2 = a^2 \qquad \ldots(2)$$

Eliminating z from (1) and (2), we get

$$V = 8xy\,(a^2 - x^2 - y^2)^{1/2} \qquad \ldots(3)$$

Differentiate partially (3) with respect to x and y both sides respectively, we get

$$\frac{\partial V}{\partial x_1} = 8y\left[x \cdot \frac{1}{2}(a^2 - x^2 - y^2)^{-1/2}(-2x) + (a^2 - x^2 - y^2)^{1/2}\right]$$

$$= 8y\left[-\frac{x^2}{(a^2 - x^2 - y^2)^{1/2}} + (a^2 - x^2 - y^2)^{1/2}\right]$$

$$= 8y\left[\frac{(a^2 - 2x^2 - y^2)}{(a^2 - x^2 - y^2)^{1/2}}\right] \qquad \ldots(4)$$

and
$$\frac{\partial V}{\partial y} = 8x\left[\frac{(a^2 - x^2 - 2y^2)}{(a^2 - x^2 - y^2)^{1/2}}\right] \qquad \ldots(5)$$

For maxima and minima, we have

$$\frac{\partial V}{\partial x_1} = 0 \text{ and } \frac{\partial V}{\partial y} = 0$$

$\Rightarrow$
$$8y\left[\frac{(a^2 - 2x^2 - y^2)}{(a^2 - x^2 - y^2)^{1/2}}\right] = 0$$

and
$$8x\left[\frac{(a^2 - x^2 - 2y^2)}{(a^2 - x^2 - y^2)^{1/2}}\right] = 0$$

$$\Rightarrow \quad a^2 - 2x^2 - y^2 = 0 \qquad \ldots(6)$$

and
$$a^2 - x^2 - 2y^2 = 0 \qquad \ldots(7)$$

Solving these, we get

$$x = \frac{a}{\sqrt{3}} \text{ and } y = \frac{a}{\sqrt{3}}$$

Differentiate again partially (4) and (5) and we get

$$r = \frac{\partial^2 V}{\partial x^2} = 8y\left[\frac{(a^2 - x^2 - y^2)^{1/2}(-4x) - (a^2 - 2x^2 - y^2)\cdot\frac{1}{2}(a^2 - x^2 - y^2)^{-1/2}(-2x)}{(a^2 - x^2 - y^2)}\right]$$

$$= 8y\left[\frac{(a^2 - x^2 - y^2)(-4x) + x(a^2 - 2x^2 - y^2}{(a^2 - x^2 - y^2)^{3/2}}\right]$$

$$= \frac{-32xy}{(a^2 - x^2 - y^2)^{1/2}} \qquad \text{(Using (6))}$$

Similarly,
$$t = \frac{\partial^2 V}{\partial y^2} = \frac{-32xy}{(a^2 - x^2 - y^2)^{1/2}}$$

and
$$s = \frac{\partial^2 V}{\partial x \partial y} = \frac{-16x^2}{(a^2 - x^2 - y^2)^{1/2}}$$

Now we have

$$rt - s^2 = \left(\frac{-32xy}{(a^2 - x^2 - y^2)^{1/2}}\right)\left(\frac{-32xy}{(a^2 - x^2 - y^2)^{1/2}}\right) + \left(\frac{-16x^2}{(a^2 - x^2 - y^2)^{1/2}}\right)^2$$

$$= \frac{256x^2(x^2 + 4y^2)}{(a^2 - x^2 - y^2)}$$

At $\left(\frac{a}{\sqrt{3}}, \frac{a}{\sqrt{3}}\right)$, $= rt - s^2$ is positive and $r < 0$, i.e., V is maximum.

The maximum value of $V = 8xyz = 8\left(\frac{a}{\sqrt{3}}\right)\left(\frac{a}{\sqrt{3}}\right)\left(\frac{a}{\sqrt{3}}\right) = \frac{8a^3}{3\sqrt{3}}$.

Example 23: Optimize $Z = x^2 + y^2 + z^2$ s.t. $4x + y^2 + 2z = 14$.

Solution: It is given that

$$Z = x^2 + y^2 + z^2 \qquad \ldots(1)$$

and
$$4x + y^2 + 2z = 14$$

$$\Rightarrow \quad g(x, y, z) = 4x + y^2 + 2z - 14 = 0 \qquad \ldots(2)$$

The Lagrangian function L is

$$L(x, y, z; \lambda) = x^2 + y^2 + z^2 + \lambda(4x + y^2 + 2z - 14) \qquad \ldots(3)$$

The necessary conditions for extreme L are

$$\left.\begin{aligned}\frac{\partial L}{\partial x} &= 2x + 4\lambda = 0\\ \frac{\partial L}{\partial y} &= 2y + 2y\lambda = 0\\ \frac{\partial L}{\partial z} &= 2z + 2\lambda = 0\\ \frac{\partial L}{\partial \lambda} &= 4x + y^2 + 2z - 14 = 0\end{aligned}\right\} \qquad \ldots(4)$$

Solving these we get $x = -2\lambda$, $z = -\lambda$ and $\lambda = -1$

$\Rightarrow$ $x = 2$, $z = 1$ and $y = \pm 2$, putting $x = -2\lambda$, $z = -\lambda$, $y = 0$ in (2), we get $\lambda = -1.4$.

Here $(2, 2, 1, -1)$, $(2, -2, 1, -1)$ and $(-2\lambda, 0, -\lambda, \lambda)$ or $(2.8, 0, 1.4, -1.4)$ are extreme points.

Differentiate again partially (4) and we get

$$\frac{\partial^2 L}{\partial x^2} = 2,\ \frac{\partial^2 L}{\partial x \partial y} = 0,\ \frac{\partial^2 L}{\partial x \partial z} = 0,\ \frac{\partial^2 L}{\partial y \partial x} = 0,\ \frac{\partial^2 L}{\partial y^2} = 2 + 2\lambda,\ \frac{\partial^2 L}{\partial y \partial z} = 2,\ \frac{\partial^2 L}{\partial z \partial x} = 0,\ \frac{\partial^2 L}{\partial z \partial y}$$

$$= 0,\ \frac{\partial^2 L}{\partial z^2} = 2,\ \frac{\partial g}{\partial x} = 4,\ \frac{\partial g}{\partial y} = 2y,\ \frac{\partial g}{\partial z} = 2.$$

The sufficient condition for the extreme point is

$$H = \begin{vmatrix} \dfrac{\partial^2 L}{\partial x^2} - k & \dfrac{\partial^2 L}{\partial x \partial y} & \dfrac{\partial^2 L}{\partial x \partial z} & \dfrac{\partial g}{\partial x} \\ \dfrac{\partial^2 L}{\partial yx \partial x} & \dfrac{\partial^2 L}{\partial y^2} - k & \dfrac{\partial^2 L}{\partial y \partial z} & \dfrac{\partial g}{\partial y} \\ \dfrac{\partial^2 L}{\partial z \partial x} & \dfrac{\partial^2 L}{\partial z \partial y} & \dfrac{\partial^2 L}{\partial z^2} - k & \dfrac{\partial g}{\partial z} \\ \dfrac{\partial g}{\partial x} & \dfrac{\partial g}{\partial y} & \dfrac{\partial g}{\partial z} & 0 \end{vmatrix} = 0$$

i.e.,

$$H = \begin{vmatrix} 2-k & 0 & 0 & 4 \\ 0 & 2+2\lambda-k & 0 & 2y \\ 0 & 0 & 2-k & 2 \\ 4 & 2y & 2 & 0 \end{vmatrix} = 0$$

or $4(2-k)[-10 + 5k - 10\lambda - 2y^2 + y^2 k] = 0$

At $(2, 2, 1, -1)$, equation (2) we have

$$(2-k)(-10 + 5k + 10 - 8 + 4k) = 0$$

$$\Rightarrow \qquad k = 2,\ 8/9$$

i.e., values of k are positive then there is a minima.

At $(2, -2, 1, -1)$, equation (2) we have

$$(2 - k)(-10 + 5k + 10 - 8 + 4k) = 0$$

$$\Rightarrow \qquad k = 2, 8/9$$

Also values of k are positive then there is a minima.

At $(2.8, 0, 1.4, -1.4)$, from (2) we have

$$(2 - k)(-10 + 5k + 14 - 0 + 0) = 0$$

$$\Rightarrow \qquad k = 2, -4/5.$$

i.e., value of k are positive and negative (neither maxima and nor minima) i.e., saddle point.

Example 24: Minimize $f(x) = \frac{1}{2}(x_1^2 + x_2^2 + x_3^2)$

s.t. $\qquad g_1(x) = x_1 - x_2 = 0$

and $\qquad g_2(x) = x_1 + x_2 + x_3 - 1 = 0$ by Lagrange multiplier method.

Solution: It is given that

Minimize
$$f(x) = \frac{1}{2}(x_1^2 + x_2^2 + x_3^2) \qquad \ldots(1)$$

s.t.
$$g_1(x) = x_1 - x_2 = 0 \qquad \ldots(2)$$

and
$$g_2(x) = x_1 + x_2 + x_3 - 1 = 0 \qquad \ldots(3)$$

The Lagrangian function L is

$$L(x_1, x_2, x_3; \lambda_1, \lambda_2) = \frac{1}{2}(x_1^2 + x_2^2 + x_3^2) + \lambda_1(x_1 - x_2) + \lambda_2(x_1 + x_2 + x_3 - 1) \qquad \ldots(4)$$

The necessary conditions for extreme of L are

$$\left.\begin{aligned} \frac{\partial L}{\partial x_1} &= x_1 + \lambda_1 + \lambda_2 = 0 \\ \frac{\partial L}{\partial x_2} &= x_2 - \lambda_1 + \lambda_2 = 0 \\ \frac{\partial L}{\partial x_3} &= x_3 + \lambda_2 = 0 \\ \frac{\partial L}{\partial \lambda_1} &= x_1 - x_2 = 0 \\ \frac{\partial L}{\partial \lambda} &= x_1 + x_2 + x_3 - 1 = 0 \end{aligned}\right\} \qquad \ldots(5)$$

Solving these, we get

$$x_1 = x_2 = x_3 = \frac{1}{3}; \quad \lambda_1 = 0, \lambda_2 = -\frac{1}{3}$$

Differentiate again partially (6) and we get

$$\frac{\partial^2 L}{\partial x_1^2} = 1, \quad \frac{\partial^2 L}{\partial x_1 \partial x_2} = 0, \quad \frac{\partial^2 L}{\partial x_1 \partial x_3} = 0, \quad \frac{\partial^2 L}{\partial x_2 \partial x_1} = 0, \quad \frac{\partial^2 L}{\partial x_2^2} = 1, \quad \frac{\partial^2 L}{\partial x_2 \partial x_3} = 0, \quad \frac{\partial^2 L}{\partial x_3 \partial x_1} = 0, \quad \frac{\partial^2 L}{\partial x_3 \partial x_2}$$

$$= 0, \quad \frac{\partial^2 L}{\partial x_3^2} = 1, \quad \frac{\partial g_1}{\partial x_1} = 1, \frac{\partial g_1}{\partial x_2} = 1, \quad \frac{\partial g_1}{\partial x_3} = 0, \quad \frac{\partial g_2}{\partial x_1} = 1, \quad \frac{\partial g_2}{\partial x_2} = 1, \quad \frac{\partial g_2}{\partial x_3} = 1.$$

The sufficient condition for extreme point is

$$H = \begin{vmatrix} 1-K & 0 & 0 & 1 & 1 \\ 0 & 1-K & 0 & -1 & 1 \\ 0 & 0 & 1-K & 0 & 1 \\ 1 & -1 & 0 & 0 & 0 \\ 1 & 1 & 1 & 0 & 0 \end{vmatrix} = 0$$

This implies we get $k = 1, 1, 1$. Here all the values of K are of same sign and positive. i.e., f is minimum at $\left(\frac{1}{3}, \frac{1}{3}, \frac{1}{3}, 0, -\frac{1}{3}\right)$.

The minimum value of $f = \frac{1}{2}\left(\frac{1}{9} + \frac{1}{9} + \frac{1}{9}\right) = \frac{1}{6}$

Example 25: State Kuhn-Tucker conditions. Use them to

Minimize $\quad f(x, y, z) = x^2 + y^2 + z^2 + 20x + 10y$

s.t.

$$x \geq 40,$$
$$x + y \geq 80,$$
$$x + y + z \geq 120.$$

Solution: Let us consider the optimization problem

Minimize $$Z = f(X) \quad \ldots(1)$$

s.t. $$g_j(X) \geq 0; j = 1, 2, 3 \quad \ldots(2)$$

where $$X = \begin{pmatrix} x \\ y \\ z \end{pmatrix}$$

From equation (2), we have

$$G_j(X, Y) = g_j(X) - y_j^2 = 0; j = 1, 2, 3 \quad \ldots(3)$$

Now define a Lagrangian function

$$L(x, y, z; \lambda_1, \lambda_2, \lambda_3; y_1, y_2, y_3) = f(X) + \sum_{J=1}^{3} \lambda_j \left[g_j(X) - y_j^2\right] \quad \ldots(4)$$

It is given that

Minimize $$f(x, y, z) = x^2 + y^2 + z^2 + 20x + 10y \quad \ldots(5)$$

s.t. $\quad x \geq 40,$

$$x + y \geq 80,$$
$$x + y + z \geq 120.$$

$$g_1 = x - 40,\ g_2 = x + y - 80,\ g_3 = x + y + z - 120 \qquad \ldots(6)$$

Using equations (4), (5) and (6), we have

$$L = x^2 + y^2 + z^2 + 20x + 10y + \lambda_1(x - 40 - y_1^2) + \lambda_2(x + y - 80 - y_2^2) + \lambda_3(x + y + z - 120 - y_3^2) \qquad \ldots(7)$$

The Kuhn-Tucker necessary conditions for minimization of L (with $g_j(X) \geq 0$) are

$$\frac{\partial L}{\partial x} = 0,\ \frac{\partial L}{\partial y} = 0,\ \frac{\partial L}{\partial z} = 0$$
$$\lambda_j g_j = 0;\quad j = 1, 2, 3$$
$$\lambda \geq 0; j = 1, 2, 3$$

Differentiate partially (7) and we get

$$\left.\begin{aligned} \frac{\partial L}{\partial x} &= 2x + 20 + \lambda_1 + \lambda_2 + \lambda_3 = 0 \\ \frac{\partial L}{\partial y} &= 2y + 10 + \lambda_2 + \lambda_3 = 0 \\ \frac{\partial L}{\partial z} &= 2z + \lambda_3 = 0 \end{aligned}\right\} \qquad \ldots(8)$$

$$\left.\begin{aligned} \lambda_1 g_1 &= \lambda_1(x - 40) = 0 \\ \lambda_2 g_2 &= \lambda_2(x + y - 80) = 0 \\ \lambda_3 g_3 &= \lambda_3(x + y + z - 120) = 0 \end{aligned}\right\} \qquad \ldots(9)$$

$$\lambda_1, \lambda_2, \lambda_3, \geq 0 \qquad \ldots(10)$$

If $\lambda_1 \neq 0, \lambda_2 \neq 0, \lambda_3 \neq 0$ from (9) we have

$$\left.\begin{aligned} (x - 40) &= 0 \\ (x + y - 80) &= 0 \\ (x + y + z - 120) &= 0 \end{aligned}\right\}$$

Solving these we get

$$x = 40, y = 40, z = 40.$$

Using values of x, y and z. From equation (6), we get

$$\lambda_1 = -10,\ \lambda_2 = -10,\ \lambda_3 = -80$$

Hence, the condition $\lambda_j \leq 0$ from equation (10) is satisfied.

Hence, the optimum solution is

$$x = y = z = 40.$$

and minimize $f = (40)^2 + (40)^2 + (40)^2 + 20(40) + 10(4)$
$= 6000.$

Example 26: Solve the following problem

Minimize $f(X) = x_1^2 + x_2^2 + x_3^2$

s.t. $g_1(X) = 2x_1 + x_2 - 5 \le 0$

$g_2(X) = x_1 + x_3 - 2 \le 0$

$g_3(X) = 1 - x_1 \le 0$

$g_4(X) = 2 - x_2 \le 0$

$g_5(X) = -x_3 \le 0.$

Solution: It is given that

Minimize $$f(X) = x_1^2 + x_2^2 + x_3^2 \quad \ldots (1)$$

s.t. $$\left.\begin{aligned} g_1(X) &= 2x_1 + x_2 - 5 \le 0 \\ g_2(X) &= x_1 + x_3 - 2 \le 0 \\ g_3(X) &= 1 - x_1 \le 0 \\ g_4(X) &= 2 - x_2 \le 0 \\ g_5(X) &= -x_3 \le 0 \end{aligned}\right\} \quad \ldots (2)$$

Now define a Lagrangian function by introducing slack variable y_j^2, we have

$$L(x_1, x_2, x_3; \lambda_1, \lambda_2, \lambda_3, \lambda_4, \lambda_5; y_1, y_2, y_3, y_4, y_5) = f(X) + \sum_{J=1}^{5} \lambda_j \, [g_j(X) + y_j^2] \quad \ldots (3)$$

From equations (1), (2) and (3), we have

$$L = x_1^2 + x_2^2 + x_3^2 + \lambda_1(2x_1 + x_2 - 5 + y_1^2) + \lambda_2(x_1 + x_3 - 2 + y_2^2) + \lambda_3(1 - x_1 + y_3^2) + \lambda_4(2 - x_2 + y_4^2) + \lambda_5(-x_3 + y_5^2) \quad \ldots (4)$$

The Kuhn-Tucker necessary conditions for minimization of L (with $g_j(X) \le 0$) are

$$\frac{\partial L}{\partial x_i} = 0; \; i = 1, 2, 3.$$

$$\lambda_j g_j = 0; \; i = 1, 2, 3, 4, 5.$$

$$\lambda_j \ge 0; \; j = 1, 2, 3, 4, 5.$$

Differentiate partially (4) and we get

$$\left.\begin{aligned} \frac{\partial L}{\partial x_1} &= 2x_1 + 2\lambda_1 + \lambda_2 - \lambda_3 = 0 \\ \frac{\partial L}{\partial x_2} &= 2x_2 + \lambda_1 - \lambda_4 = 0 \\ \frac{\partial L}{\partial x_3} &= 2x_3 + \lambda_2 - \lambda_5 = 0 \end{aligned}\right\} \quad \ldots (5)$$

$$\left.\begin{aligned} \lambda_1(2x_1 + x_2 - 5) &= 0 \\ \lambda_2(x_1 + x_3 - 2) &= 0 \\ \lambda_3(1 - x_1) &= 0 \\ \lambda_4(2 - x_2) &= 0 \\ \lambda_5(-x_3) &= 0 \end{aligned}\right\} \quad \ldots (6)$$

$$\lambda_j \ge 0; \; j = 1, 2, 3, 4, 5 \quad \ldots (7)$$

Let $\lambda_3 \neq 0$ and $\lambda_4 \neq 0$, we have

$$x_1 = 1, x_2 = 2, x_3 = 0 \text{ and } \lambda_5 = 0.$$

If we take $x_1 = 1, x_2 = 2, x_3 = 0$ then equation (6) is satisfied.

Now from equation (6) we have

$$2x_1 + x_2 - 5 \neq 0$$

and $$x_1 + x_2 - 2 \neq 0$$

so $$\lambda_1 = 0 \text{ and } \lambda_2 = 0.$$

From equation (5) we have

$$2 - \lambda_3 = 0 \Rightarrow \lambda_3 = 2$$
$$4 - \lambda_4 = 0 \Rightarrow \lambda_4 = 4$$
$$\lambda_2 - \lambda_5 = 0 \Rightarrow \lambda_2 = \lambda_5$$

Hence, the optimum solution is $x_1 = 1, x_2 = 2, x_3 = 0, \lambda_1 = \lambda_2 = \lambda_5 = 0, \lambda_3 = 2$ and $\lambda_4 = 4$ and minimize $f = (1)^2 + (2)^2 + (0)^2 = 5$.

EXERCISE 2.2

1. Find the extreme values of $x^2 + y^2 + z^2$ when $ax + by + cz = p$.
2. Find the maximum and minimum values of $u = x + y + z$ s.t. $\frac{a}{x} + \frac{b}{y} + \frac{c}{z} = 1$.
3. Find the extreme value of $x^p y^q z^r$ s.t. $\frac{a}{x} + \frac{b}{y} + \frac{c}{z} = 1$.
4. Find the volume of the largest rectangular parallel piped that can be inscribed in the ellipsoid $\frac{x^2}{a^2} + \frac{y^2}{b^2} + \frac{z^2}{c^2} = 1$.
5. Find the maxima and minima of $x^2 + y^2 + z^2$ s.t. $ax^2 + by^2 + cz^2 = 1$ and $lx + my + nz = 0$.
6. Using: (i) Direct method and (ii) Lagrange multiplier method to solve the following problem:
   Minimize $f = 2x_1^2 + 2x_2^2 + x_3^2 + 2x_1x_2 - 8x_1 - 6x_2 - 4x_3 + 2x_1x_3 + 9$
   s.t. $x_1 + x_2 + 2x_3 = 3$.
7. Using Lagrange multiplier method to solve the following problems:
   (i) Minimize $f(x_1, x_2) = 3x_1^2 + x_2^2 + 2x_1x_2 + 6x_1 + 2x_2$
   s.t. $2x_1 - x_2 = 4$.
   (ii) Optimize $f(x_1, x_2) = 6x_1x_2$
   s.t. $2x_1 + x_2 = 10$.
   Also state whether the stationary point is a maxima or minima.
   (iii) Minimize $f(x, y) = kx^{-1}y^{-2}$
   s.t. $g(x, y) = x^2 + y^2 - a^2 = 0$.
   (iv) Maximize $f(x_1, x_2) = 6x_1 + 8x_2 - x_1^2 - x_2^2$
   s.t. $4x_1 + 3x_2 = 16$
   and $3x_1 + 5x_2 = 15$

8. Using the Kuhn-Tucker conditions to solve the following problem:

   (i) Maximize $f(x_1, x_2) = -x_1^2 + x_2 + 2x_1$

   s.t. $2x_1 + 3x_2 \le 6$

   $2x_1 + x_2 \le 4$

   and $x_1, x_2 \ge 0$.

   (ii) Maximize $f(x_1, x_2) = 8x_1^2 + 2x_2^2$

   s.t. $x_1^2 + x_2^2 \le 9$

   and $x_1 \ge 0$.

   (iii) Minimize $f(x, y) = -x_1^2 - 2x_2^2 + 2x_1 + 3x_2$

   s.t. $x_1 + 3x_2 \le 6$

   $5x_1 + 2x_2 \le 10$

   and $x_1, x_2 \ge 0$.

   (iv) Maximize $f(x_1, x_2) = -x_1^2 - x_2^2 + x_1x_2 + 8x_1 + 4x_2$

   s.t. $2x_1 + 3x_2 \le 24$

   $-5x_1 + 12x_2 \le 24$

   $x_2 \le 5$.

ANSWERS

1. Minimum value of u is $\dfrac{P^2}{a^2+b^2+c^2}$.

2. $u = \left(\sqrt{a}+\sqrt{b}+\sqrt{c}\right)^2$.

3. $\dfrac{a^p\, b^q\, c^r}{p^p\, q^q\, r^r}(p+q+r)^{p+q+r}$.

4. $V = \dfrac{8abc}{3\sqrt{3}}$.

5. $\dfrac{l^2}{au-1}+\dfrac{m^2}{bu-1}+\dfrac{n^2}{cu-1}=0$.

6. $\left(\dfrac{4}{3},\dfrac{7}{9},\dfrac{4}{9}\right)$.

7. (i) $(1, -2)$ minimize $f = 5$. (ii) $\left(\dfrac{5}{2},\dfrac{5}{2}\right)$ maximize $f = 75$.

   (iii) $\left(\dfrac{a}{\sqrt{3}},\dfrac{\sqrt{2}}{\sqrt{3}}a\right)$. (iv) $\left(\dfrac{35}{11},\dfrac{12}{11}\right)$ maximize $f = 281$.

8. (i) $\left(\dfrac{2}{3},\dfrac{14}{9}\right)$ maximize $f = \dfrac{22}{9}$ (ii) $\left(2,\sqrt{5}\right)$ maximize $f = 42$

   (iii) $\left(1,\dfrac{3}{4}\right)$ minimize $f = \dfrac{17}{8}$ (iv) No solution.

3 Chapter

Linear Programming Problem – I (Formulation)

3.1 INTRODUCTION

A linear programming problem (LPP) is an important optimization technique. It forms the base for the development of many other optimization techniques. A linear programming problem is therefore, can be considered as the backbone of optimization techniques. This technique can be used to solve a constrained problem when it is required to find the maximum or minimum value of a linear function under certain constraints which are also linear in nature. In this chapter, we shall consider practical applications of a linear programming problem.

3.2 LINEAR PROGRAMMING PROBLEM

Let $f(X)$ and $g(X)$ be linear functions $i = 1, 2, 3, \ldots\ldots, m$, where $X \in E_n$ i.e., $X = (x_1, x_2, \ldots\ldots, x_n)$, then a problem is defined as:

Maximize or Minimize) $\quad Z = f(X) \quad \ldots(1)$

Subject to (s.t.)

$$g_i(X) \leq \text{ or } = \text{ or } \geq b,\ i = 1, 2, 3, \ldots\ldots, m \quad \ldots(2)$$

and $\quad X = (x_1, x_2, \ldots\ldots\ldots\ldots, x_n) \geq 0 \quad \ldots(3)$

is called a linear programming problem with m constraints in n variables. $f(X)$ in (1) is called objective function, $g_i(X)$ in (2) are called constraints and X_i's ≥ 0 are called non-negativity restrictions. In *matrix* notation LPP can be written as

$$\text{Maximize or Minimize) } Z = f(X) = C^T X$$

s.t.

$$AX \leq \text{ or } = \text{ or } \geq b$$

and $\quad X \geq 0$

where $\mathbf{C} = [c_1, c_2 \ldots.c_n]^T$, $\mathbf{b} = [b_1, b_2 \ldots.b_m]^T$, $\mathbf{X} = [x_1, x_2 \ldots\ldots x_n]^T$ and $\mathbf{A} = [a_{ij}]_{m \times n}$.

3.3 FORMULATION OF LINEAR PROGRAMMING MODEL

We follow the steps given below to formulate a linear programming problem:

Step 1. Identify the decision variables.

Step 2. Recognize the function to be maximized or minimized.

Step 3. Identify the restriction on decision variables and put them in the form of inequalities.

SOLVED EXAMPLES

Example 1: A firm manufactures 3 products *A*, *B* and *C*. The profits are Rs. 3, Rs. 2 and Rs. 4 respectively. The firm has 2 machines and below is the required processing time in minutes for each machine on each product:

Machine	*Product*		
	A	*B*	*C*
M_1	4	3	5
M_2	2	2	4

Machines M_1 and M_2 have 2,000 and 2,500 machine minutes respectively to work. The firm must manufacture 100 A's, 200 B's and 50 C's, but not more than 150 A's. Set up a linear programming problem to maximize profit.

Solution: Let x_1, x_2 and x_3 be the number of product *A*, *B* and *C* respectively produced by the firm.

The total time required to produce these products on machines M_1 and M_2 are $4x_1 + 3x_2 + 5x_3$ and $2x_1 + 2x_2 + 4x_3$ minutes respectively.

Since the time available with the machines M_1 and M_2 are 2,000 and 2,500 minutes respectively, so we have

$$4x_1 + 3x_2 + 5x_3 \le 2{,}000$$

and
$$2x_1 + 2x_2 + 4x_3 \le 2{,}500$$

Also the firm must manufacture 100 A's, 200 B's and 50 C's but not more than 150 A's, we have

$$100 \le x_1 \le 150,\ x_2 \ge 200 \text{ and } x_3 \ge 50.$$

Also the profit (in Rs.)
$$Z = 3x_1 + 2x_2 + 4x_3.$$

Hence, the problem formulated as linear programming problem is:

Maximize
$$Z = 3x_1 + 2x_2 + 4x_3.$$

Subject to
$$4x_1 + 3x_2 + 5x_3 \le 2{,}000,$$
$$2x_1 + 2x_2 + 4x_3 \le 2{,}500,$$
$$100 \le x_1 \le 150,$$
$$x_2 \ge 200,$$

and
$$x_3 \ge 50.$$

Example 2: A tyre factory produces three types of tyres T_1, T_2, T_3. Three different types of chemicals say C_1, C_2, C_3 are required for production. One T_1 tyre needs 2 units of C_1, 3 units of C_3; one T_2 tyre needs 3 units of C_1, 2 units of C_2 and 2 units of C_3; and one T_3 tyre needs 5 units of C_2 and 4 units of C_3. The factory has only a stock of 20 units of C_1, 25 units of C_2 and 30 units of C_3. Further the profit from the sale of one tyre T_1 is Rs. 6, one tyre T_2 is Rs 10 and of one tyre T_3 is Rs 8. Assuming that the factory can sell all that it produces, formulate a linear programming problem to minimize its profits.

Solution: Let the tyre factory produce x_1 tyres of type T_1, x_2 tyres of type T_2 and x_3 tyres of type T_3. Putting the given data in the following tabular form:

	Tyre T_1	*Tyre T_2*	*Tyre T_3*	*Total chemical available*
Chemical C_1	2	3	0	20
Chemical C_2	0	2	5	25
Chemical C_3	3	2	4	30
Profit in Rs. (per tyre)	6	10	8	

The total profit Z of tyre factory is

$$Z = 6\,x_1 + 10\,x_2 + 8\,x_3.$$

The total quantity of chemical C_1 is

$$2\,x_1 + 3\,x_2 \text{ units.}$$

Since the tyre factory has a stock of 20 units of chemical C_1, so we have

$$2\,x_1 + 3\,x_2 \le 20.$$

Similarly, we have

$$2\,x_2 + 5\,x_3 \le 25$$

and

$$3\,x_1 + 2\,x_2 + 4\,x_3 \le 30.$$

Since the units of the product produced cannot be negative, so we have

$$x_1, x_2, x_3 \ge 0.$$

Hence, the linear programming problem is formulated as:

$$\text{Maximize } Z = 6\,x_1 + 10\,x_2 + 8\,x_3$$

Subject to

$$2\,x_1 + 3\,x_2 \le 20,$$
$$2\,x_2 + 5\,x_3 \le 25,$$
$$3\,x_1 + 2\,x_2 + 4\,x_3 \le 30,$$

and

$$x_1, x_2, x_3 \ge 0.$$

Example 3: A firm can produce two products A and B during the given period of time. Each of these product required four different operations, viz. grinding, turning, assembling and testing. The requirement in hours per unit of manufacturing of these products is as given:

Product	*Grinding*	*Turning*	*Assembling*	*Testing*	*Profit in Rs.*
A	1	3	4	5	3
B	2	1	3	4	2
Available capacities (in hours)	30	60	200	200	

Formulate the problem as a linear programming model to maximize the profit assuming that the firm can sell all the items that it produces at the prevailing market price.

Solution: Let x_1 and x_2 be the units of type A and B respectively produced by the firm.

The total grinding time required to produce the products is

$$x_1 + 2\,x_2.$$

Since capacity of grinding is 30 hours per period, so we have

$$x_1 + 2\,x_2 \leq 30$$

Similarly, considering the times for turning, assembling and testing, we have

$$3x_1 + x_2 \leq 60,$$

$$4x_1 + 3x_2 \leq 200,$$

and $$5x_1 + 4x_2 \leq 200$$

Since the units of the product produced cannot be negative, so we have

$$x_1, x_2 \geq 0.$$

Also the profit (in Rs.) $\quad Z = 3x_1 + 2x_2.$

Hence, the problem formulated as linear programming problem is:

Maximize $\quad Z = 3x_1 + 2x_2$

Subject to

$$x_1 + 2\,x_2 \leq 30,$$

$$3\,x_1 + x_2 \leq 60,$$

$$4\,x_1 + 3\,x_2 \leq 200,$$

$$5\,x_1 + 4\,x_2 \leq 200,$$

and $$x_1, x_2 \geq 0.$$

Example 4: A factory produces two products P_1 and P_2. Each of the products P_1 requires 2 hours of moulding, 3 hours of grinding and 4 hrs for polishing and each of the products P_2 requires 4 hrs for moulding, 2 hrs of grinding and 2 hrs for polishing. The moulding machine can work for 20 hrs, grinding machine for 24 hrs and polishing machine is available for 13 hrs. The profit is Rs 5 per unit of P_1 and Rs 3 per unit for P_2. Assume the factory can sell all it produces, formulate the problem as a linear programming problem to maximize the profit.

Solution: Let the factory produce x_1 units of product P_1 and x_2 units of product P_2. We can put the given data in the following a tabular form:

Product	*Moulding*	*Grinding*	*Polishing*	*Profit*
P_1	2	3	4	5
P_2	4	2	2	3
Total time available	20	24	13	

The total profit on selling x_1 units of product P_1 and x_2 units of product P_2 is (in Rs)

$$Z = 5x_1 + 3x_2$$

The total time required for moulding is

$$2x_1 + 4x_2 \text{ hours.}$$

Since the moulding machine is available for not more than 20 hours, so we have

$$2x_1 + 4x_2 \leq 20.$$

Similarly, the restrictions for grinding and polishing machine, we have

$$3x_1 + 2x_2 \leq 24$$

and $$4x_1 + 2x_2 \leq 13.$$

Since the units of the product produced cannot be negative, so we have

$$x_1, x_2 \geq 0.$$

Hence, the problem formulated as linear programming problem is:

Maximize $$Z = 5x_1 + 3x_2$$

Subject to $$2x_1 + 4x_2 \leq 20,$$

$$3x_1 + 2x_2 \leq 24,$$

$$4x_1 + 2x_2 \leq 13,$$

and $$x_1, x_2 \geq 0.$$

Example 5: A firm manufactures two types of products *A* and *B* and sells them at a profit of Rs. 2 and Rs. 3 respectively. Each product is processed on two machines *M* and *N*. Type *A* requires one minute of processing time on *M* and two minutes on *N*, type *B* requires one minute on *M* and one minute on *N*. The machine *M* is available for not more than 6 hours 40 minutes while machine *N* is available for 10 hours during any working day. Formulate the problem as a linear programming problem.

Solution: Let the firm produce product of type *A* in x_1 units and product of type *B* in x_2 units. Putting the given data in a tabular form:

Machine	*Time of processing (minutes per unit)*		*Time available (minutes)*
	A	*B*	
M	1	1	400
N	2	1	600
Profit per unit (in Rs.)	2	3	

The total profit on selling x_1 units of type *A* and x_2 units of type *B* is (in Rs)

$$Z = 2\,x_1 + 3\,x_2.$$

The total time required on machine *M* is

$$x_1 + x_2 \text{ minutes.}$$

Since the machine *M* is available for not more than 6 hours and 40 minutes i.e., 400 minutes, so we have

$$x_1 + x_2 \leq 400.$$

Similarly, the total time required on machine *N* and availability of machine *N*, we have

$$2\,x_1 + x_2 \leq 600.$$

Since the units of the product produced cannot be negative, so we have

$$x_1, x_2 \geq 0.$$

Hence, the problem formulated as linear programming problem is:

Maximize $\quad Z = 2x_1 + 3x_2$

Subject to $\quad x_1 + x_2 \leq 400,$

$$2\,x_1 + x_2 \leq 600,$$

and $\quad x_1, x_2 \geq 0.$

Example 6: A diet is to contain at least 4000 units of carbohydrates, 500 units of fat and 300 units of protein. Two foods *A* and *B* are available. One unit of food *A* contains 10 units of carbohydrates, 20 units of fat and 15 units of protein. One unit of food *B* contains 25 units of carbohydrates, 10 units of fat and 20 units of protein. Formulate the problem as a linear programming problem so as to find the minimized cost for a diet that consists of a mixture of these two foods and also meets the minimum nutrition requirements.

Solution: Let the diet consist of x_1 and x_2 units of food *A* and *B* respectively. Putting the given data in a tabular form:

Food	*Carbohydrates*	*Fat*	*Protein*	*Cost in Rs.*
A	10	20	15	2
B	25	10	20	4
Requirement (least)	4000	500	300	

In a diet, carbohydrates, fat and protein units are

$$10\,x_1 + 25\,x_2,\ 20\,x_1 + 10\,x_2 \text{ and } 15\,x_1 + 20\,x_2.$$

While they are required at least 4000, 500 and 300 units respectively, so we have

$$10\,x_1 + 25\,x_2 \geq 4000$$

$$20\,x_1 + 10\,x_2 \geq 500$$

and $\quad 15\,x_1 + 20\,x_2 \geq 300.$

Since the daily diet cannot contain negative quantity of any food items, so we have

$$x_1, x_2 \geq 0.$$

Also the profit (in Rs.) $\quad Z = 2x_1 + 4x_2.$

Hence, the problem formulated as linear programming problem is:

Maximize $\quad Z = 2x_1 + 4x_2.$

Subject to $\quad 10\,x_1 + 25\,x_2 \geq 4000,$

$$20\,x_1 + 10\,x_2 \geq 500,$$

$$15\,x_1 + 20\,x_2 \geq 300,$$

and $\quad x_1, x_2 \geq 0.$

Example 7: According to the medical experts it is necessary for an adult to consume at least 75 grams of proteins, 85 grams of fats and 300 grams of carbohydrates daily. The following table gives the analysis of the food items readily available in the market with their respective costs.

Food Type	*Food value (in grams) per 100 grams*			*Cost in Rs. per kg*
	Proteins	*Fats*	*Carbohydrate*	
A	18	15	-	3
B	16	4	7	4
C	4	20	2.5	2
D	5	8	40	1.5
Minimum daily requirement	75	85	300	

Formulate a linear programming problem for an optimum diet.

Solution: Let x_1, x_2, x_3 and x_4 kg units daily diet consist of food *A*, *B*, *C* and *D* respectively.

The total amount of proteins in the daily diet is

$$(180\,x_1 + 160\,x_2 + 40\,x_3 + 50\,x_4)$$

Since the minimum daily requirement of proteins is 75 grams, so we have

$$180\,x_1 + 160\,x_2 + 40\,x_3 + 50\,x_4 \geq 75.$$

Similarly, considering the total amounts of fats and carbohydrates in the diet, we have

$$150\,x_1 + 40\,x_2 + 200\,x_3 + 80\,x_4 \geq 85$$

and $$70\,x_2 + 250\,x_3 + 400\,x_4 \geq 300.$$

Since the daily diet cannot contain negative quantity of any food items, so we have

$$x_1, x_2, x_3, x_4 \geq 0.$$

The total cost per day is

$$Z = 3\,x_1 + 4\,x_2 + 2\,x_3 + 1.5\,x_4$$

Hence, the problem formulated as linear programming problem is:

Minimize $$Z = 3\,x_1 + 4\,x_2 + 2\,x_3 + 1.5\,x_4$$

Subject to $$180\,x_1 + 160\,x_2 + 40\,x_3 + 50\,x_4 \geq 75,$$

$$150\,x_1 + 40\,x_2 + 200\,x_3 + 80\,x_4 \geq 85,$$

$$70\,x_2 + 250\,x_3 + 400\,x_4 \geq 300,$$

and $$x_1, x_2, x_3, x_4 \geq 0.$$

Example 8: A city hospital has the following minimal daily requirements for nurses:

Period	*Clock time (24 hr day)*	*Minimum number of nurses required*
1	6 a.m – 10 a.m	2
2	10 a.m – 2p.m	7
3	2 p.m – 6 p.m	15
4	6 p.m – 10 p.m	8
5	10 p.m – 2 a.m	20
6	2 a.m – 6 a.m	6

Nurses report to the hospital at the beginning of each period and work for 8 consecutive hours. The hospital wants to determine the minimum number of nurses available for each period. Formulate this as a linear programming problem by setting up appropriate constraints and objective function.

Solution: Let $x_1, x_2 \ldots\ldots\ldots x_6$ be the number of nurses reporting on duty at 6 a.m., 10 a.m., 2 p.m., 6 p.m., 10 p.m., and 2 a.m. respectively (i.e., at the beginning of alone six periods).

Since x_1 nurses reporting on duty at 6 a.m. will work upto 2 p.m. while x_2 nurses report on duty at 10 a.m. will work upto 6 p.m. Thus, the number of nurses available between 10 a.m. to 2 p.m. is $x_1 + x_2$. Similarly, $x_2 + x_3$, $x_3 + x_4$, $x_5 + x_6$, and $x_6 + x_1$ number of nurses remain available between 10 a.m. to 2 p.m., 2 p.m. to 6 p.m., 6 p.m. to 10 p.m., 10 p.m. to 2 a.m. and 2 a.m. to 6 a.m. respectively. Total number of nurses reporting on duty in 24 hours is

$$x_1 + x_2 + x_3 + x_4 + x_5 + x_6$$

Hence, the problem formulated as linear programming problem is:

Minimize $$z = x_1 + x_2 + x_3 + x_4 + x_5 + x_6$$

Subject to
$$x_1 + x_2 \geq 7,$$
$$x_2 + x_3 \geq 15,$$
$$x_3 + x_4 \geq 8,$$
$$x_4 + x_5 \geq 20,$$
$$x_5 + x_6 \geq 6,$$
$$x_6 + x_1 \geq 2,$$

and $$x_i \geq 0,\ i = 1, 2, 3 \ldots\ldots. 6.$$

EXERCISE 3.1

1. A goldsmith manufactures necklaces and bracelets. The total number of necklaces and bracelets that he can handle per day is at most 24. It takes one hour to make a bracelet and half an hour to make a necklace. It is assumed that he can work for a maximum of 16 hours a day. Further the profit on a bracelet is Rs. 300 and the profit on a necklace is Rs. 100. Formulate this problem as a linear programming problem so as to maximize the profit.
2. A furniture firm manufactures chairs and tables. Each requiring the use of three machines *A*, *B* and *C*. Production of one chair requires 2 hours on machine *A*, 1 hour on machine *B* and 1 hour on machine *C*. Each table requires 1 hour each on machines *A* and *B* and 3 hours on machine *C*. The profit realized by selling one chair is Rs. 30 while for a table is Rs. 60. The total time available per week on machine *A* is 70 hours, on machine *B* is 40 hours and on machine *C* is 90 hours. How many chairs and tables should be manufactured per week so as to maximize the profit? Formulate a mathematical model for the problem.
3. A factory produces two products *A* and *B*. To manufacture one unit of product *A*, a machine has to work 3/2 hours and a craftsman has to work for 2 hours. To manufacture one unit of product *B*, the machine has to work for 3 hours and the craftsman for one hour. In week the factory can available of 80 hours of machine time and 70 hours of craftsman time. The profit on the sale each unit of *A* and *B* is of Rs. 10 and Rs 8 respectively. If the manufacturer can sell all the items produced, how many of each should be produced to get maximum profit per week? Formulate the problem as a linear programming problem.
4. A toy company manufactures two types of dolls; an ordinary doll *A* and a deluxe doll *B*. Each type of doll *B* takes twice as long to produce one of type *A*. It is given that the

company would have time to make a maximum of 2000 dolls per day if it produces only the ordinary version. The supply of plastic is sufficient to produce 1500 dolls per day (both A and B combined). The deluxe version requires a fancy dress of which there are only 600 pieces per day available. If the company makes profit of Rs. 3 and Rs. 5 per doll respectively on doll A and doll B, formulate the problem as a linear programming problem to maximize the profit.

5. A resourceful home decorator manufactures two types of lamps say A and B. Both lamps go through two technicians, first a cutter, second a finisher. These products are as given:

Technicians types	*Time of processing (hours per piece)*		*Time requires (in hours)*
	Type A (x_1 lamps)	*Type B (x_2 lamps)*	
Cutter	2	1	104
Finisher	1	6	76
Profit in Rs.	6	11	

Assuming that he can sell all that he produces, how many of each type O lamps should be manufactured per month to obtain the best returns? Formulate a linear programming problem for this problem.

6. A manufactures of medicines is to prepare a production plan for medicines M and N. There are sufficient ingredients available to make 20,000 bottles of medicine M and 40,000 bottles medicine N. But there are only 45,000 bottles into which either of the medicine can be filled. Further it takes 3 hours to prepare enough material to fill 1000 bottles of medicine N and there are 66 hours available for this operation. This profit is Rs. 8 per bottle for medicine M and Rs. 7 per bottle for medicine N. Formulate this problem as a linear programming problem.

7. A company manufactures two kinds of leather purses, L_1 and L_2. L_1 is the high quality purse and L_2 is of lower quality. The rates of each of these purses L_1 and L_2 earn profit of Rs. 4 and Rs. 3 respectively. Each purse of type requires twice as much time as purse of type L_2, and if all purses are of type L_2. The company could make 1000 purses per day. The supply of leather is sufficient for only 800 purses per day (both L_1 and L_2 combined). Purse L_1 requires a fancy buckle and only 400 buckles per day are available. There are 700 buckles available for purse L_2. What should be the daily production of each type of purse to get the maximum profit? Formulate the problem as a linear programming problem.

8. A farmer has 100 acre farm. He can sell tomatoes, lettuce or radishes he can raise. The price he can obtain is Rs 1.00 per kg for radishes. The average yield per acre is 2000 kg of tomatoes, 3000 heads of lettuce and 1000 kg of radishes. Fertilizer is available at Rs 50 per kg and the amount required per acre is 100 kg each for tomatoes and lettuce and 50 kg for radishes. Labour required for sowing, cultivating and harvesting per acre is 5 man-days for tomatoes and radishes, and 6 man-days for lettuce. A total of 400 man-days of labour are available at Rs 20 per man-day. Formulate this problem as a linear programming problem to maximize the farmer's total profit.

9. A firm can produce three types of cloth, say: A, B and C. Three kinds of wool are require for it, say red wool, green wool and blue wool. One unit length of type A cloth needs 2 yards of red wool and 3 yards of blue wool; one unit length of type B cloth needs 3 yards of red wool, 2 yards of green wool and z yards for blue wool and one unit length of type

c cloth needs 5 yards of green wool and 4 yards of blue wool. The firm has only a stack of 8 yards of red wool, 10 yards for green wool and 15 yards for blue wool. It is assumed that the income obtained from one unit length of type *A* cloth is Rs. 3.00, of type *B* cloth is Rs 5.00 and of type *C* is Rs 4.00. Formulate this problem as a linear programming model to maximize the income from the finished cloth.

10. A dietician decides a certain minimum intake of vitamins *A*, *B* and *C* for family. The minimum daily needs of the vitamins *A*, *B*, *C* are 30, 20, 16 units respectively. For the supply of these, the dietician depends on two types of foods *X* and *Y*. The first one gives 7, 5, 2 units per gram of vitamins *A*, *B*, *C* respectively. The second one gives 2, 4, 8 units per gram of these vitamins respectively. The first food costs Rs 2 per gram and the second Re 1 per gram. How much grams of each food should the family buy everyday to keep the food expense at a minimum? Formulate a linear programming problem for this problem.
11. The objective of a diet problem is to ascertain the quantities of certain foods that should be eaten to meet certain nutritional requirement at a minimum cost. The consideration is limited to milk, green vegetables and eggs, and to vitamins *A*, *B*, *C*. The number of milligrams of each of these vitamins contained within a unit of each food and their daily minimum requirement along with the cost of each food is given in the table below:

Vitamin	*Litre of milk*	*Kg. of vegetables*	*Dozens of eggs*	*Minimum daily requirement*
A	1	1	10	1 mg
B	100	10	10	50 mg
C	10	100	10	10 mg
Cost in Rs.	20	10	8	

Formulate a linear programming problem for this diet problem.

12. A company has two grades inspector, 1 and 2, who are to be assigned for a quality control inspection. It is required that at least 2000 pieces BE inspected per to 8 hours day. Grade 1 inspector can check the pieces at the rate of 50/hour with the accuracy of 97% grade 2 inspector can check pieces at rate of 40/hour with the accuracy of 95%. The wage rate of grade 1 inspector is 4.50/hour and that of grade 2 is Rs 2.50/hour. Each time an error is made by the inspector, the cost to company is one rupee. The company has available for the inspection job, 10 grade 1 and 5 grade 2 inspectors. Formulate the problem to minimize the total cost of inspection.
13. The manager of an oil refinery must decide on the optimal mix of 2 possible blending processes of which the inputs and outputs per production run are as follows:

Process	*Input*		*Output*	
	Crude A	*Crude B*	*Gasoline X*	*Gasoline Y*
1	6	4	6	9
2	5	6	5	5

The maximum amounts available of crudes *A* and *B* are 500 units and 400 units respectively. Market demand shows that at least 300 units of gasoline *X* and 260 units of gasoline *Y* must be produced. The profits per production run from process 1 and 2 are Rs 40 and Rs 50 respectively. Formulate the linear programming problem for maximizing the profit.

14. A manufacturer produces three models (1, 2 and 3) of a certain product. He uses two types of raw materials (*A* and *B*) of which 4000 and 6000 units respectively are available. The raw material requirements per unit of the three models are given below:

Raw material	*Requirement per unit of given model*		
	1	2	3
A	2	3	5
B	4	2	7

The labour time for each unit of model 1 is twice that of model 2 and three times of that model 3. The entire model force of the factory can the equivalent of 2500 units of model 1. A market survey indicates that the minimum demand of three models are 500, 500 and 350 units respectively. However, the ratio of the number of units produced must be equal to 3:2:5. Assume that the profit of per unit of model 1, 2 and 3 are rupees 60, 40 and 100 respectively. Formulate the problem as a linear programming problem, in order to determine the number of units of each product which will maximize profit.

ANSWERS

1. Maximize $Z = 100\, x_1 + 300\, x_2$
   Subject to $x_1 + 2x_2 \leq 32$
   $2x_1 + x_2 \leq 24$
   and the non-negative restrictions $x_1 \geq 0,\, x_2 \geq 0$.
2. Maximize $Z = 10\, x_1 + 8\, x_2$,
   Subject to $1.5\, x_1 + 3\, x_2 \leq 80$,
   $2\, x_1 + x_2 \leq 70$,
   and the non-negative restrictions $x_1 \geq 0,\, x_2 \geq 0$.
3. Maximize $Z = 30\, x_1 + 60\, x_2$,
   Subject to $2\, x_1 + x_2 \leq 70$,
   $x_1 + x_2 \leq 40$,
   $x_1 + 3x_2 \leq 90$,
   and the non-negative restrictions $x_1 \geq 0,\, x_2 \geq 0$.
4. Maximize $Z = 3x_1 + 5x_2$
   Subject to $x_1 + 2x_2 \leq 2000$,
   $x_1 + x_2 \leq 1500$,
   $x_2 \leq 600$,
   and the non-negative restrictions $x_1 \geq 0,\, x_2 \geq 0$.
5. Maximize $Z = 6x_1 + 11x_2$
   Subject to $2x_1 + x_2 \leq 104$,
   $x_1 + 2x_2 \leq 76$,
   and the non-negative restrictions $x_1 \geq 0,\, x_2 \geq 0$.

6. Maximize Z = $8000x_1 + 7000x_2$,
   Subject to $x_1 + x_2 \leq 45$,
   $3x_1 + x_2 \leq 66$,
   $x_1 \leq 20$,
   $x_2 \leq 40$
   and the non-negative restrictions $x_1 \geq 0$, $x_2 \geq 0$.
7. Maximize $Z = 4x_1 + 3x_2$
   Subject to $2x_1 + x_2 \leq 1000$,
   $x_1 + x_2 \leq 800$,
   $x_1 \leq 400$,
   $x_2 \leq 700$,
   and the non-negative restrictions $x_1 \geq 0$, $x_2 \geq 0$.
8. Maximize Z = $1850x_1 + 2080x_2 + 1875x_3$
   Subject to $x_1 + x_2 + x_3 \leq 100$,
   $5x_1 + 6x_2 + 5x_3 \leq 400$,
   and the non-negative restrictions $x_1 \geq 0$, $x_2 \geq 0$, $x_3 \geq 0$.
9. Maximize $Z = 3x_1 + 5x_2 + 4x_3$
   Subject to $2x_1 + 3x_2 \leq 8$,
   $2x_2 + 5x_3 \leq 10$,
   $3x_1 + 2x_2 + 4x_3 \leq 15$
   and the non-negative restrictions $x_1 \geq 0$, $x_2 \geq 0$, $x_3 \geq 0$.
10. Minimize $Z = 2x_1 + x_2$
    Subject to $7x_1 + 2x_2 \geq 30$,
    $5x_1 + 4x_2 \geq 20$,
    $2x_1 + 8x_2 \geq 16$,
    and the non-negative restrictions $x_1 \geq 0$, $x_2 \geq 0$.
11. Minimize $Z = 20x_1 + 10x_2 + 8x_3$
    Subject to $x_1 + x_2 + 10x_3 \geq 1$,
    $100x_1 + 10x_2 + 10x_3 \geq 50$,
    $10x_1 + 100x_2 + 10x_3 \geq 10$,
    and the non-negative restrictions $x_1 \geq 0$, $x_2 \geq 0$, $x_3 \geq 0$.

12 Minimize $Z = 48x_1 + 36x_2$
   Subject to $5x_1 + 4x_2 \geq 25$,
   $x_1 \leq 10$,
   $x_2 \leq 5$,
   and the non-negative restrictions $x_1 \geq 0$, $x_2 \geq 0$.

13. Maximize $Z = 40x_1 + 50x_2$
    Subject to $6x_1 + 5x_2 \leq 500$,
    $4x_1 + 6x_2 \leq 400$,
    $6x_1 + 5x_2 \geq 300$,
    $9x_1 + 5x_2 \geq 260$,
    and the non-negative restrictions $x_1 \geq 0$, $x_2 \geq 0$.

14. Maximize $Z = 60x_1 + 40x_2 + 100x_3$

Subject to $2x_1 + 3x_2 + 5x_3 \leq 4000,$

$4x_1 + 2x_2 + 7x_3 \leq 6000,$

$6x_1 + 3x_2 + 2x_3 \leq 15{,}000,$

$2x_1 = 3x_2,\ 5x_2 = 2x_3,$

and the non-negative restrictions $x_1 \geq 500,\ x_2 \geq 500,\ x_3 \geq 375.$

4 Chapter
Linear Programming Problem – II (Graphical Method)

4.1 INTRODUCTION

In this chapter our objective is to solve the linear programming problem by sketching the graph of constraints. The graphical method is applicable when linear programming problem is given in two variables and constraints are in the form of inequality. An inequality constraint in two variables represents the set of points on one side of the line including the points on the line obtained by replacing the inequality sign of the constraint by the equality sign. This set is known as the set of feasible solutions and denoted by S_F. In graphical method we find the point (s) of S_F where the objective function is optimum.

4.2 IMPORTANT TERMINOLOGY

4.2.1 Neighbourhood

The neighbourhood of a point X_O in E_n is the set S, $S = \{X \in E_n: |X - X_O| < \varepsilon\}$

(i) It is an interval if $n = 1$.
(ii) Interior point of a circle if $n = 2$.

4.2.2 Bounded Set

A set $S \subset E_n$ is said to be bounded from below if there exists a $Y \in E_n$ such that $Y \leq X$ for all $X \in E_n$ i.e., every component y_j of $Y = (y_1, y_2, \ldots\ldots, y_n)^T$ is less than or equal to corresponding component x_j of $X = (x_1, x_2, \ldots\ldots, x_n)^T$. A set $S \subset E_n$ is said to be bounded from above if there exists a $Y \in E_n$ such that $Y \geq X$ for all X in E_n. A set which is both bounded from above and below is called a bounded set.

4.2.3 Convex Linear Combination (CLC)

A point X is called CLC of m points $X_1, X_2, \ldots\ldots, X_m$ if there exist scalars $\lambda_i \geq 0$, $i = 1, 2, \ldots..,$ m such that $X = \lambda_1 X_1 + \lambda_2 X_2 + \ldots.. + \lambda_m X_m$, $\lambda_1 + \lambda_2 + \ldots.. + \lambda_m = 1$.

For $m = 2$, $X = \lambda_1 X_1 + \lambda_2 X_2$, λ_i's ≥ 0 and $\lambda_1 + \lambda_2 = 1$ or $X = \lambda X_2 + (1 - \lambda)X_1$, $0 \leq \lambda \leq 1$ and $\lambda \geq 0$.

4.2.4 Hyper Plane

The set $S = \{X \in E_n: C^T X = \alpha\}$, where $C = (c_1, c_2, \ldots\ldots, c_n)^T$, $X = (x_1, x_2, \ldots\ldots, x_n)^T$ and α a constant is called a hyper plane in E_n.

4.2.5 Half Space

The set $S = \{(x_1, x_2, \ldots\ldots, x_n): c_1x_1 + c_2x_2 + \ldots. + c_nx_n \leq \text{ or } \geq \alpha\}$ is called half space in E_n.

4.2.6 Convex Set

A set $S \subset E_n$ is said to be a convex set if CLC of every two points in S belongs to S. That is, for all X_1 and X_2 in S the point $X = (1 - \lambda)X_1 + \lambda X_2$, $0 \leq \lambda \leq 1$ is again in S. A point X of a convex set S which cannot be expressed as a CLC of two distinct points X_1 and X_2 of S is known as a vertex of S.

Example 1: Let us consider the set $S = \{(x, y): 0 \leq x \leq 2, 0 \leq y \leq 3\}$

Solution: (i) S is bounded below by (0, 0)
(ii) S is bounded above by (2, 3).

Thus S which is the set of all points on and inside the rectangle $0 \leq x \leq 2$, $0 \leq y \leq 3$ is a bounded set. Points on rectangle are called boundary points of the set and points lying inside the set are called interior points. Points in E_2 outside the rectangle are called exterior points of S.

(iii) Set S is a closed set as it contains all its boundary points.

Example 2: Let us consider the set $S = \{(x, y): x^2 + y^2 < 2\}$

Solution: (i) S is an open set
(ii) S is a bounded set.

Example 3: Find CLC of $X_1 = (1, 3)^T$ and $X_2 = (3, 1)^T$. Does

(i) (3, 2) belong to CLC of X_1 and X_2?
(ii) (2, 2) belong to CLC of X_1 and X_2?

Solution: Let X be the CLC of X_1 and X_2, then

$$\begin{aligned} X &= (1 - \lambda)X_1 + \lambda X_2 \\ &= (1-\lambda)\begin{pmatrix}1\\3\end{pmatrix} + \lambda\begin{pmatrix}3\\1\end{pmatrix} \\ &= \begin{pmatrix}1-\lambda+3\lambda\\3(1-\lambda)+\lambda\end{pmatrix} \\ &= \begin{pmatrix}1+2\lambda\\3-2\lambda\end{pmatrix},\ 0 \leq \lambda \leq 1 \end{aligned}$$

Thus, required CLC is $S = \{X: X = (1 + 2\lambda, 3 - 2\lambda)^T, 0 \le \lambda \le 1\}$. If we take $\lambda = 1/2$, then $X = (2, 2)^T$ is the middle point of the line segment joining X_1 and X_2.

(i) If possible let $(3, 2) \in S$, then $3 = 1 + 2\lambda \Rightarrow \lambda = 1$ and $3 - 2\lambda = 3 - 2 = 1 \neq 2$. Thus (3, 2) does not belong to the CLC of X_1 and X_2.

(ii) For $(2, 2) \in S$, then

$$2 = 1 + 2\lambda \quad \Rightarrow \quad \lambda = 1/2 \qquad \ldots(1)$$

and

$$2 = 3 - 2\lambda \quad \Rightarrow \quad \lambda = 1/2 \qquad \ldots(2)$$

Thus both (1) and (2) are satisfied for $\lambda = 1/2$ which also satisfies the condition $0 \le \lambda \le 1$.

Example 4: Let $X_1 = (-1, 2, 0)$, $X_2 = (1, 2, 0)$ and $X_3 = (1, 0, 2)$ be three points in E_3.

(i) Is $X = (1, 1, 1)$ a CLC of X_1, X_2 and X_3?

(ii) Is $X = (1/2, - 1/2, 3/4)$ a CLC of X_1, X_2 and X_3?

Solution:

(i) For $X = (1, 1, 1)$ to be a CLC of X_1, X_2 and X_3.

$$(1, 1, 1) = \lambda_1(-1, 2, 0) + \lambda_2 (1, 2, 0) + \lambda_3 (1, 0, 2),\ \lambda_i\text{'s} \ge 0 \text{ and } \lambda_1 + \lambda_2 + \lambda_3 = 1$$
$$= (-\lambda_1 + \lambda_2 + \lambda_3, 2\lambda_1 + 2\lambda_2, 2\lambda_3)$$

$$\therefore \quad 1 = -\lambda_1 + \lambda_2 + \lambda_3 \qquad \ldots(1)$$
$$1 = 2\lambda_1 + 2\lambda_2 \qquad \ldots(2)$$
$$1 = 2\lambda_3 \qquad \ldots(3)$$

Solving (1), (2) and (3), we get

$$\lambda_1 = 1/4, \lambda_2 = 1/4 \text{ and } \lambda_3 = 1/2.$$

Thus, λ_i's ≥ 0, $\forall\, i = 1, 2, 3$ and $\lambda_1 + \lambda_2 + \lambda_3 = (1/4) + (1/4) + (1/2) = 1$. Hence (1, 1, 1) is a CLC of X_1, X_2 and X_3.

(ii) For $X = (1/2, - 1/2, 3/4)$ to be a CLC of X_1, X_2 and X_3.

$$(1/2, -1/2, 3/4) = \lambda_1(-1, 2, 0) + \lambda_2 (1, 2, 0) + \lambda_3 (1, 0, 2),\ \lambda_i\text{'s} \ge 0 \text{ and } \lambda_1 + \lambda_2 + \lambda_3 = 1$$
$$= (-\lambda_1 + \lambda_2 + \lambda_3, 2\lambda_1 + 2\lambda_2, 2\lambda_3)$$

$$\therefore \quad 1/2 = -\lambda_1 + \lambda_2 + \lambda_3 \qquad \ldots(4)$$
$$-1/2 = 2\lambda_1 + 2\lambda_2 \qquad \ldots(5)$$
$$3/4 = 2\lambda_3 \qquad \ldots(6)$$

Solving (4), (5) and (6), we get

$\lambda_1 = -3/16$, $\lambda_2 = -1/16$ and $\lambda_3 = 3/8$. Therefore $X = (1/2, - 1/2, 3/4)$ is not a CLC of X_1, X_2 and X_3.

Theorem 1: A closed hyper plane in E_n is a convex set.

Proof: Let us consider a closed hyper plane in E_n as

$S = \{X \in E_n : C^TX = \alpha\}$, where $C = (c_1, c_2, \ldots.., c_n)^T$, $X = (x_1, x_2, \ldots.., x_n)^T$ and α a constant. S is closed as all its boundary points are in S.

To prove that it is convex, consider any two points $X_1, X_2 \in S$. Then $X_1, X_2 \in S$ imply that

$$C^TX_1 = \alpha \text{ and } C^TX_2 = \alpha \qquad \ldots(1)$$

Let $X = (1 - \lambda)X_1 + \lambda X_2$, $0 \le \lambda \le 1$ be a CLC of X_1 and X_2. Then

$$C^TX = C^T\{(1 - \lambda)X_1 + \lambda X_2\}$$
$$= (1 - \lambda)\, C^T X_1 + \lambda\, C^TX_2$$
$$= (1 - \lambda)\, \alpha + \lambda\alpha = \alpha$$

Thus $C^T X = \alpha$ and hence $X \in S$. Hence, CLC of any two points $X_1, X_2 \in S$ is again in S. Therefore S is a convex set.

Theorem 2: The intersection of finite number of convex sets in E_n is again a convex set.

Proof: Suppose $S_1, S_2, \ldots\ldots, S_n$ are n convex sets in E_n. Let $S = S_1 \cap S_2 \cap S_3, \ldots\ldots\ldots\ldots, S_n$. Let $X_1, X_2 \in S$ then $X_1, X_2 \in S_i$, $i = 1, 2, \ldots\ldots\ldots, n$. Let $X = (1 - \lambda)X_1 + \lambda X_2$, $0 \le \lambda \le 1$ be a CLC of X_1 and X_2. As $X_1, X_2 \in S_i$ and S_i, $i = 1, 2, \ldots\ldots n$ are convex sets, $X \in S_i$, $i = 1, 2, \ldots\ldots n$ and hence $X \in S$. Thus, we have proved that CLC of any two points in S is again in S and hence S, which is the intersection of S_i, $i = 1, 2, \ldots\ldots, n$ is a convex set.

Now we state the following theorems without proof:

Theorem 3: The optimum of the objective function $Z = f(X)$ of a linear programming problem always occurs at some vertex of the set of feasible solutions (S_F) provided S_F is non-empty and bounded.

Theorem 4: Let S be a set which is non-empty closed, convex and bounded, then it has at least one vertex and every point of S is a convex linear combination (CLC) of its vertices.

4.3 GRAPHICAL METHOD

In graphical method we use: (i) Corner Point Method and (ii) Iso Profit Method (Iso Value Line Method) to find the optimal solution.

4.4 CORNER POINT METHOD

In this method we use the following steps:

(i) Sketch the set of feasible solutions (S_F)
(ii) Obtain the co-ordinates of its corners (vertices)
(iii) Find the value of the objective function at all these vertices
(iv) Write the solution where objective function is optimum.

Prefer the corner point method to find the graphical optimal solution when the set of feasible solution (S_F) is bounded.

4.5 ISO PROFIT METHOD (ISO VALUE LINE METHOD)

In this method first we draw some parallel lines with different values of the objective function moving in the direction in which the objective function is tending to increase or decrease according as the problem is maximization or minimization. Continuing to draw the parallel lines in desired direction until the last line (i) passes through a vertex of S_F where optimum solution is obtained and any further movement of the parallel line pushes it out of S_F. (ii) Or the parallel line surpasses the last vertex in the desired direction and still containing the points of S_F with no bound on objective function value, then the solution is unbounded.

Example 5: Solve the following LPP by graphical method

Maximize $\qquad Z = 5x_1 + 3x_2$

s.t. $3x_1 + 5x_2 \le 15$
$5x_1 + 2x_2 \le 10$
and $x_1, x_2 \ge 0$

Solution: The given linear programming problem is

Maximize $Z = 5x_1 + 3x_2$ ...(1)

s.t. $3x_1 + 5x_2 \le 15$ ...(2)

$5x_1 + 2x_2 \le 10$ ...(3)

and $x_1, x_2 \ge 0.$ ...(4)

Now the problem is to find a point $X_0 = (x_0, y_0)$ in the first quadrant satisfying (2) and (3) where $Z = 5x_1 + 3x_2$ has maximum value. We sketch the inequalities in (2), (3) and (4) to get the region OABC (set of feasible solutions) as shown in the given figure. We shall find the solution of this linear programming problem using both the Iso profit method and the corner point method.

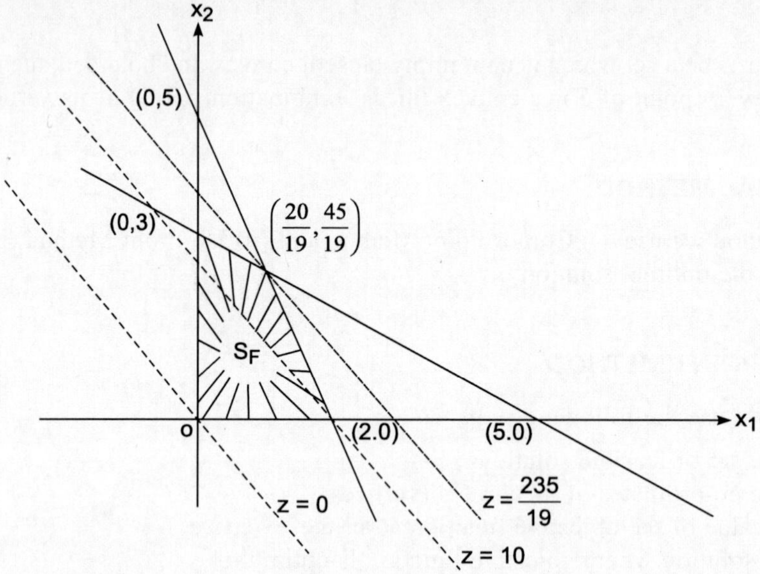

Let us first solve the problem using Iso profit method (Iso value line method). Let us consider the parallel lines that correspond to different values of Z say $Z = 0, 2$. We see that Z increases as the movement of the parallel lines is in the direction of arrow as shown in the figure. The last such line passes through the point $B(20/19, 45/19)$ with $Z = 5(20/19) + 3(45/19) = 235/19$.

Thus, optimum solution is Max $Z = 235/19$, when $x_1 = 20/19$, $x_2 = 45/19$.

Also we find the optimum solution using corner point method. For that we find the coordinates of points O, A, B and C which are vertices (corner points) of S_F and calculate Z values.

Points	*Z*
$O(0, 0)$	$Z = 5(0) + 3(0) = 0$
$A(2, 0)$	$Z = 5(2) + 3(0) = 10$
$B(20/19, 45/19)$	$Z = 5(20/19) + 3(45/19) = 235/19$ (Maximum)
$C(0, 3)$	$Z = 5(0) + 3(3) = 9$

Thus, optimum solution is Max $Z = 235/19$, when $x_1 = 20/19$, $x_2 = 45/19$.

Example 6: Solve the following LPP by graphical method

Max $$Z = 4x_1 + 5x_2$$

s.t. $$x_1 + x_2 \geq 1,$$
$$-2x_1 + x_2 \leq 1$$
$$4x_1 - 2x_2 \leq 1$$

and $$x_1, x_2 \geq 0.$$

Solution: The given linear programming problem is

Max $$Z = 4x_1 + 5x_2 \quad \text{...(1)}$$

s.t. $$x_1 + x_2 \geq 1 \quad \text{...(2)}$$
$$-2x_1 + x_2 \leq 1 \quad \text{...(3)}$$
$$4x_1 - 2x_2 \leq 1 \quad \text{...(4)}$$

and $$x_1, x_2 \geq 0. \quad \text{....(5)}$$

The region of feasible solutions (S_F) which is the intersection of the sets, represented by (2), (3), (4) and (5) is shown in the following figure.

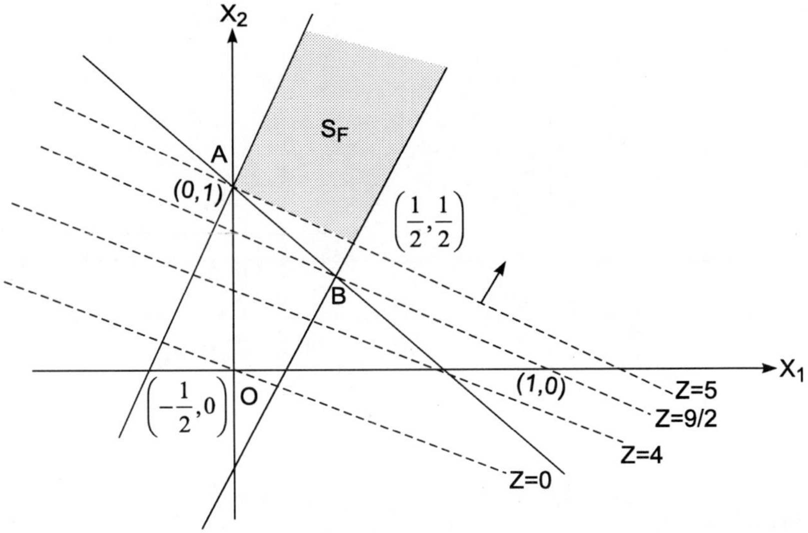

Here S_F is unbounded. We use Iso-profit technique to find the solution of the problem. For this we draw the parallel lines across S_F for various values of Z say $Z = 0, 4, 5$ etc. This shows that Z increases in the direction as shown by arrow ($\rightarrow$) having value 5 at vertex A which is the last vertex in the direction of increasing Z.

The further parallel movement of the line still passes across the feasible region and thus there exists no upper bound on Z-value. Thus the given problem has unbounded solution.

Example 7: Solve the following LPP by graphical method

Max $$Z = 4x_1 + 4x_2$$

s.t. $$x_1 + 2x_2 \leq 10,$$
$$6x_1 + 6x_2 \leq 36$$
$$x_1 \leq 6$$

and $$x_1, x_2 \geq 0.$$

Solution: The given linear programming problem is

$$\text{Max} \quad Z = 4x_1 + 4x_2 \quad \ldots(1)$$

$$\text{s.t.} \quad x_1 + 2x_2 \leq 10 \quad \ldots(2)$$

$$6x_1 + 6x_2 \leq 36 \quad \ldots(3)$$

$$x_1 \leq 6 \quad \ldots(4)$$

$$\text{and} \quad x_1, x_2 \geq 0. \quad \ldots(5)$$

The region of feasible solutions (S_F) which is the intersection of the sets, represented by (2), (3), (4) and (5) is shown in figure.

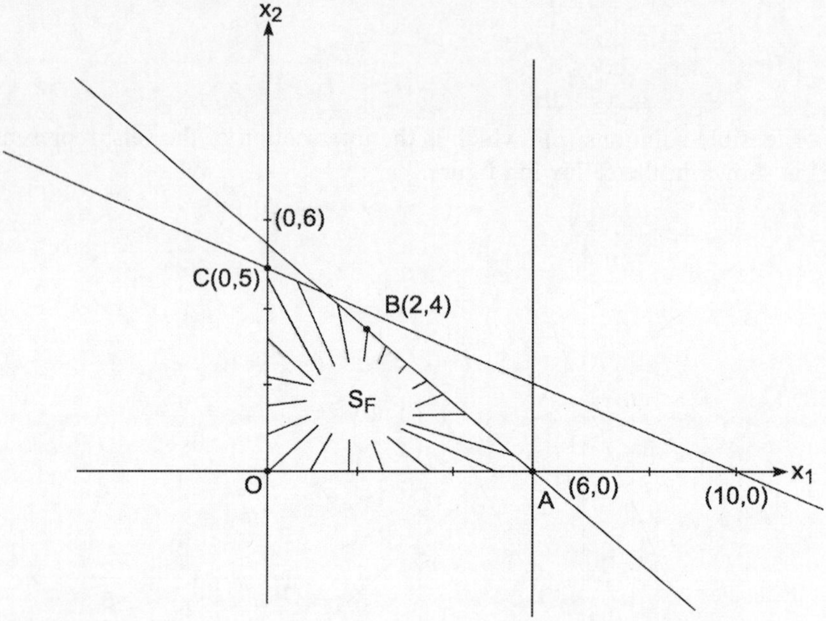

We shall find the optimum solution by corner point method. We find the co-ordinates of the vertices O, A, B and C of S_F as $O(0, 0)$, $A(6, 0)$, $B(2, 4)$ and $C(0, 5)$. We calculate the value of Z at these points as:

At vertex O, $x_1 = 0$, $x_2 = 0$ $\Rightarrow$ $Z = 4(0) + 4(0) = 0$.

At vertex A, $x_1 = 6$, $x_2 = 0$ $\Rightarrow$ $Z = 4(6) + 4(0) = 24$.

At vertex B, $x_1 = 2$, $x_2 = 4$ $\Rightarrow$ $Z = 4(2) + 4(4) = 24$.

At vertex C, $x_1 = 0$, $x_2 = 5$ $\Rightarrow$ $Z = 4(0) + 4(5) = 20$.

Thus, (i) Max $Z = 24$ when $x_1 = 6$, $x_2 = 0$ and also (ii) Max $Z = 24$ when $x_1 = 2$, $x_2 = 4$ is another optimal solution, called alternate optimal shown of the problem. In fact optimum solution will occur at every point on the line segment joining points BA.

EXERCISE 4.1

1. Identify the vertices of the following sets
   (i) $S_1 = \{x \in R: 0 \leq x < 2\}$
   (ii) $S_2 = \{x \in R: 0 < x \leq 2\}$

(iii) $S_3 = \{x \in R: 0 \le x \le 2\}$
(iv) $S_4 = \{(x, y) \in E_2: x^2 + y^2 \le 2\}$
(v) $S_5 = \{X: X = (1 - \lambda)X_1 + \lambda X_2, 0 \le \lambda \le 1, \lambda \le 0\}$

2. Prove that the set $S = \{X \in E_2 \,|X - X_0| \le 5, X_0$ is a fixed point in $E_2\}$ is a convex set.
3. Prove that intersection of two convex sets is also a convex set.
4. Prove that a closed half space in E_n is a convex set.

Solve the following linear programming problem by graphical method

5. Max $Z = -2x_1 + 2x_2$
s.t. $-x_1 + x_2 \le 1$
$x_1 + x_2 \le 2$
and $x_1, x_2 \ge 0.$
6. Maximize $Z = -x_1 + 2x_2$
s.t. $-x_1 + x_2 \le 1,$
$x_1 + x_2 \le 2$
and $x_1, x_2 \ge 0.$
7. Min $Z = -5x_1 + 10x_2$
s.t. $-x_1 + x_2 \le 1$
$x_1 + x_2 \le 2$
and $x_1, x_2 \ge 0.$
8. Max $Z = 10x_1 - 15x_2$
s.t. $-x_1 + x_2 \le 1$
$x_1 - 4x_2 \le 2$
and $x_1, x_2 \ge 0.$
9. Min $Z = 20x_1 + 10x_2$
s.t. $x_1 + 2x_2 \le 40$
$3x_1 + x_2 \ge 30$
$4x_1 + 3x_2 \ge 60$
and $x_1, x_2 \ge 60.$
10. Max $Z = 3x_1 + 5x_2$
s.t. $x_1 + 2x_2 \le 2000$
$x_1 + x_2 \le 1500$
$x_2 \le 600.$
and $x_1, x_2 \ge 0$
11. Max $Z = x_1 + x_2$
s.t. $x_1 - x_2 \ge 0$
$-3x_1 + x_2 \ge 3$
and $x_1, x_2 \ge 0$
12. Max $Z = x_1 + x_2$
s.t. $5x_1 + 10x_2 \le 50$
$x_1 + x_2 \ge 1$
$x_2 \le 4$
and $x_1, x_2 \ge 0$
13. Max $Z = 6x_1 - 2x_2$
s.t. $2x_1 - x_2 \le 2$

$$x_1 \le 3$$

and $\quad x_1, x_2 \ge 0.$

14. Max $\quad Z = 3x_1 + 2x_2$

 s.t. $\quad x_1 - x_2 \le 1$

 $\quad x_1 + x_2 \ge 3$

 and $\quad x_1, x_2 \ge 0$

15. Max $\quad Z = 100x_1 + 300x_2$

 s.t. $\quad x_1 + 2x_2 \le 32$

 $\quad x_1 + x_2 \le 24$

 and $\quad x_1, x_2 \ge 0.$

16. A company makes two kinds of leather belts. Belt *A* is a high quality belt, and belt *B* is of lower quality. The respective profits are Rs 4.00 and 3.00 per belt. Each belt of type *A* requires twice as the much time as the belt of type *B* and if all belts were of type *B*, the company could make 1000 per day. The supply of leather is sufficient for only 800 belts per day (Both *A* and *B* combined). Belt *A* requires a fancy buckle and only 400 per day are available. There are only 700 buckles a day available for belt *B*. Determine the optimal product mix.
17. A soft drink plant has two bottling machines *A* and *B*. It produces and sells 8-ounce and 16-ounce bottles. The following data is available:

Machine	8-ounce	16-ounce
A	100/minute	40/minute
B	60/minute	75/minute

 The machines can be run 8 hrs. per day, 5 days per week. Weekly production of the drinks cannot exceed 3,00,000 ounces and the market can absorb 25,000 eight-ounce bottles and 7000 sixteen-ounce bottles per week. Profit on these bottles is 15 paise and 25 paise per bottle respectively. The planner wishes to maximize his profit subject to all the production and marketing restrictions. Formulate it as a linear programming problem and solve graphically.

ANSWERS

1. (i) $x = 0$.
 (ii) $x = 2$.
 (iii) $x = 0$ and 2.
 (iv) All boundary points of a circle $x^2 + y^2 = 2$.
 (v) $X = X_1$.
5. Infinite number of solution.
6. $x_1 = 1/2$, $x_2 = 3/2$, max $Z = 5/2$.
7. $x_1 = 2$, $x_2 = 0$, Min $Z = -10$.
8. Unbounded solution.
9. $x_1 = 6$, $x_2 = 12$, Min $Z = 240$.
10. $x_1 = 1000$, $x_2 = 500$, Max $Z = 5500$.
11. No feasible solution.

12. Infinite number of solution.
13. $x_1 = 3$, $x_2 = 4$, Max $Z = 10$.
14. Unbounded solution.
15. $x_1 = 0$, $x_2 = 16$, Max $Z = 4800$.
16. $x_1 = 200$, $x_2 = 600$, Max $Z = 2600$.
17. $x_1 = 25000$, $x_2 = 6250$, Max $Z = 5312.50$.

5 Chapter

Simplex and Revised Simplex Method

5.1 INTRODUCTION

Our objective in this chapter is to solve linear programming problems in general, when graphical method fails to yield the solution. To accomplish this task we have to develop the theory for a method called the simplex method. Any linear programming problem can be solved using the simplex method. Using simplex method, we shall be in a position to investigate whether the given LPP has a unique solution, no solution, more than one solution (alternate optimal solution) or unbounded solution. Revised simplex method has also been discussed.

5.2 STANDARD (CANONICAL) FORM OF A LINEAR PROGRAMMING PROBLEM

Mathematical programming problem,

Maximize or Minimize $Z = f(X)$ ...(1)

Subject to (s. t.)

$$g_i(X) \leq \text{ or } = \text{ or } \geq b,\ i = 1, 2, 3, \ldots\ldots, m \qquad \ldots(2)$$

and $$X = (x_1, x_2, \ldots\ldots\ldots, x_n)^T \geq 0 \qquad \ldots(3)$$

$$b = (b_1, b_2, \ldots\ldots, b_m)^T$$

is a linear programming problem if the objective function $f(X)$ in (1) and all the constraints in (2) are linear with non-negativity conditions in (3) are satisfied. The standard form of LPP given by equations (1), (2) and (3) is

Maximize or Minimize $Z = C^T X$

s. t. $AX = b$

where $\mathbf{C} = [c_1, c_2 \ldots\ldots c_n]^T$, $\mathbf{b} = [b_1, b_2 \ldots\ldots b_m]^T$, $\mathbf{X} = [x_1, x_2 \ldots\ldots x_n]^T \geq 0$ and $\mathbf{A} = [a_{ij}]_{m \times n}$.

Thus, a linear programming problem is said to be in its standard (canonical) form if all the following four conditions are satisfied:

(i) The objective function must be either maximization or minimization.

(ii) All constraints should hold with equality sign.
(iii) Right hand side of all constraints is non-negative.
(iv) All the variables involved in the problems are non-negative.

Every linear programming problem can be converted into standard form.

Example 1: Write the standard (canonical) form of the following LPP:

$$\text{Max.} \quad Z = 2x_1 + 3x_2 - 2x_3 \qquad \ldots(1)$$
$$\text{s.t.} \quad x_1 + 5x_2 - x_3 \leq 15 \qquad \ldots(2)$$
$$2x_1 - x_2 + x_3 \leq -10 \qquad \ldots(3)$$
$$x_1 + 2x_3 = 10 \qquad \ldots(4)$$
$$x_1, x_2 \geq 0, \ x_3 \text{ unrestricted in sign.}$$

Solution: Adding x_4 in equation (2), we get

$$x_1 + 5x_2 - x_3 + x_4 = 15 \qquad \ldots(5)$$

Here the variable x_4 added to '$\leq$' type constraints to get equality sign is called slack variable.
Multiply (3) by (–1) to get

$$-2x_1 + x_2 - x_3 \geq 10 \qquad \ldots(6)$$

Now subtract x_5 from LHS of equation (6) to get

$$-2x_1 + x_2 - x_3 - x_5 = 10 \qquad \ldots(7)$$

Here the variable x_5 subtracted to '$\geq$' type constraints to get equality sign is called surplus variable.

As x_3 is unrestricted in sign it can be written as difference of two positive variables x_{31} and x_{32}, i.e.,

$$x_3 = x_{31} - x_{32}, \text{ where } x_{31}, x_{32} \geq 0.$$

Now replacing x_3 by $x_{31} - x_{32}$ in equations (1), (4), (5) and (7) we get

$$\text{Max.} \quad Z = 2x_1 + 3x_2 - 2(x_{31} - x_{32})$$
$$\text{s.t.} \quad x_1 + 5x_2 - (x_{31} - x_{32}) + x_4 = 15$$
$$-2x_1 + x_2 - (x_{31} - x_{32}) - x_5 = 10$$
$$x_1 + 2(x_{31} - x_{32}) = 10$$
$$x_1, x_2, x_{31}, x_{32} \geq 0.$$

Here all the above four conditions of standard (canonical) form are satisfied and hence is the required standard form.

5.3 SIMPLEX METHOD

Let us consider the linear programming problem in the standard form as:

$$\text{Max } Z = f(X) = c_1x_1 + c_2x_2 + c_3x_3 + \ldots\ldots\ldots + c_nx_n$$
$$a_{11}x_1 + a_{12}x_2 + \ldots\ldots\ldots + a_{1n}x_1 = b_1$$
$$a_{21}x_1 + a_{22}x_2 + \ldots\ldots\ldots + a_{2n}x_2 = b_2$$
$$\cdots \quad \cdots \quad \cdots \quad \cdots \quad \cdots$$
$$\cdots \quad \cdots \quad \cdots \quad \cdots \quad \cdots$$
$$\cdots \quad \cdots \quad \cdots \quad \cdots \quad \cdots$$
$$\cdots \quad \cdots \quad \cdots \quad \cdots \quad \cdots$$
$$a_{m1}x_1 + a_{m2}x_2 + \ldots\ldots\ldots\ldots a_{mn}x_n = b_m$$

$$X = (x_1, x_2, \ldots\ldots\ldots, x_n)^T \geq 0 \qquad \ldots(1)$$

or
$$\text{Max } Z = C^T X$$

s.t.
$$AX = b$$

where $\mathbf{X} = [x_1, x_2 \ldots\ldots x_n]^T \geq 0$, $\mathbf{C} = [c_1, c_2 \ldots\ldots c_n]^T$, $\mathbf{b} = [b_1, b_2 \ldots\ldots b_m]^T$ and $\mathbf{A} = [a_{ij}]_{m \times n} = (A_1, A_2, \ldots, A_n)$.

Now we consider one solution by putting $(n - m)$ unknowns say $x_{m+1}, x_{m+2}, \ldots\ldots\ldots\ldots\ldots\ldots$, x_n equal to zero in equation (1), the resulting system after putting in matrix form can be written as

$$\text{Max } Z = C_B^T X_B \qquad \ldots(2)$$

s.t.
$$BX_B = b \qquad \ldots(3)$$

$$X_B \geq 0 \qquad \ldots(4)$$

Here components of X_B i.e., $x_1, x_2, \ldots, x_m$ are called *basic variables* and variables x_{m+1}, $x_{m+2}, \ldots\ldots\ldots\ldots\ldots, x_n$ which have been equated to zero are called *non-basic variables*.

where $B = (A_1, A_2, \ldots\ldots\ldots, A_m) = \begin{pmatrix} a_{11} & a_{12} & \cdots & a_{1m} \\ a_{21} & a_{22} & \cdots & a_{2m} \\ \cdots & \cdots & \cdots & \cdots \\ \cdots & \cdots & \cdots & \cdots \\ a_{m1} & a_{m2} & \cdots & a_{mm} \end{pmatrix}$, $X_B = \begin{pmatrix} x_1 \\ x_2 \\ \cdots \\ \cdots \\ x_m \end{pmatrix}$, $b = \begin{pmatrix} b_1 \\ b_2 \\ \cdots \\ \cdots \\ b_m \end{pmatrix}$ $C_B = \begin{pmatrix} c_1 \\ c_2 \\ \cdots \\ \cdots \\ c_m \end{pmatrix}$

If B is non-singular, then from (3) we have

$$X_B = B^{-1}\, b \qquad \ldots(5)$$

B is called *basis matrix* and X_B is called corresponding *basic feasible solution*. Columns of B from a basis for E_m. Thus any of A_j, $j = m + 1, m + 2, \ldots\ldots\ldots, n$; can be written as

$$\begin{aligned}
A_{m+1} &= \alpha_1^{m+1} A_1 + \alpha_2^{m+1} A_2 + \ldots\ldots\ldots\ldots\ldots\ldots + \alpha_m^{m+1} A_m \\
A_{m+2} &= \alpha_1^{m+2} A_1 + \alpha_2^{m+2} A_2 + \ldots\ldots\ldots\ldots\ldots\ldots + \alpha_m^{m+2} A_m \\
\cdots &\quad \cdots \qquad \cdots \qquad \cdots \qquad \cdots \\
\cdots &\quad \cdots \qquad \cdots \qquad \cdots \qquad \cdots \\
A_n &= \alpha_1^{n} A_1 + \alpha_2^{n} A_2 + \ldots\ldots\ldots\ldots\ldots\ldots\ldots\ldots + \alpha_m^{n} A_m
\end{aligned} \qquad \ldots(6)$$

Combining all the equations in (6), we can write

$$A_j = \alpha_1^j A_1 + \alpha_2^j A_2 + \ldots\ldots\ldots\ldots\ldots\ldots\ldots\ldots\ldots + \alpha_m^j A_m$$

$$= (A_1, A_2, \ldots\ldots, A_m) \begin{pmatrix} \alpha_1^j \\ \alpha_2^j \\ \cdots \\ \cdots \\ \alpha_m^j \end{pmatrix} = B\, \alpha^j$$

or
$$\alpha^j = B^{-1} A_j \qquad \ldots(7)$$

where $\alpha^j = \begin{pmatrix} \alpha_1^j \\ \alpha_2^j \\ \cdots \\ \cdots \\ \alpha_m^j \end{pmatrix} = (\alpha_1^j, \alpha_2^j, \ldots\ldots\ldots\ \alpha_m^j)^T$ and $(\alpha_1^j, \alpha_2^j, \ldots\ldots\ldots\ \alpha_m^j$ are m components of vector α^j.

In the simplex method we move from one BFS to another one, maintaining the feasibility of the solution in such a way that the value of the objective function improves. For this we should have another set of m basic variables which can be obtained by replacing one of the existing basic variables by some non-basic variable. This amounts the same thing as replacing a basic column of the existing basic matrix B by one of the non-basic columns $A_j, j = m + 1, m + 2, \ldots\ldots\ldots, n$, to have a new basis matrix say B_1.

$$B = (A_1, A_2, \ldots, A_{r-1}, \overset{\uparrow}{A_r}, A_{r+1}, \ldots, A_m),\ A = (A_1, A_2, \ldots, A_r, \ldots, A_m, A_{m+1} \ldots, \overset{\downarrow}{A_j}, \ldots, A_m)$$

Thus, $B_1 = (A_1, A_2, \ldots\ldots, A_{r-1}, A_j, A_{r+1}, \ldots\ldots, A_m)$, where A_r of B is replaced by A_j of A, A_j is one of $A_{m+1}, A_{m+2}, \ldots\ldots\ldots, A_n$. Since B is the basis matrix in E_m. Therefore $A_j = \alpha_1^j A_1 + \alpha_2^j A_2 + \ldots\ldots\ldots + \alpha_r^j A_r + \ldots\ldots\ldots + \alpha_m^j A_m$. If $\alpha_r^j \neq 0$, then we get

$$A_r = \frac{1}{\alpha_r^j}(-\alpha_1^j A_1 - \alpha_2^j A_2 - \ldots\ldots - \alpha^j_{r-1} A_{r-1} + A_j - \alpha^j_{r+1} A_r \ldots\ldots - \alpha_m^j A_m) \qquad \ldots(8)$$

The original basis matrix B gave

$$x_1A_1 + x_2A_2 + \ldots\ldots + x_rA_r + \ldots\ldots + x_mA_m = b \qquad \ldots(9)$$

Putting the value of A_r from (8) in equation (9), we get

$$\left(x_1 - \alpha_1^j \frac{x_r}{\alpha_r^j}\right)A_1 + \ldots\ldots + \left(x_{r-1} - \alpha_{r-1}^j \frac{x_r}{\alpha_r^j}\right)A_{r-1} + \frac{x_r}{\alpha_r^j}A_j + \left(x_{r+1} - \alpha_{r+1}^j \frac{x_r}{\alpha_r^j}\right)A_{r+1} + \ldots\ldots + \left(x_m - \alpha_m^j \frac{x_r}{\alpha_r^j}\right)A_m = b \qquad \ldots(10)$$

Equation (10) shows that when A_r leaves and A_j enters then the new basic solution is $x_1', x_2', \ldots, x_r', \ldots\ldots, x_m'$, where

$$x_1' = \left(x_1 - \alpha_1^j \frac{x_r}{\alpha_r^j}\right), \ldots\ldots\ldots, x_{r-1}' = \left(x_{r-1} - \alpha_{r-1}^j \frac{x_r}{\alpha_r^j}\right), x_r' = \frac{x_r}{\alpha_r^j}, x_{r+1}' = \left(x_{r+1} - \alpha_{r+1}^j \frac{x_r}{\alpha_r^j}\right) \ldots\ldots\ldots x_m' = \left(x_m - \alpha_m^j \frac{x_r}{\alpha_r^j}\right) \qquad \ldots(11)$$

With m basic variables and n-m non-basic variable satisfying the system $AX = b$, we should choose the leaving column A_r in such a way that the basic solution given in equation (11) becomes feasible and hence a BFS. To ensure it, we choose r such that

$$\frac{x_{B_r}}{\alpha_r^j} = \min_i \left(\frac{x_{B_i}}{\alpha_i^j}, \alpha_i^j \succ 0\right) = \theta_j\ (say) \qquad \ldots(12)$$

This determines r, i.e. which column A_r (variable x_r) should leave in order that column A_j (i.e., variable x_j) enters the basis (j is yet to be determined). This is leaving variable rule which ensures the feasibility of the new basic solution. Thus equation (11) along with (12) is a BFS. The new value Z' of the objective function corresponding to this BFS is

$$Z' = c_1x_1' + c_2x_2' + c_3x_3' + \ldots\ldots + c_jx_r' + \ldots\ldots + c_mx_m' \qquad \ldots(13)$$

From equations (2) and (13), on simplification we get

$$Z' - Z = \theta_j(c_j - Z_j), \quad \ldots(14)$$

where $Z_j = c_1\alpha_1^{\,j} + c_2\alpha_2^{\,j} + \ldots\ldots + c_{r-1}\alpha^j_{r-1} + c_r\alpha^j_r + c_{r+1}\alpha^j_{r+1} + \ldots\ldots + c_m\alpha_m^{\,j} = C_B^{\,T}\alpha^j$ ...(15)

and $$\alpha^j = (\alpha_1^j, \alpha_2^j, \ldots\ldots, \alpha_m^j)^T \quad \ldots(16)$$

For maximization problem $\theta_j(c_j - Z_j)$ must be most positive. Thus for maximum change we select that column A_j of coefficient matrix A for which $\theta_j\,(c_j - Z_j)$ is most positive. But this rule is not convenient as we have to calculate θ_j for all non-basic columns. To simplify this we select that A_j as entering column for which $c_j - Z_j$ is most positive. The rule based on $c_j - Z_j$ value may increase one or two simplex iterations but saves a lot of calculations later on. Thus the variable x_j with most positive $c_j - Z_j$ value should enter the basis for the improvement of the value of the objective function. We may call it the entering variable rule.

5.4 CONDITION FOR OPTIMAL SOLUTION

Let us consider a linear programming problem

$$\text{Max } Z = C^T X \quad \ldots(1)$$

s.t. $$AX = b, \quad \ldots(2)$$

$$X \geq 0. \quad \ldots(3)$$

where $A = [a_{ij}]_{m \times n}$, $X = (x_1, x_2, \ldots, x_n)^T$, $C = (c_1, c_2, \ldots, c_n)^T$, $b = (b_1, b_2, \ldots, b_m)^T$.

If for a BFS X_B, of an *LPP* in (1), $c_j - Z_j \leq 0$ for all j; $j = 1, 2, \ldots\ldots, n$, then X_B is the optimal basic feasible solution of (1).

5.5 ALGORITHM FOR SIMPLEX METHOD

Step 1: We express the given linear programming problem in standard (canonical) form to use slack, surplus and artificial variables.

Step 2: Get the starting basic feasible solution.

Step 3: After recognizing starting BFS, put the problem in simplex format.
Where, $\Delta_j = c_j - Z_j = c_j - C_B^T\alpha^j$, Δ_j is called relative cost of variable x_j. In particular $\Delta_1 = c_1 - Z_1 = c_1 - C_B^{\,T}\alpha^1$. In C_B column we write the costs of basic variables. In basic variable column all the basic variables are written in definite order. In this column x_i stands for the ith basic variable. Its column will have i^{th} entry 1 & all other entries zero. In X_B column, we write the values of basic variables i.e., $X_B = B^{-1}\,b$. In the initial simplex table α^j is equal to j^{th} column A_j of the coefficient matrix A (i.e., $\alpha^j = A_j$).

Step 4: Start simplex iterations if the conditions:

(i) Identify matrix is the sub-matrix of the coefficient matrix.
(ii) For basic variables $c_j - Z_j$ value is zero.

are satisfied. If all Δ_j are non-positive (i.e., $\Delta_j \leq 0$) then the starting simplex table gives the optimal solution. If not then go to step 5.

Costs of all variables involved in the problem →				c_1 c_2 c_3 ... ... c_j ... ... c_n	Minimum Ratio
All variables involved in the problem →				x_1 x_2 x_3 ... ... x_j ... ... x_n	$\frac{X_{B_i}}{\alpha_i^j}$
Table Number	Costs of Basic Variables $\mathbf{C_B}$	Basic Variables	Value of Basic Variables $\mathbf{X_B}$	α^1 α^2 α^3 ... ... α^j ... ... α^n	
	C_{B1}	x_1	X_{B1}	α_1^1 α_1^2 α_1^3 ... ... α_1^j ... ... α_1^n	...
	C_{B2}	x_2	X_{B2}	α_2^1 α_2^2 α_2^3 ... ... α_2^j ... ... α_2^n	...
1	...	...	...	... ...	...
	...	...	...	... ...	...
	...	...	...	... ...	...
	C_{Bm}	x_m	X_{Bm}	α_m^1 α_m^2 α_m^3 ... ... α_m^j ... ... α_m^n	...
	$c_j - Z_j$			Δ_1 Δ_2 Δ_3 ... ... Δ_j ... ... Δ_m	

Step 5: To get new BFS, one of the non-basic variables should enter the basis and one of existing basic variable should leave the basis. In simplex method, first we decide about the entering variable.

Entering variable rule: The variable x_j with most positive $c_j - Z_j$ value row will enter the basis.

Leaving variable rule: Entering variable rule fixes the column α^j. Now for leaving variable calculate the ratio of basic variable values x_{Bi} to corresponding entries α_i^j i.e., $x_{Bi}/\alpha_i{}^j$, $\alpha_i{}^j \geq 0$ and take the minimum ratio. Let this minimum ratio be for $i = r^{\text{th}}$ then x_r will leave the basis. Thus, when x_j enters, x_r leaves the basis.

Step 6: Mark the entry at the intersection of entering variable column and leaving variable row and enrectangle it. This entry is called the pivot (key) entry and the row containing the pivot entry is called the pivot row. To prepare the next simplex table bring 1 in the position corresponding to the pivot entry and zeros above and below this 1, using elementary row operations with the help of the pivot row.

Step 7: Repeat step 5 and 6 until $\Delta_j > 0$ and one of the following cases arise:

Case I: If all the entries in the entering column α^j are non-positive but the entry Δ_j is positive then the solution of the problem is **unbounded.**

Case II: If $\Delta_j \leq 0$ for all j, then the solution is optimum. Table thus obtained will be the optimal table and simplex iterations will stop. Further in the optimum table for some non-basic variable x_j if

(i) $\Delta_j = 0$, then there exists an **alternate optimal solution.**

(ii) $\Delta_j < 0$ for all j, optimal solution is **unique.**

Note. We put the starting basic variables in all problems towards the end in the simplex table.

SOLVED EXAMPLES

Example 2: Solve the following linear programming problem using the simplex method:

Max. $Z = 2x_1 - x_2 + x_3$

s.t. $3x_1 + x_2 + x_3 \le 60$ ...(1)

$x_1 - x_2 + 2x_3 \le 10$ ...(2)

$-x_1 - x_2 + x_3 \ge -20$ ...(3)

$x_1, x_2, x_3 \ge 0$

Solution: In the given problem multiplying (3) with (–1) to make its RHS positive and adding slack variables x_4, x_5 and x_6 to get

Max. $Z = 2x_1 - x_2 + x_3 + 0.x_4 + 0.x_5 + 0.x_6$

s.t. $3x_1 + x_2 + x_3 + x_4 + 0.x_5 + 0.x_6 = 60$

$x_1 - x_2 + 2x_3 + 0.x_4 + x_5 + 0.x_6 = 10$

$x_1 + x_2 - x_3 + 0.x_4 + 0.x_5 + x_6 = 20$

$x_1, x_2, x_3, x_4, x_5, x_6 \ge 0.$

This is the canonical (standard) form of the given LPP. Here $A \begin{bmatrix} 3 & 1 & 1 & 1 & 0 & 0 \\ 1 & -1 & 2 & 0 & 1 & 0 \\ 1 & 1 & -1 & 0 & 0 & 1 \end{bmatrix}$.

Column of x_4, x_5 and x_6 give the identity matrix $\begin{bmatrix} 1 & 0 & 0 \\ 0 & 1 & 0 \\ 0 & 0 & 1 \end{bmatrix}$. Thus x_4, x_5 and x_6 are the starting basic variables and $\begin{bmatrix} 1 & 0 & 0 \\ 0 & 1 & 0 \\ 0 & 0 & 1 \end{bmatrix}$ is the basis matrix. For simplex BFS is obtained by putting $x_1 = x_2 = x_3 = 0$ so that $x_4 = 60$, $x_5 = 10$ and $x_6 = 20$ will be the basic feasible solution with x_4, x_5 and x_6 as basic variables and x_1, x_2 and x_3.

		Cost →		2	–1	1	0	0	0	Minimum ratio
		Variable →		x_1	x_2	x_3	x_4	x_5	x_6	
Table number	$\mathbf{C_B}$	Basic variable	$\mathbf{X_B}$	α^1	α^2	α^3	α^4	α^5	α^6	$\dfrac{X_{B_i}}{\alpha_i^j}$
	0	x_4	60	3	1	1	1	0	0	
1	0	x_5	10	1	–1	2	0	1	0	
	0	x_6	20	1	1	–1	0	0	1	
		$c_j - Z_j$		2	–1	1	0	0	0	

In the following table we see that x_1 is the variable with most positive entry $\Delta_1 = 2$ in $c_j - Z_j$ row and hence will enter the basis. This fixes α^1 column. For leaving variable the minimum ratio $x_{Bi}/\alpha_i^{\,j}$ is for x_5 row and hence x_5 will leave the basis. Thus when x_1 enters, x_5 leaves the basis. Continue simplex iterations.

Cost →				2	−1	1	0	0	0	Minimum ratio
Variable →				x_1	x_2	x_3	x_4	x_5	x_6	
Table number	$\mathbf{C_B}$	Basic variable	$\mathbf{X_B}$	α^1	α^2	α^3	α^4	α^5	α^6	$\frac{X_{B_i}}{\alpha_i^j}$
1	0	x_4	60	3	1	1	1	0	0	$\frac{60}{3} = 20$
	0	x_5	10	[1]	−1	2	0	1	0	$\frac{10}{1} = 20$ →
	0	x_6	20	1	1	−1	0	0	1	$\frac{20}{1} = 20$
	$c_j - Z_j$			2↑	−1	1	0	0	0	
2	0	x_4	30	0	4	−5	1	−3	0	$\frac{30}{4} = \frac{15}{2}$
	2	x_1	10	1	−1	2	0	1	0	---
	0	x_6	10	0	[2]	−3	0	−1	1	$\frac{10}{2} = 5$ →
	$c_j - Z_j$			0	1↑	−3	0	−2	0	
3	0	x_4	10	0	0	1	1	−1	−2	
	2	x_1	15	1	0	1/2	0	1/2	1/2	
	−1	x_2	5	0	1	−3/2	0	−1/2	1/2	
	$c_j - Z_j$			0	0	−3/2	0	−3/2	−1/2	

In above table 3, $\Delta_j \le 0$ for all j, therefore solution is optimal. Hence, optimal solution is $x_1 = 15$, $x_2 = 5$, $x_3 = 0$ and Max. $Z = 25$.

Example 3: Solve the following linear programming problem using simplex method

Min $Z = x_1 - 3x_2 + 2x_3$

s.t.

$$3x_1 - x_2 + 3x_3 \le 7$$
$$-2x_1 + 4x_2 \le 12$$
$$-4x_1 + 3x_2 + 8x_3 \le 10$$

and $x_1, x_2, x_3 \ge 0$

Solution: Converting the minimization problem into maximization, we have

$$\text{Max } Z' = -x_1 + 3x_2 - 2x_3, \text{ where } Z' = -Z$$

The problem can be written in the standard form as:

$$\text{Max} \quad Z' = -x_1 + 3x_2 - 2x_3$$

$$\text{s.t.} \quad 3x_1 - x_2 + 3x_3 + x_4 = 7$$
$$-2x_1 + 4x_2 + x_5 = 12$$
$$-4x_1 + 3x_2 + 8x_3 + x_6 = 10$$

and $x_1, x_2, x_3, x_4, x_5, x_6 \geq 0$, where x_4, x_5 and x_6 are slack variables. Here $\begin{bmatrix} 3 & -1 & 3 & 1 & 0 & 0 \\ -2 & 4 & 0 & 0 & 1 & 0 \\ -4 & 3 & 8 & 0 & 0 & 1 \end{bmatrix}$.

Column of x_4, x_5 and x_6 give the identity matrix $\begin{bmatrix} 1 & 0 & 0 \\ 0 & 1 & 0 \\ 0 & 0 & 1 \end{bmatrix}$. Thus x_4, x_5 and x_6 are the starting basic variables and $\begin{bmatrix} 1 & 0 & 0 \\ 0 & 1 & 0 \\ 0 & 0 & 1 \end{bmatrix}$ is the basis matrix. By taking x_1, x_2 and x_3 as non-basic variables (i.e., $x_1 = 0$, $x_2 = 0$ and $x_3 = 0$), we write the corresponding basic solution with $x_4 = 7$, $x_5 = 12$ and $x_6 = 10$ in the form of the following simplex table and there after use simplex iterations till to get optimal solution:

	Cost →			-1	3	-2	0	0	0	Minimum ratio
	Variable →			x_1	x_2	x_3	x_4	x_5	x_6	
Table number	$\mathbf{C_B}$	Basic variable	$\mathbf{X_B}$	α^1	α^2	α^3	α^4	α^5	α^6	$\dfrac{X_{B_i}}{\alpha_i^j}$
	0	x_4	7	3	-1	3	1	0	0	-----
1	0	x_5	12	-2	**[4]**	0	0	1	0	$\dfrac{12}{4} = 3 \rightarrow$
	0	x_6	10	-4	3	8	0	0	1	$\dfrac{10}{3} = 3.33$
		$c_j - Z_j$		-1	$3\uparrow$	-2	0	0	0	
	0	x_4	10	**[5/2]**	0	3	1	1/4	0	$\dfrac{10}{5/2} = 4 \rightarrow$
2	3	x_2	3	$-1/2$	1	0	0	1/4	0	-----
	0	x_6	1	$-5/2$	0	8	0	$-3/4$	1	-----
		$c_j - Z_j$		$1/2\uparrow$	0	-2	0	$-3/4$	0	

3	–1	x_1	4	1	0	6/5	2/5	1/10	0	
	3	x_2	5	0	1	3/5	1/5	3/10	0	
	0	x_6	11	0	0	11	1	–1/2	1	
	$c_j - Z_j$			0	0	–13/5	–1/5	–4/5	0	

In above Table 3, $\Delta_j \leq 0$ for all j, therefore solution is optimal. Thus optimal solution is $x_1 = 4$, $x_2 = 5$, $x_3 = 0$ and Min $Z = -Z' = -11$.

Example 4: Solve the following linear programming problem using the simplex method

Max. $\quad Z = 3x_1 + 2x_2$

Subject to (s.t.) $\quad x_1 + x_2 \leq 4$

$x_1 - x_2 \leq 2$

and $\quad x_1, x_2 \geq 0$

Solution: The problem can be written in the standard form as:

Max. $\quad Z = 3x_1 + 2x_2 + 0.x_3 + 0.x_4$

Subject to $\quad x_1 + x_2 + x_3 + 0.x_4 = 4$

$x_1 - x_2 + 0.x_3 + x_4 = 2$

and $x_1, x_2, x_3, x_4 \geq 0$, x_3 and x_4 are slack variables.

Here $A = \begin{bmatrix} 1 & 1 & 1 & 0 \\ 1 & -1 & 0 & 1 \end{bmatrix}$. Columns of x_3 and x_4 give the identity matrix $\begin{bmatrix} 1 & 0 \\ 0 & 1 \end{bmatrix}$. Thus x_3 and x_4 are the starting basic variables and $\begin{bmatrix} 1 & 0 \\ 0 & 1 \end{bmatrix}$ is the basis matrix. By taking x_1 and x_2 as non-basic variables (i.e., $x_1 = 0$ and $x_2 = 0$), we write the corresponding basic solution with $x_3 = 4$ and $x_4 = 2$ in the form of the following simplex table and there after use simplex iterations till to get optimal solution:

Cost →				3	2	0	0	Minimum ratio
Variable →				x_1	x_2	x_3	x_4	
Table number	$\mathbf{C_B}$	Basic variable	$\mathbf{X_B}$	α^1	α^2	α^3	α^4	$\dfrac{X_{B_i}}{\alpha_i^j}$
1	0	x_3	4	1	1	1	0	$\frac{4}{1} = 4$
	0	x_4	2	[1]	–1	0	1	$\frac{2}{1} = 2 \rightarrow$
	$c_j - Z_j$			3↑	2	0	0	

2	0	x_3	2	0	[2]	1	– 1	$\frac{2}{2} = 1 \rightarrow$
	3	x_1	2	1	– 1	0	1	---
	$c_j - Z_j$			0	5↑	0	– 3	
3	2	x_2	1	0	1	1/2	–1/2	
	3	x_1	3	1	0	1/2	1/2	
	$c_j - Z_j$			0	0	– 5/2	– 1/2	

In the above Table 3, $\Delta_j \leq 0$ for all j, therefore the solution is optimal. Hence, optimal solution is $x_1 = 3$, $x_2 = 1$ and Max. $Z = 11$.

Example 5: Solve the following linear programming problem using the simplex method

Max. $Z = 4x_1 + 6x_2$

s.t. $2x_1 + 3x_2 \leq 6$

$x_1 - x_2 \leq 1$

and $x_1, x_2 \geq 0$

Solution: The problem can be written in the standard form as:

Max. $Z = 4x_1 + 6x_2 + 0.x_3 + 0.x_4$

Subject to $2x_1 + 3x_2 + x_3 + 0.x_4 = 6$

$x_1 - x_2 + 0.x_3 + x_4 = 1$

and $x_1, x_2, x_3, x_4 \geq 0$, x_3 and x_4 are slack variables. Here $A = \begin{bmatrix} 2 & 3 & 1 & 0 \\ 1 & -1 & 0 & 1 \end{bmatrix}$. Columns of x_3 and x_4 give the identity matrix $\begin{bmatrix} 1 & 0 \\ 0 & 1 \end{bmatrix}$. Thus x_3 and x_4 are the starting basic variables and $\begin{bmatrix} 1 & 0 \\ 0 & 1 \end{bmatrix}$ is the basis matrix. By taking x_1 and x_2 as non-basic variables (i.e., $x_1 = 0$ and $x_2 = 0$), we write the corresponding basic solution with $x_3 = 6$ and $x_4 = 1$ in the form of the following simplex table and thereafter use simplex iterations till we get optimal solution:

Cost →				4	6	0	0	Minimum ratio $\frac{X_{B_i}}{\alpha_i^j}$
Variable →				x_1	x_2	x_3	x_4	
Table number	$\mathbf{C_B}$	Basic variable	$\mathbf{X_B}$	α^1	α^2	α^3	α^4	
1	0	x_3	6	2	[3]	1	0	$\frac{6}{3} = 2 \rightarrow$
	0	x_4	1	1	–1	0	1	….
	$c_j - Z_j$			4	6↑	0	0	

	6	x_2	2	2/3	1	1/3	0	
2	0	x_4	3	5/3	0	1/3	1	
	$c_j - Z_j$			0	0	– 2	0	

In above Table 2, $\Delta_j \leq 0$ for all j, therefore the solution is optimal. Hence, optimal solution is $x_1 = 0$, $x_2 = 2$ and Max. $Z = 12$.

Now if in above Table 2, enter $\Delta_1 = 0$ in $c_j - Z_j$ row, in x_1 column indicates the existence of alternate optimal solution. So if x_1 is made to enter the basic the new solution obtained will again be optimal solution.

Cost →				4	6	0	0	Minimum ratio $\frac{X_{B_i}}{\alpha_i^j}$
Variable →				x_1	x_2	x_3	x_4	
Table number	$\mathbf{C_B}$	Basic variable	$\mathbf{X_B}$	α^1	α^2	α^3	α^4	
2	6	x_2	2	2/3	1	1/3	0	$\frac{2}{2/3} = 3$
	0	x_4	3	[5/3]	↑0	1/3	1	$\frac{3}{5/3} = \frac{9}{5}$ →
	$c_j - Z_j$			0↑	0	– 2	0	
3	6	x_2	4/5	0	1	1/5	– 2/5	
	4	x_1	9/5	1	0	1/5	3/5	
	$c_j - Z_j$			0	0	– 2	0	

In above Table 3, $\Delta_j \leq 0$ for all j, therefore again the solution is optimal. Thus optimal solution is $x_1 = 0$, $x_2 = 2$ and Max. $Z = 12$.

Example 6: Solve the following linear programming problem using the simplex method

Max. $\quad Z = -x_1 + 4x_2$

s.t. $\quad x_1 - x_2 \leq 1$

$\quad -2x_1 + x_2 \leq 2$

and $\quad x_1, x_2 \geq 0$

Solution: The problem can be written in the standard form as:

Max. $\quad Z = -x_1 + 4x_2 + 0.x_3 + 0.x_4$

Subject to $\quad x_1 - x_2 + x_3 + 0.x_4 = 1$

$\quad -2x_1 + x_2 + 0.x_3 + x_4 = 2$

and $x_1, x_2, x_3, x_4 \geq 0$, x_3 and x_4 are slack variables.

Here $A = \begin{bmatrix} 1 & -1 & 1 & 0 \\ -2 & 1 & 0 & 1 \end{bmatrix}$. Columns of x_3 and x_4 give the identity matrix $\begin{bmatrix} 1 & 0 \\ 0 & 1 \end{bmatrix}$. Thus x_3 and x_4 are the starting basic variables and $\begin{bmatrix} 1 & 0 \\ 0 & 1 \end{bmatrix}$ is the basis matrix. By taking x_1 and x_2 as non-basic variables (i.e., $x_1 = 0$ and $x_2 = 0$), we write the corresponding basic solution with $x_3 = 1$ and $x_4 = 2$ in the form of the following simplex table and thereafter use simplex iterations till we get the optimal solution:

		Cost →		–1	4	0	0	Minimum ratio $\frac{X_{B_i}}{\alpha_i^j}$
		Variable →		x_1	x_2	x_3	x_4	
Table number	$\mathbf{C_B}$	Basic variable	$\mathbf{X_B}$	α^1	α^2	α^3	α^4	
1	0	x_3	1	1	–1	1	0	....
	0	x_4	2	– 2	[1]	0	1	$\frac{2}{1} = 1 \rightarrow$
		$c_j - Z_j$		–1	4↑	0	0	
2	0	x_3	3	–1	0	1	1	
	4	x_2	2	–2	1	0	1	
		$c_j - Z_j$		7↑	0	0	– 4	

In above Table 2, x_1 enters as $\Delta_1 = 7$, but there is no variable to leave the basis because of all entry in α_i^1 column being negative. Hence, the solution is unbounded.

5.6 TO SOLVE AN LPP USING ARTIFICIAL VARIABLE TECHNIQUES

When in a linear programming problem all the constraints are not of '≤' type, then to get the identity matrix as the submatrix of the coefficient matrix we add artificial variables. To solve such an LPP, we shall discuss the following methods.

5.6.1 Big-M Method

The following steps are used to solve a linear programming problem:

Step 1: First we convert the given linear programming problem in the standard form using slack, surplus and artificial variables.

Step 2: Modify the objective function by adding (–M) times the sum of artificial variables involved in the problem, in case of a maximization problem, where M is a large positive number.

Step 3: Construct the initial simplex table and use simplex iterations after bringing it in requisite format till the optimal solution is obtained.

Note that the large positive value of M in Big-M method will tend to pass the artificial variables out of the basis in simplex iterations, if the solution of the problem exists. However, if the artificial variable continues with positive value in the optimal simplex table, the given problem has no solution. Further no solution situation for a linear programming problem will arise only in the artificial variable technique when it is solved by the simplex method. Again whenever, there is a tie among leaving variables with one as artificial variable then an artificial variable is preferred, because our aim is to bring them to zero. An artificial variable is never considered for re-entry into the basis once it leaves the basis. We drop its column and the column entries are not calculated in subsequent simplex tables unless B^{-1} is needed in the optimal table.

Example 7: Solve the following linear programming problem using Big-M method

Max. $Z = -x_1 + x_2$

s.t. $x_1 + x_2 \leq 1$

$2x_1 + 3x_2 \leq 6$

and $x_1, x_2 \geq 0$

Solution: The problem can be written in the standard form as:

Max. $Z = -x_1 + x_2 + 0.x_3 + 0.x_4$

s.t. $x_1 + x_2 + x_3 = 1$ ...(1)

$2x_1 + 3x_2 - x_4 = 6$

and $x_1, x_2, x_3, x_4 \geq 0$, x_3 slack and x_4 is surplus variable.

Modified problem is

Max. $Z = -x_1 + x_2 + 0.x_3 + 0.x_4 - M x_5$

s.t. $x_1 + x_2 + x_3 = 1$ ...(2)

$2x_1 + 3x_2 - x_4 + x_5 = 6$

and $x_1, x_2, x_3, x_4, x_5 \geq 0$, x_5 is artificial variable.

Optimal solution of (2) with $x_5 = 0$ is also an optimal solution of (1). By taking x_1, x_2 and x_4 as non-basic variables (i.e., $x_1 = 0$, $x_2 = 0$ and $x_4 = 0$), we write the corresponding basic solution with $x_3 = 1$ and $x_5 = 6$ in the form of the following simplex table and thereafter use simplex iterations till we get the optimal solution:

		Cost →		-1	1	0	0	$-M$	Minimum ratio $\frac{X_{B_i}}{\alpha_i^j}$
		Variable →		x_1	x_2	x_4	x_3	x_5	
Table number	C_B	Basic variable	X_B	α^1	α^2	α^4	α^3	α^5	
1	0	x_3	1	1	[1]	0	1	0	$\frac{1}{1} = 1 \rightarrow$
	$-M$	x_5	6	2	3	-1	0	1	$\frac{6}{3} = 2$
		$c_j - Z_j$		-1	1	$-M$	0	0	
2	1	x_2	1	1	1	0	1	1	
	$-M$	x_5	3	-1	0	-1	0	1	
		$c_j - Z_j$		$-2-M$	0	$-M$	$-1-3M$	0	

This is the optimal table in which artificial variable x_5 is present with positive value $x_5 = 3$. Thus the problem has no solution.

Example 8: Solve the following linear programming problem using Big-M method

$$\text{Max} \quad Z = 4x_1 + 5x_2 - 3x_3$$

$$\text{s.t.} \quad x_1 + x_2 + x_3 = 10$$

$$x_1 - x_2 \geq 1$$

$$2x_1 + 3x_2 + x_3 \leq 40$$

$$\text{and} \quad x_1, x_2, x_3 \geq 0$$

Solution: The problem can be written in the standard form as:

$$\text{Max.} \quad Z = 4x_1 + 5x_2 - 3x_3 + 0.x_5 + 0.x_7$$

$$\text{s.t.} \quad x_1 + x_2 + x_3 = 10$$

$$x_1 - x_2 - x_5 = 1$$

$$2x_1 + 3x_2 + x_3 + x_7 = 40$$

and $x_1, x_2, x_3, x_5, x_7 \geq 0$, x_5 slack and x_7 is surplus variable.

Modified problem is

$$\text{Max.} \quad Z = 4x_1 + 5x_2 - 3x_3 - Mx_4 + 0.x_5 - M x_6 + 0.x_7$$

$$\text{s.t.} \quad x_1 + x_2 + x_3 + x_4 = 10$$

$$x_1 - x_2 - x_5 + x_6 = 1$$

$$2\,x_1 + 3x_2 + x_3 + x_7 = 40$$

and $\quad x_1, x_2, x_3, x_4, x_5, x_6, x_7 \geq 0$, x_4 and x_6 are artificial variables.

By taking x_1, x_2, x_3 and x_5 as non-basic variables (i.e., $x_1 = 0$, $x_2 = 0$, $x_3 = 0$ and $x_5 = 0$), we write the corresponding basic solution with $x_4 = 10$, $x_6 = 1$ and $x_7 = 40$ in the form of the following simplex table and thereafter use simplex iterations till we get the optimal solution:

Cost →				4	5	–3	0	$-M$	$-M$	0	Minimum ratio
Variable →				x_1	x_2	x_3	x_5	x_4	x_6	x_7	
Table number	C_B	Basic variable	X_B	α^1	α^2	α^3	α^5	α^4	α^6	α^7	$\dfrac{X_{B_i}}{\alpha_i^j}$
1	$-M$	x_4	10	1	1	1	0	1	0	0	$\frac{10}{1} = 10$
					[2]						
	$-M$	x_6	1	[1]	–1	0	–1	0	1	0	$\frac{1}{1} = 1 \rightarrow$
	0	x_7	40	2	3	1	0	0	0	1	$\frac{40}{2} = 20$
	$c_j - Z_j$			$4+2M\uparrow$	5	$-3+M$	$-M$	0	0	0	

2	$-M$	x_4	9	0		1	1	1		0	$\frac{9}{2} = 4.5 \rightarrow$
	4	x_1	1	1	–1	0	–1	0		0	-----
	0	x_7	38	0	5	1	2	0		1	$\frac{38}{5} = 7.6$
	$c_j - Z_j$			0	9+2M↑	–3+M	M+4	0		0	
3	5	x_2	9/2	0	1	1/2	1/2			0	
	4	x_1	11/2	1	0	1/2	–1/2			0	
	0	x_7	31/2	0	0	–3/2	–1 /2			1	
	$c_j - Z_j$			0	0	–15/2	–1/2			0	

In above table 3, $\Delta_j \leq 0$ for all j, therefore solution is optimal. Thus, optimal solution is $x_1 = 11/2$, $x_2 = 9/2$, $x_3 = 0$ and Max $Z = 89/2$.

5.6.2 Two-Phase Method

Here we solve the problem in two phases.

Phase I: In phase I, introduce objective function of maximization form with cost zero for given variables and (–1) for artificial variables and solve it using the simplex method rules to get the optimal table of phase I.

Note that if the artificial variable is present with non-zero value in the optimal table of phase I, then given problem has no solution. The purpose of using phase I is to get the starting BFS for phase II.

Phase II: In phase II, we start with the optimal solution of phase I. Include the given objective function of the problem and bring the simplex table in proper format. This will give initial simplex table of phase II. Now apply simplex iterations until the optimality conditions are satisfied. This gives the optimal table of phase II and the optimal solution given by this optimal table will be the optimal solution of the given problem.

Example 9: Solve the following linear programming problem using two-phase method

Max $\quad Z = 3x_1 + 2x_2 - 3x_3$

s.t. $\quad 2x_1 + x_2 + x_3 \leq 2$

$\quad 3x_1 + 4x_2 - x_3 \geq 12$

and $\quad x_1, x_2, x_3 \geq 0$

Solution: **Phase I:** In phase I, introduce objective function of the maximization form with cost zero for given variables and (–1) for artificial variables.

Max $\quad Z' = -x_6$

s.t. $2x_1 + x_2 + x_3 + x_4 = 2$

$3x_1 + 4x_2 - x_3 - x_5 + x_6 = 12$

and $x_1, x_2, x_3, x_4, x_5, x_6 \geq 0$

By taking x_1, x_2, x_3 and x_5 as non-basic variables (i.e., $x_1 = 0$, $x_2 = 0$, $x_3 = 0$ and $x_5 = 0$), we write the corresponding basic solution with $x_4 = 2$ and $x_6 = 12$ in the form of the following simplex table and thereafter use simplex iterations till we get the optimal solution:

		Cost →		0	0	0	0	0	–1	Minimum ratio
		Variable →		x_1	x_2	x_3	x_5	x_4	x_6	
Table number	$\mathbf{C_B}$	Basic variable	$\mathbf{X_B}$	α^1	α^2	α^3	α^5	α^4	α^6	$\frac{X_{B_i}}{\alpha_i^j}$
1	0	x_4	2	2	[1]	1	0	1	0	$\frac{2}{1} = 2 \rightarrow$
	–1	x_6	12	3	4	–1	–1	0	1	$\frac{12}{4} = 3$
		$c_j - Z_j$		3	4↑	–1	–1	0	0	
2	0	x_2	2	2	1	1	0	1	0	
	–1	x_6	4	–5	0	–5	–1	–4	1	
		$c_j - Z_j$		–5	0	–5	–1	–4	0	

In above table 2, $\Delta j \leq 0$ for all j, but the artificial variable x_6 is present with positive value ($x_6 = 4$). Hence, the given problem has no solution.

Example 10: Solve the following linear programming problem using two-phase method

Max $Z = 3x_1 - x_2$

s.t. $2x_1 + x_2 \geq 2$

$x_1 + 3x_2 \leq 2$

$x_2 \leq 4$

and $x_1, x_2 \geq 0$

Solution: **Phase I:** In phase I, introduce objective function of the maximization form with cost zero for given variables and (–1) for artificial variables.

Max $Z' = -x_4$

s.t. $2x_1 + x_2 - x_3 + x_4 = 2$

$x_1 + 3x_2 + x_5 = 2$

$x_2 + x_6 = 4$

and $x_1, x_2, x_3, x_4, x_5, x_6 \geq 0$

By taking x_1, x_2 and x_3 as non-basic variables (i.e., $x_1 = 0$, $x_2 = 0$ and $x_3 = 0$), we write the corresponding basic solution with $x_4 = 2$, $x_5 = 2$ and $x_6 = 4$ in the form of the following simplex table and thereafter use simplex iterations till we get the optimal solution:

		Cost →		0	0	0	−1	0	0	Minimum ratio
		Variable →		x_1	x_2	x_3	x_4	x_5	x_6	
Table number	$\mathbf{C_B}$	Basic variable	$\mathbf{X_B}$	α^1	α^2	α^3	α^4	α^5	α^6	$\frac{X_{B_i}}{\alpha_i^j}$
1	−1	x_4	2	[2]	1	−1	1	0	0	$\frac{2}{2}=1\rightarrow$
	0	x_5	2	1	3	0	0	1	0	$\frac{2}{1}=2$
	0	x_6	4	0	1	0	0	0	1	…
		$c_j - Z_j$		2↑	1	−1	0	0	0	
2	0	x_1	1	1	1/2	−1/2		0	0	
	0	x_5	1	0	5/2	1/2		1	0	
	0	x_6	4	0	1	0		0	1	
		$c_j - Z_j$		0	0	0		0	0	

In above table 2, $\Delta j \leq 0$ for all j, and hence is the optimal table of phase I.

Phase II: In phase II, consider the objective function of the given problem, omit the artificial variable column and use the optimal table of phase I to get

		Cost →		3	−1	0	0	0	Minimum ratio
		Variable →		x_1	x_2	x_3	x_5	x_6	
Table number	$\mathbf{C_B}$	Basic variable	$\mathbf{X_B}$	α^1	α^2	α^3	α^5	α^6	$\frac{X_{B_i}}{\alpha_i^j}$
2	3	x_1	1	1	1/2	−1/2	0	0	…
	0	x_5	1	0	5/2	[1/2]	1	0	$\frac{1}{1/2}=2\rightarrow$
	0	x_6	4	0	1	0	0	1	…
		$c_j - Z_j$		0	−5/2	3/2↑	0	0	
3	3	x_1	2	1	3	0	1	0	
	0	x_3	2	0	5	1	2	0	
	0	x_6	4	0	1	0	0	1	
		$c_j - Z_j$		0	−10	0	−3	0	

In above table 3, $\Delta j \leq 0$ for all j, therefore solution is optimal. Thus, the optimal solution is $x_1 = 2$, $x_2 = 0$ and Max $Z = 6$.

5.7 ALGORITHM FOR REVISED SIMPLEX METHOD

Step 1: First we convert the given linear programming problem in the standard form such that the identity matrix is the sub-matrix of the coefficients matrix A.

Step 2: Identify the basic variables, corresponding costs and the basis matrix B and find B^{-1}. Initially $B = I$ (identity matrix) and thus $B^{-1} = \text{I}$.

Step 3: Obtain Δj, $j = m + 1, m + 2, \ldots\ldots\ldots\ldots, n$ for the feasible basis xj, $j = 1, 2, \ldots\ldots\ldots\ldots, m$ using $\Delta_j = c_j - Z_j = c_j - C_B^T B^{-1} A_j$, $C_B^T B^{-1}$ is called the simplex multiplier. The variable x_j with most positive Δj value will enter the basis.

Step 4: Obtain the entries $\alpha_1^j, \alpha_2^j, \ldots\ldots\ldots\ldots\ldots\ldots, \alpha_m^j$ (i.e., vector $\overset{j}{\alpha}$) and consider $\min\limits_i \left(\frac{x_i}{\alpha_i^j}, \alpha_i^j \succ 0 \right) = \frac{x_r}{\alpha_r^j}$. Then variable x_r will leave the basis.

Step 5: Put the above computations done in step 2, 3 and 4 in the following table:

Iteration number	C_B	Basic variable	$X_B = C_B^T B^{-1}$	B^{-1}	Non-basic variables x_{m+1} x_{m+2} … … x_n	Minimum ratio $\frac{X_{B_i}}{\alpha_i^j}$
1	C_{B1} … … C_{Bm}	x_1 …. …. x_m	X_{B1} … … X_{Bm}			
		$c_j - Z_j$				

Step 6: After finding the leaving and entering variable, enrectangle the entry at the intersection of entering and leaving variable. This is called pivot entry and leaving variable row is called pivot row. Interchange the position and leaving and entering variables for the next simplex table.

Use the same elementary row operations on existing B^{-1} and X_B column which are needed to bring 1 in the position of pivot entry and zero below and above entry 1 thus obtained. These computations will give new basic variables, new C_B, new X_B and new B^{-1}.

Step 7: Now make the next simplex iteration table for computations done in step 6 with new basic variables, new C_B, new X_B and new B^{-1}.

Step 8: Repeat the above steps till $\Delta_j > 0$ for non-basic variables to get the optimal solution. Solution of the table thus obtained will be optimal solution and write it.

Note that step 6 provides with a method to find new X_B and new B^{-1} for the next table.

Example 11: Solve the following linear programming problem using the revised simplex method

Max $\quad Z = 6x_1 - 2x_2 + 3x_3$

s.t. $\quad 2x_1 - x_2 + 2x_3 \le 2$

$\quad x_1 + 4x_3 \le 4$

and $\quad x_1, x_2, x_3 \ge 0$

Solution: The problem can be written in the standard form as

Max. $Z = 6x_1 - 2x_2 + 3x_3 + 0 \cdot x_4 + 0 \cdot x_5$

s.t. $2x_1 - x_2 + 2x_3 + x_4 + 0 \cdot x_5 = 2$

$x_1 + 4x_3 + 0 \cdot x_4 + x_5 = 4$

$x_1, x_2, x_3, x_4, x_5 \geq 0$, x_4 and x_5 are slack variables.

Iteration I: Here coefficient matrix $A = (A_1, A_2, A_3, A_4, A_5) = \begin{pmatrix} 2 & -1 & 2 & 1 & 0 \\ 1 & 0 & 4 & 0 & 1 \end{pmatrix}$. Basic variables are x_4 and x_5. $B = \begin{pmatrix} 1 & 0 \\ 0 & 1 \end{pmatrix}$, $B^{-1} = \begin{pmatrix} 1 & 0 \\ 0 & 1 \end{pmatrix}$, $b = \begin{pmatrix} 2 \\ 4 \end{pmatrix}$ and $C^T = (6, -2, 3, 0, 0)$.

Basic feasible solution $X_B = B^{-1}b = (2, 4) = (2, 4)$.

Calculate for non-basic variables (x_1, x_2, x_3)

$$\Delta_j = c_j - Z_j = c_j - C_B^T B^{-1} A_j$$

$$C_B^T B^{-1} = (0, 0) \begin{pmatrix} 1 & 0 \\ 0 & 1 \end{pmatrix} = (0, 0)$$

$$C_B{}^T B^{-1} A_j = (0, 0) \begin{pmatrix} 2 & -1 & 2 \\ 1 & 0 & 4 \end{pmatrix} = (0, 0, 0)$$

$$\Delta_j = c_j - Z_j = (6, -2, 3) - (0, 0, 0) = (6, -2, 3).$$

Here $\Delta_j = 6$ is the most positive. So we calculate α^1.

$$x_1 \text{ column vector } \alpha^1 = B^{-1} A_1 = \begin{pmatrix} 1 & 0 \\ 0 & 1 \end{pmatrix} \begin{pmatrix} 2 \\ 1 \end{pmatrix} = \begin{pmatrix} 2 \\ 1 \end{pmatrix}.$$

The above computational work can be put in a tabular form in Table 1.

Table 1

Iterations number	C_B	Basic variable	X_B	B^{-1}		Non-basic variables x_1	x_2	x_3	Minimum ratio $\frac{X_{B_i}}{\alpha_i^j}$
1	0	x_4	2	1	0	[2]			$\frac{2}{2} = 1 \rightarrow$
	0	x_5	4	0	1	1			$\frac{4}{1} = 4$
	$c_j - Z_j$					$6\uparrow$	-2	3	

Here entry $\Delta j = 6$ is most positive for x_1 column. So x_1 enters the basis in place of x_4.

Iteration II: Now use the same elementary row operations on column X_B and B^{-1} of table 1 which are needed to bring zero above the pivot entry 1 and interchange the position of entering and leaving variable.

New basic variables are x_1 and x_5 with new $C_B^T = (6, 0)$, new $B^{-1} = \begin{pmatrix} 1/2 & 0 \\ -1/2 & 1 \end{pmatrix}$ and new $X_B = (1, 3)$.

Calculate for non-basic variables (x_4, x_2, x_3)

$$\Delta_j = c_j - Z_j = c_j - C_B^T B^{-1} A_j$$

$$C_B^T B^{-1} = (6, 0) \begin{pmatrix} 1/2 & 0 \\ -1/2 & 1 \end{pmatrix} = (3, 0)$$

$$C_B^T B^{-1} A_j = (3, 0) \begin{pmatrix} 0 & -1 & 2 \\ 1 & 0 & 4 \end{pmatrix} = (0, -3, 6)$$

$$\Delta j = c_j - Z_j = (0, -2, 3) - (0, -3, 6) = (0, 1, -3).$$

Here $\Delta_j = 1$ is most positive. So we calculate α^2.

$$x_2 \text{ column vector } \alpha^2 = B^{-1} A_2 = \begin{pmatrix} 1/2 & 0 \\ -1/2 & 1 \end{pmatrix} \begin{pmatrix} -1 \\ 0 \end{pmatrix} = \begin{pmatrix} -1/2 \\ 1/2 \end{pmatrix}.$$

These computations we put in a tabular form in Table 2.

Table 2

Iterations number	$\mathbf{C_B}$	Basic variable	$\mathbf{X_B}$	B^{-1}		Non-basic variables x_4	x_2	x_3	Minimum ratio $\frac{X_{B_i}}{\alpha_i^j}$
1	6	x_1	1	1/2	0		–1/2		
	0	x_5	3	–1/2	1		[1/2]		
		$c_j - Z_j$				0	1↑	–3	

Here entry $\Delta_j = 1$ is most positive for x_2 column. So x_2 enters the basis in place of x_5.

Iteration III: Now use the same elementary row operations on column X_B and B^{-1} of Table 2 which are needed to bring 1 in the position of pivot entry and zero above the pivot entry 1 and interchange the position of entering and leaving variable.

New basic variables are x_1 and x_2 with new $C_B^T = (6, -2)$, new $B^{-1} \begin{pmatrix} 0 & 1 \\ -1 & 2 \end{pmatrix}$ and new $X_B = (4, 6)$.

Calculate for non-basic variables (x_5, x_4, x_3)

$$\Delta_j = c_j - Z_j = c_j - C_B^T B^{-1} Aj$$

$$C_B^T B^{-1} = (6, -2) \begin{pmatrix} 0 & 1 \\ -1 & 2 \end{pmatrix} = (2, 2)$$

$$C_B^T B^{-1} A_j = (2, 2)\begin{pmatrix} 0 & 1 & 2 \\ 1 & 0 & 4 \end{pmatrix} = (2, 2, 12)$$

$$\Delta_j = c_j - Z_j = (0, 0, 3) - (2, 2, 12) = (-2, -2, -9). \text{ Here } \Delta_j \le 0.$$

These computations we put in a tabular form in Table 3.

Table 3

Iterations number	$\mathbf{C_B}$	Basic variable	$\mathbf{X_B}$	B^{-1}		Non-basic variables			Minimum ratio
						x_4	x_5	x_3	$\frac{X_{B_i}}{\alpha_i^j}$
1	6	x_1	4	0	1				
	–2	x_2	6	–1	2				
	$c_j - Z_j$					–2	–2	–9	

In Table 3, $\Delta_j \le 0$ for all non-basic variables x_j. Therefore, solution obtained from Table 3 is optimal. Hence, optimal solution is $x_1 = 4$, $x_2 = 6$, $x_3 = 0$ and Max. $Z = 12$.

EXERCISE 5.1

1. Write a short note on the simplex method.
2. Write the procedure to solve an LPP by using Big-M and two-phase methods.

   Solve the following linear programming problems using simplex method and as well as the revised simplex method:

3. Max $Z = -4x_1 + x_2$

   s.t. $-x_1 + x_2 \le 1$

   $x_1 - 2x_2 \le 2$

   and $x_1, x_2 \ge 0$

4. Max $Z = 7x_1 + 5x_2$

   s.t. $-x_1 - 2x_2 \ge -6$

   $4x_1 + 3x_2 \le 12$

   and $x_1, x_2 \ge 0.$

5. Max $Z = 5x_1 + 3x_2$

   s.t. $3x_1 + 5x_2 \le 15$

   $5x_1 + 2x_2 \le 10$

   and $x_1, x_2 \ge 0.$

6. Max $Z = 2x_1 + x_2$

   s.t. $x_1 + 2x_2 \le 10$

   $x_1 + x_2 \le 6$

   $x_1 - x_2 \le 2$

   $x_1 - 2x_2 \le 1$

   and $x_1, x_2 \ge 0.$

7. Max $Z = 3x_1 + 5x_2 + 4x_3$

   s.t. $2x_1 + 3x_2 \le 8$

$2x_2 + 5x_3 \leq 10$
$3x_1 + 2x_2 + 4x_3 \leq 15$
and $x_1, x_2, x_3 \geq 0.$

8. Max $Z = 3x_1 + 2x_2 + 5x_3$
s.t. $x_1 + 2x_2 + x_3 \leq 430$
$3x_1 + 2x_3 \leq 460$
$x_1 + 4x_2 \leq 420$
and $x_1, x_2, x_3 \geq 0.$

9. Max $Z = 3x_1 + 4x_2$
s.t. $x_1 - x_2 \leq 1$
$-x_1 + x_2 \leq 2$
and $x_1, x_2 \geq 0.$

10. Max $Z = 4x_1 + 10x_2$
s.t. $2x_1 + x_2 \leq 50$
$2x_1 + 5x_2 \leq 100$
$2x_1 + 3x_2 \leq 90$
and $x_1, x_2 \geq 0.$

Solve the following linear programming problems using Big-M method and as well as the revised simplex method:

11. Min $Z = 4x_1 + 8x_2 + 3x_3$
s.t. $x_1 + x_2 \leq 2$
$2x_1 + x_3 \geq 5$
and $x_1, x_2, x_3 \geq 0.$

12. Max $Z = x_1 + 2x_2 + 3x_3 - x_4$
s.t. $x_1 + 2x_2 + 3x_3 = 15$
$2x_1 + x_2 + 5x_3 = 20$
$x_1 + 2x_2 + x_3 + x_4 = 10$
and $x_1, x_2, x_3, x_4 \geq 0.$

13. Max $Z = -x_1 - x_2$
s.t. $3x_1 + 2x_2 \geq 30$
$-2x_1 + 3x_2 \geq -30$
$x_1 + x_2 \leq 5$
and $x_1, x_2 \geq 0.$

Solve the following linear programming problem using two-phase method and as well as the revised simplex method:

14. Max $Z = 5x_1 + 8x_2$
s.t. $3x_1 + 2x_2 \geq 3$
$x_1 + 4x_2 \geq 4$
$x_1 + x_2 \leq 5$
and $x_1, x_2 \geq 0.$

15. Min $Z = x_1 - 2x_2 - 3x_3$
s.t. $-2x_1 + x_2 + 3x_3 = 2$
$2x_1 + 3x_2 + 4x_3 = 1$
and $x_1, x_2, x_3 \geq 0.$

ANSWERS

3. $x_1 = 0, x_2 = 1$ and Max. $Z = 1$.
4. $x_1 = 3, x_2 = 0$ and Max $Z = 21$.
5. $x_1 = 20/19, x_2 = 45/19$ and Max $Z = 235/19$.
6. $x_1 = 4, x_2 = 2$ and Max $Z = 10$.
7. $x_1 = 89/41, x_2 = 50/41, x_3 = 62/41$ and Max $Z = 765/41$.
8. $x_1 = 0, x_2 = 100, x_3 = 230$ and Max $Z = 1350$.
9. Unbounded solution.
10. $x_1 = 4, x_2 = 1$ and Min $Z = 11$.
11. $x_1 = 5/2, x_2 = 0, x_3 = 0$ and Min $Z = 10$.
12. $x_1 = x_2 = x_3 = 5/2, x_4 = 0$ and Max $Z = 15$.
13. No feasible solution.
14. $x_1 = 0, x_2 = 5$ and Max $Z = 40$.
15. No feasible solution.

6 Chapter

Duality in Linear Programming

6.1 INTRODUCTION

The concept of duality is very useful to explain that corresponding to every linear programming problem there is another linear programming problem. Solutions of these two linear programming problems are closely associated with each other. To be very specific the solution of any of these problems can be obtained from the solution of another. It is easier to solve that one which has smaller number of constraints. Thus, duality gives us the freedom to choose that linear programming problem out of the two which has smaller number of constraints and then to get the solution of another if needed.

As it is clear from the earlier chapters, that when we solve a linear programming problem using graphical method, then constraints of the problem must be in the inequality form and while solving it by the simplex method we have to bring the problem in standard form. Likewise when we solve a linear programming problem using duality then we have to put the problem in the form called standard primal form (canonical form) of the linear programming problem which we define as follows.

6.2 STANDARD PRIMAL FORM OR CANONICAL FORM

A linear programming problem is said to be in the standard primal form (canonical form), if it be either

(i) A maximization problem with all the constraints of '≤' type and variables involved are non-negative.

or

(ii) A minimization problem with all the constraints of '≥' type and variables involved are non-negative.

Example 1: Max. $Z = 2x_1 - 3x_2$

s.t. $-x_1 + x_2 \geq 1$

$x_1 + x_2 \geq 2$

and $x_1, x_2 \geq 0$. ...(1)

Solution: In matrix notation (1) can be written as

$$\begin{aligned} \text{Max.} \quad & Z = C^T X \\ \text{s.t.} \quad & AX \le b \\ & X \ge 0 \end{aligned}$$

where $C = (2, -3)^T$, $X = (x_1, x_2)^T$, $b = (1, 2)^T$ and $A = \begin{pmatrix} -1 & 1 \\ 1 & 1 \end{pmatrix}$.

Note that in the standard primal form, there is no restriction on the sign of b_i's, whereas in the standard form of the LPP while solving by the simplex method we require all b_i's ≥ 0. Further every LPP can be expressed in the standard primal (canonical) form.

6.3 DUALITY

Consider the linear programming problems

$$\begin{aligned} \text{Max. } Z &= C^T X \\ \text{s.t } AX &\le b \\ X &\ge 0 \end{aligned} \qquad \ldots(1)$$

and

$$\begin{aligned} \text{Min. } Z &= C^T X \\ \text{s.t. } AX &\ge b \\ X &\ge 0 \end{aligned} \qquad \ldots(2)$$

as given in (1) and (2) above. The dual of the linear programming problem in (1) is defined as

$$\begin{aligned} \text{Min. } Z_D &= b^T Y \\ \text{s.t. } A^T Y &\ge C \\ Y &\ge 0 \end{aligned} \qquad \ldots(3)$$

The dual of the linear programming problem given in (2) is defined as

$$\begin{aligned} \text{Max. } Z_D &= b^T Y \\ \text{s.t. } A^T Y &\le C \\ Y &\ge 0 \end{aligned} \qquad \ldots(4)$$

where $C = (c_1, c_2,\ldots,c_n)^T$, $Y = (y_1, y_2,\ldots,y_m)^T$, $A = [a_{ij}]_{m\times n}$ and $b = (b_1, b_2,\ldots,b_m)^T$.

Note that C^T is a row matrix, b and Y are column matrices (vectors). The linear programming problem (1) is called primal and (3) as its dual problem and so is for problem in (2) and (4). The m constraints in (3) and (4) are called dual constraints. The variable y_i is called the dual variable associated with i^{th} primal constraint. Note that dual of a maximization becomes a minimization problem and vice-versa. If some constraints are given in equality form in the primal problem then while writing the dual treats them as the corresponding variables in the dual problem will be unrestricted in sign. Conversely if the primal variable is unrestricted in sign in the given problem, then the corresponding dual constraint will hold with equality sign.

Theorem 1: The dual of the dual is primal.

Proof: Let us consider a LPP

$$\begin{aligned} \text{Max} \quad & Z(X) = C^T X \\ \text{s.t.} \quad & AX \le b \end{aligned} \qquad \ldots(1)$$

$$X \geq 0.$$

where $\quad C = (c_1, c_2, \ldots, c_n)^T, X = (x_1, x_2, \ldots, x_n)^T, A = [a_{ij}]_{m \times n}, b = (b_1, b_2, \ldots, b_m)^T.$

Dual of (1) is

$$\begin{aligned} \text{Min} \quad & Z(Y) = b^T Y \\ \text{s.t.} \quad & A^T Y \geq C \\ & Y \geq 0. \end{aligned}$$

$$\begin{aligned} \text{Or Max} \quad & [-Z(Y)] = -b^T Y \\ \text{s.t.} \quad & -A^T Y \leq -C \qquad \ldots(2) \\ & Y = (y_1, y_2, \ldots, y_m)^T \geq 0, y_i\text{'s are dual variables of (1)} \end{aligned}$$

Dual of (2) is

$$\begin{aligned} \text{Min} \quad & [-Z(W)] = -C^T W \\ \text{s.t.} \quad & (-A^T)^T W \geq (-b^T)^T \\ & W \geq 0 \end{aligned}$$

$$\begin{aligned} \text{Or Max} \quad & Z(W) = C^T W \\ \text{s.t.} \quad & AW \leq b \qquad \ldots(3) \\ & W \geq 0 \end{aligned}$$

Problem in (3) is the same as in (1), thus proving that the dual of the dual is the primal (i.e., original LPP).

Example 2: Find the dual of the following LPP

$$\begin{aligned} \text{Max} \quad & Z = 2x_1 - x_2 + 5x_3 \\ \text{s.t.} \quad & 2x_1 + x_2 + x_3 \leq 7 \\ & 3x_1 - x_3 \leq 4 \\ & x_1 - x_2 + x_3 \leq 3 \\ \text{and} \quad & x_1, x_2, x_3 \geq 0. \end{aligned}$$

Solution: The given problem can be written in the standard primal form as

$$\text{Max} \quad Z = C^T X = (2, -1, 5)\,(x_1, x_2, x_3)^T$$

$$\text{s.t.} \quad \begin{pmatrix} 2 & 1 & 1 \\ 3 & 0 & -1 \\ 1 & -1 & 1 \end{pmatrix} \begin{pmatrix} x_1 \\ x_2 \\ x_3 \end{pmatrix} \leq \begin{pmatrix} 7 \\ 4 \\ 3 \end{pmatrix}$$

$$\text{and} \quad x_1, x_2, x_3 \geq 0$$

where $C = (2, -1, 5)^T, X = (x_1, x_2, x_3)^T, b = (7, 4, 3)^T.$

The dual is

$$\begin{aligned} \text{Min} \quad & Z_D = b^T Y \\ \text{s.t.} \quad & A^T Y \geq C \\ \text{and} \quad & Y \geq 0. \end{aligned}$$

$$\text{or Min} \quad Z_D = (7, 4, 3)(y_1, y_2, y_3)^T = 7y_1 + 4y_2 + 3y_3$$

s.t.
$$\begin{pmatrix} 2 & 3 & 1 \\ 1 & 0 & -1 \\ 1 & -1 & 1 \end{pmatrix}\begin{pmatrix} y_1 \\ y_2 \\ y_3 \end{pmatrix} \geq \begin{pmatrix} 2 \\ -1 \\ 5 \end{pmatrix}$$

or
$$2y_1 + 3y_2 + y_3 \geq 2$$
$$y_1 - y_3 \geq -1$$
$$y_1 - y_2 + y_3 \geq 5$$

and
$$y_1, y_2, y_3 \geq 0.$$

Aliter (Short-cut method): The given linear programming problem is

Max
$$Z = 2x_1 - x_2 + 5x_3$$

s.t.
$$2x_1 + x_2 + x_3 \leq 7, y_1$$
$$3x_1 + 0.x_2 - x_3 \leq 4, y_2$$
$$x_1 - x_2 + x_3 \leq 3, y_3$$

and
$$x_1, x_2, x_3 \geq 0.$$

In the given problem, there are three constraints. Suppose associate three dual variables y_1, y_2, y_3 with these constraints. Then dual is

Min
$$Z_D = 7y_1 + 4y_2 + 3y_3$$

s.t.
$$2y_1 + 3y_2 + y_3 \geq 2$$
$$y_1 + 0.y_2 - y_3 \geq -1$$
$$y_1 - y_2 + y_3 \geq 5$$
$$y_1, y_2, y_3 \geq 0.$$

From the above problem we see that to get the objective function of the dual we multiply the right hand side of the primal constraints with corresponding y_i's. Max Z is replaced by Min Z_D. Multiply the x_1, x_2 and x_3 columns in coefficient matrix with corresponding y_i's to get the constraints of the dual problem and the right hand side of the dual constraints are the corresponding coefficients of primal variables in the objective function. Signs '≤' becomes '≥' and x_i's ≥ 0 become y_i's ≥ 0.

Example 3: Find the dual of the following LPP

Max
$$Z = 13x_1 + 24x_2$$

s.t.
$$x_1 + 7x_2 \leq 1$$
$$5x_1 + x_2 \geq 2$$

and
$$x_1, x_2 \geq 0.$$

Solution: The given linear programming problem can be written as

Max
$$Z = 13x_1 + 24x_2$$

s.t.
$$x_1 + 7x_2 \leq 1, y_1$$
$$-5x_1 - x_2 \leq -2, y_2$$

and
$$x_1, x_2 \geq 0.$$

In the given problem, there are two constraints. Suppose associate two dual variables y_1, y_2 with these constraints. Dual of this problem is

Min
$$Z_D = y_1 - 2y_2$$

s.t.
$$y_1 - 5y_2 \geq 13$$

$$7y_1 - y_2 \geq 24$$

and $$y_1, y_2 \geq 0.$$

From the above problem we see that to get the objective function of the dual we multiply the right hand side of the primal constraints with corresponding y_i's. Max Z is replaced by Min Z_D. Multiply the x_1 and x_2 columns in coefficient matrix with corresponding y_i's to get the constraints of the dual problem and the right hand side of the dual constraints are the corresponding coefficients of primal variables in objective function. Signs '≤' becomes '≥' and x_i's ≥ 0 become y_i's ≥ 0.

Example 4: Find the dual of the following linear programming problem

Min $$Z = x_1 + x_2 + 2x_3$$

s.t.
$$x_1 + 2x_2 + x_3 \leq 2$$
$$2x_1 - x_2 + 3x_3 = 5$$
$$x_1 + 4x_2 - x_3 \geq 1$$
$$x_1, x_2 \geq 0 \text{ and } x_3 \text{ unrestricted in sign.}$$

Solution: The given linear programming problem can be written as

Min $$Z = x_1 + x_2 + 2x_3$$

s.t.
$$-x_1 - 2x_2 - x_3 \geq -2\ ,\ y_1$$
$$2x_1 - x_2 + 3x_3 = 5\ ,\ y_2$$
$$x_1 + 4x_2 - x_3 \geq 1\ ,\ y_3$$
$$x_1, x_2 \geq 0 \text{ and } x_3 \text{ unrestricted in sign.}$$

In the given problem, there are three constraints. Suppose associate three dual variables y_1, y_2, y_3 with these constraints. Dual of this problem is

Max $$Z_D = -2y_1 + 5y_2 + y_3$$

s.t.
$$-y_1 + 2y_2 + y_3 \leq 1$$
$$-2y_1 - y_2 + 4y_3 \leq 1$$
$$-y_1 + 3y_2 - y_3 = 2$$
$$y_1, y_3 \geq 0,\ y_2 \text{ unrestricted in sign.}$$

From the above problem we see that to get the objective function of the dual we multiply the right hand side of the primal constraints with corresponding y_i's. Min Z is replaced by Max Z_D. Multiply the x_1, x_2 and x_3 columns in the coefficient matrix with corresponding y_i's to get the constraints of the dual problem and the right hand side of the dual constraints are the corresponding coefficients of primal variables in objective function. Here in the primal problem the second constraint is given in equality form, we treat it as '≥' and take the corresponding variable y_2 in the dual problem unrestricted in sign. Also the primal variable x_3 is unrestricted in sign, then the corresponding third dual constraint holds with equality sign. Signs '≥' becomes '≤' and x_i's ≥ 0 become y_i's ≥ 0.

Example 5: Verify for the following linear programming problem, that dual of the dual is primal (i.e., given problem).

Max $$Z = 4x_1 - 9x_2$$

s.t.
$$-x_1 + x_2 \leq 3$$
$$x_1 + x_2 \leq 2$$

and $$x_1, x_2 \geq 0.$$

Solution: The given problem can be written as

Max $$Z = (4, -9)\ (x_1, x_2)^T = C^T X$$

s.t. $$\begin{pmatrix} -1 & 1 \\ 1 & 1 \end{pmatrix}\begin{pmatrix} x_1 \\ x_2 \end{pmatrix} \le \begin{pmatrix} 3 \\ 2 \end{pmatrix}$$

and $$x_1, x_2 \ge 0 \qquad ...(1)$$

where C = $(4, -9)^T$, $X = (x_1, x_2)^T$, $A = \begin{pmatrix} -1 & 1 \\ 1 & 1 \end{pmatrix}$, $b = (3, 2)^T$

The dual of (1) is

Min $$Z_D = b^T Y$$

s.t. $$A^T Y \ge (C^T)^T$$

and $$Y \ge 0$$

or $$\text{Min } Z_D = (3, 2)(y_1, y_2)^T = y_1 + 2y_2$$

s.t. $$\begin{pmatrix} -1 & 1 \\ 1 & 1 \end{pmatrix}\begin{pmatrix} y_1 \\ y_2 \end{pmatrix} \ge \begin{pmatrix} 4 \\ -9 \end{pmatrix}$$

and $$y_1, y_2 \ge 0$$

or Min $$Z_D = 3y_1 + 2y_2$$

s.t. $$-y_1 + y_2 \ge 4$$

$$y_1 + y_2 \ge -9 \qquad ...(2)$$

and $y_1, y_2 \ge 0$; y_i's are dual variables.

Dual of equation (2) is

Max $$Z = C^T W$$

s.t. $$(A^T)^T W \le (b^T)^T$$

and $$W \ge 0$$

Or $$\text{Max } Z = (4, -9)(w_1, w_2)^T = 2w_1 - 3w_3$$

s.t. $$\begin{pmatrix} -1 & 1 \\ 1 & 1 \end{pmatrix}\begin{pmatrix} w_1 \\ w_2 \end{pmatrix} \le \begin{pmatrix} 3 \\ 2 \end{pmatrix}$$

and $w_1, w_2 \ge 0$; w_i's are dual variables of (2)

where $C^T = (4, -9)$, $W = (w_1, w_2)^T$, $(A^T)^T = \begin{pmatrix} -1 & 1 \\ 1 & 1 \end{pmatrix}$, $(b^T)^T = (3, 2)$

Or Max $$Z = 4w_1 - 9w_2$$

s.t. $$-w_1 + w_2 \le 3$$

$$w_1 + w_2 \le 2$$

and $$w_1, w_2 \ge 0.$$

Which is the same problem as in (1) and thus, dual of the dual is primal is verified.

6.4 ALGORITHM FOR THE DUAL SIMPLEX METHOD

We observe that in the simplex method, in starting simplex table feasibility conditions are satisfied, i.e., we start with a basic feasible solution and move towards the optimal solution. In the dual simplex method we proceed the other way, i.e., in the starting simplex table optimality criterion is satisfied but the basic solution is infeasible. In the dual simplex method, starting with this infeasible solution, we move towards the feasible solution maintaining optimality criterion. Optimality criteria being satisfied is evidently visible without any calculations made as $Z_j = 0, \forall j$ in the starting simplex table.

Step 1: Consider the given linear programming problem:

Max or Min $\qquad Z = f(X) = C^T X$

s.t. $\qquad AX = b$

$X = (x_1, x_2, \ldots, x_n)^T \geq 0$, where $A = [a_{ij}]_{m \times n}$ is the coefficient matrix containing identity matrix as its submatrix and one or more b_i's are negative.

Step 2: Writing the problem in the form of simplex table and check that optimality conditions ($c_j - Z_j \leq 0$ for maximize problem $\forall j$) are satisfied. But the basic solution of the table is infeasible.

Step 3: Use the following rules of leaving and entering variable to restore the feasibility:

(i) *Leaving variable rule:* The variables x_r of the row r which is most negative among the basic variables leaves the basis i.e., leaving variable $x_r = \lim\limits_{i} \{x_i ; x_i < 0\}$.

(ii) *Entering variable rule:* Look for the negative entries in the row of leaving variable, i.e., r^{th} row and take their ratio with corresponding entries in $c_j - Z_j$ row and select the entering variable x_k for column k for which this ratio in absolute sign is minimum, i.e.,

$$\text{Min}\left\{\left|\frac{c_j - Z_j}{\alpha_r^j}\right|, \alpha_r^j \langle 0\right\} = \theta, \text{ say for } j = k,\ r \text{ is fixed.}$$

Step 4: Continue step 3 till all the basic variables become non-negative. Table thus obtained is the optimal table and solution given by this table is the optimal solution.

Note 1: This rule ensures that from the existing non-feasible solution we shall move towards feasibility.

Note 2: This rule ensures that optimality conditions will continue to hold good. If all α_r^j are non-negative, (i.e., $\alpha_r^j \langle 0; \forall j$) then the given problem has no feasible solution.

Note 3: If the primal has no solution, then its dual also has no solution. If the primal has unbounded solution, then the dual has no feasible solution or unbounded solution.

Note 4: The problem which we solve can be considered the primal problem and the solution of its dual is given by the simplex multiplier $C_B{}^T B^{-1}$. Vector $C_B{}^T$ (row matrix) is the vector of costs of basic variables in order in the optimal table and B^{-1} is available below the starting basic variables in the optimal table.

Example 6: Solve the following linear programming problem by the dual simplex method:

Max $\qquad Z = -x_1 - 6x_2 - 18x_3$

s.t. $$-x_1 - 3x_3 \leq -3$$
$$-x_2 - 2x_3 \leq -5$$
and $$x_1, x_2, x_3 \geq 0.$$

Solution: The given problem can be written in the standard primal form as

Max. $$Z = -x_1 - 6x_2 - 18x_3 + 0.x_4 + 0.x_5$$
s.t. $$-x_1 - 3x_3 + x_4 = -3$$
$$-x_2 - 2x_3 + x_5 = -5$$

and $x_1, x_2, x_3, x_4, x_5 \geq 0$, x_4 and x_5 are slack variables. The starting dual simplex table is

	Cost →			-1	-6	-18	0	0	Minimum ratio
	Variable →			x_1	x_2	x_3	x_4	x_5	$\frac{X_{B_i}}{\alpha_i^j}$
Table number	C_B	Basic variable	X_B	α^1	α^2	α^3	α^4	α^5	
1	0	x_4	-3	-1	0	-3	1	0	–
	0	$\leftarrow x_5$	-5	0	[-1]	-2	0	1	$\min\left(\left\lvert\frac{-6}{-1}\right\rvert, \left\lvert\frac{-18}{-2}\right\rvert\right) = 6$ for x_2
		$c_j - Z_j$		-1	$-6\uparrow$	-18	0	0	
2	0	$\leftarrow x_4$	-3	[-1]	0	-3	1	0	$\min\left(\left\lvert\frac{-1}{-1}\right\rvert, \left\lvert\frac{-6}{-3}\right\rvert\right) = 1$ for x_1
	-6	x_2	5	0	1	2	0	-1	–
		$c_j - Z_j$		$-1\uparrow$	0	-6	0	-6	
3	-1	x_1	3	1	0	3	-1	0	
	-6	x_2	5	0	1	2	0	-1	
		$c_j - Z_j$		-2	0	-3	-1	-6	

In table 3, all the basic variables become non-negative. The optimum solution is $x_1 = 3$, $x_2 = 5$, $x_3 = 0$ and Max $Z = -33$.

Example 7: Solve the following linear programming problem using the dual simplex method:

Min $$Z = 2x_1 + 3x_2 + 2x_3$$
s.t. $$x_1 + 2x_2 + 3x_3 \geq 2$$
$$2x_1 - x_2 + x_3 \geq 3$$
and $$x_1, x_2, x_3 \geq 0.$$

Solution: Converting the minimization problem into maximization problem, we have

Max $$Z = -2x_1 - 3x_2 - 2x_3, \text{ where } Z' = -Z$$

The problem can be written in the standard form as:

Max $$Z = -2x_1 - 3x_2 - 2x_3$$
s.t. $$-x_1 - 2x_2 - 3x_3 + x_4 = -2$$
$$-2x_1 - x_2 - x_3 + x_5 = -3$$

and $x_1, x_2, x_3, x_4, x_5 \geq 0$, x_4 and x_5 are slack variables.

Observe that the dual simplex method is applicable when $c_j - Z_j$ are all non-positive.The starting dual simplex table is

	Cost →			−2	−3	−2	0	0	Minimum ratio
	Variable →			x_1	x_2	x_3	x_4	x_5	$\frac{X_{B_i}}{\alpha_i^j}$
Table number	C_B	Basic variable	X_B	α^1	α^2	α^3	α^4	α^5	
1	0	x_4	−2	−1	−2	−3	1	0	–
	0	← x_5	−3	**[−2]**	1	−1	0	1	$\min\left(\left\|\frac{-2}{-2}\right\|, \left\|\frac{-5}{-1}\right\|\right) = 1$ for x_1
		$c_j - Z_j$		−2↑	−3	−2	0	0	–
2	0	← x_4	−1/2	0	−5/2	**[−5/2]**	1	−1/2	$\min\left(\left\|\frac{-4}{-5/2}\right\|, \left\|\frac{-1}{-5/2}\right\| \left\|\frac{-1}{-1/2}\right\|\right) = \frac{2}{5}$ for x_3
	−2	x_1	3/2	1	−1/2	1/2	0	−1/2	
		$c_j - Z_j$		0	−4	−1↑	t	−1	
3	−2	x_3	1/5	0	1	1	−2/5	1/5	
	−2	x_1	7/5	1	−1	0	1/5	−3/5	
		$c_j - Z_j$		0	−3	0	−2/5	−4/5	

In table 3, all the basic variables become non-negative. The optimum solution is $x_1 = 7/5$, $x_2 = 0$, $x_3 = 1/5$ and Min $Z = 16/5$.

EXERCISE 6.1

1. Write a short note on duality.

   Find the dual of the following linear programming problem:

2. Min $Z = 3x_1 + 2x_2$

   s.t. $2x_1 + 3x_2 \geq 2$

   $x_1 + x_2 \geq 1$

   and $x_1, x_2 \geq 0.$

3. Min $Z = 2x_1 + 2x_2 + 4x_3$

   s.t. $2x_1 + 3x_2 + 5x_3 \geq 2$

   $3x_1 + x_2 + 4x_3 \leq 3$

   $x_1 + 4x_2 + 6x_3 \leq 5$

   and $x_1, x_2, x_3 \geq 0.$

4. Min $Z = 3x_1 - 2x_2 + 4x_3$

   s.t. $3x_1 + 5x_2 + 4x_3 \geq 7$

   $6x_1 + x_2 + 3x_3 \geq 4$

   $7x_1 - 2x_2 - x_3 \leq 10$

$x_1 - 2x_2 + 5x_3 \geq 3$
$4x_1 + 7x_2 - 2x_3 \geq 2$
and $x_1, x_2, x_3 \geq 0.$

5. Max $Z = x_1 + 3x_2$
s.t. $3x_1 + 2x_2 \leq 6$
$3x_1 + x_2 = 4$
and $x_1, x_2 \geq 0.$

6. Min $Z = x_1 + x_2 + x_3$
s.t. $x_1 - 3x_2 + 4x_3 = 5$
$x_1 - 2x_2 \leq 3$
$2x_2 - x_3 \geq 4$
and $x_1, x_3 \geq 0$, x_3 is unrestricted in sign.

7. Write the algorithm for dual simplex method.
Solve the following linear programming problems by the dual simplex method:

8. Min $Z = x_1 + 4x_2 + 3x_4$
s.t. $x_1 + 2x_2 - x_3 + x_4 \geq 3$
$-2x_1 - x_2 + 4x_3 + x_4 \geq 2$
and $x_1, x_2, x_3, x_4 \geq 0.$

9. Max $Z = -x_1 - 4x_2 - 3x_4$
s.t. $x_1 + 2x_2 - x_3 + x_4 \geq 3$
$-2x_1 - x_2 + 4x_3 + x_4 \geq 2$
and $x_1, x_2, x_3, x_4 \geq 0.$

10. Write the dual of the following linear programming problem
Min $Z = 2y_1 + 3y_2 + 5y_3$
s.t. $y_1 + 2y_2 + 3y_3 \geq 2$
$2y_1 - y_2 + y_3 \geq 3$
and $y_1, y_2, y_3 \geq 0.$
From the solution of the dual, find the solution of the given (primal) solution.

ANSWERS

2. Max $Z = 3y_1 + y_2$
s.t. $2y_1 + y_2 \leq 3$
$3y_1 + y_2 \leq 1$
and $y_1, y_2 \leq 0.$

3. Max $Z = 2y_1 - 3y_2 - 5y_3$
s.t. $2y_1 - 3y_2 - y_3 \leq 2$
$3y_1 - y_2 - 4y_3 \leq 2$
$5y_1 - 7y_2 - 6y_3 \leq 4$
and $y_1, y_2, y_3 \geq 0.$

4. Max $Z = 7y_1 + 4y_2 - 10y_3 + 3y_4 + 2y_5$
s.t. $3y_1 + 6y_2 - 7y_3 + y_4 + y_5 \leq 3$

$5y_1 + y_2 + 2y_3 - 2y_4 + 7y_5 \le -2$
$4y_1 + 3y_2 + y_3 + 5y_4 - 2y_5 \le 4$
and $y_1, y_2, y_3, y_4, y_5 \ge 0.$

5. Min $Z = 6y_1 + 4y_2$
s.t. $3y_1 + 3y_2 \ge 1$
$2y_1 + y_2 \ge 3$
$y_1 + 5y_2 \ge 0$
$y_1 \ge 0$ and y_2 is unrestricted in sign.

6. Max $Z = 5y_1 + 3y_2 + 4y_3$
s.t. $y_1 - y_2 \le 1$
$-3y_1 + 2y_2 + 2y_3 \ge 1$
$4y_1 - y_3 \le 1$
$y_2, y_3 \ge 0$ and y_1 is unrestricted in sign.

8. $x_1 = 7, x_2 = 0, x_3 = 4, x_4 = 0$ and Min $Z = 7$.

9. $x_1 = 7, x_2 = 0, x_3 = 4, x_4 = 0$ and Max $Z = -7$.

10. Max $Z_D = 2x_1 + 3x_2$
s.t. $x_1 + 2x_2 \le 2$
$2x_1 - x_2 \le 3$
$3x_1 + x_2 \le 5$
and $x_1, x_2 \ge 0.$
$x_1 = 8/5, x_2 = 1/5$ and Max $Z_D = 19/5$; $y_1 = 8/5, y_2 = 1/5, y_3 = 0$ and Min $Z = 19/5$.

7 Chapter
Post-Optimality Analysis or Sensitivity Analysis

7.1 INTRODUCTION

In solving a linear programming problem after getting its optimal solution, sometimes it is desired to make some changes in some of the parameters of the problem. There are two ways to find the solution of the new problem. One is to solve the changed problem starting from the scratch, as the original problem has been solved and other way is to solve it using the optimal table already obtained. The second option saves time and labour and that is from where emerges the idea of sensitivity or post-optimal analysis. Some of the changes in the given problem which we shall consider are:

(i) Change in a_{ij}'s i.e., technological coefficient (the coefficients of x_i's in coefficient matrix A).
(ii) Change in c_i's i.e., costs (the coefficients of x_i's in the objective function).
(iii) Change in b_j's i.e., availabilities (the right hand side of constraints).
(iv) Addition of a new variable.
(v) Addition of a new constraint.
(vi) Deletion of a constraint.
(vii) Deletion of a variable.

7.2 SENSITIVITY ANALYSIS

To start with, let us first recapitulate the form of the LPP and the notations used.
The LPP is

Max
$$Z = c_1x_1 + c_2 x_2 + \ldots + c_nx_n = C^TX$$
$$AX = b$$
$$X = (x_1, x_2,\ldots, x_n)^T \geq 0 \qquad \ldots(1)$$

where $A = [a_{ij}]_{m\times n} = (A_1, A_2,\ldots, A_n)$ is coefficient matrix with column $A_1, A_2,\ldots, A_n$, $A_j = (a_{1j}, a_{2j}\ldots, a_{mj})^T$ is the column of coefficients of variable x_j in coefficient matrix A.
Any simplex table of (1) will have m basic variables denoted as $x_1, x_2,\ldots, x_m$. The coefficients of these variables in the coefficient matrix A, taken together as columns, will constitute basis matrix

$B = (A_1, A_2, \ldots, A_m)$ and the corresponding solution $X_B = B^{-1}b$ is the basic solution given by any simplex table. C_B^T is a row vector of the costs of basic variables which are the coefficients of these variables in the objective function written in the order in which the basic variables appear in the simplex table. $C_B^T B^{-1}$ is called the simplex multiplier.

Any column in the simplex table say j^{th} column is $\alpha^j = B^{-1}A_j$. For initial simplex table $\alpha^j = A_j$ as $B = I$ (the identity matrix). Any entry in $c_j - Z_j$ row below variable x_j is $\Delta_j = c_j - Z_j = c_j - C_B^T B^{-1}A_j$. The results just discussed, we shall use in sensitivity analysis.

7.2.1 Change in Technological Coefficient a_{ij}'s of a Variable x_j

If we make some changes in the coefficients of a variable then two cases arise:

(i) The variable whose column is to be changed is a basic variable.
(ii) The variable whose column is to be changed is a non-basic variable.

In both the cases feasibility conditions are not disturbed as $X_B = B^{-1}b$ is independent of A_j but the optimality may be disturbed as $c_j - Z_j = c_j - C_B^T B^{-1}A_j$ depends on A_j. We shall illustrate this through an example.

7.2.2 Change in c_i's i.e., Costs (the Coefficients of x_i's in the Objective Function)

Here again two cases arise:

(i) c_i is to be changed is the cost of a non-basic variable.
(ii) c_i is to be changed is the cost of a basic variable.

In both these cases feasibility of new problem remains unchanged but the optimality may change. For case (i) only one entry $c_j - Z_j$ is to be evaluated. If $c_j - Z_j \leq 0$ then the solution of the existing problem remains optimal for the new problem also and write it down. In case (ii) calculate $c_j - Z_j$ for all non-basic variables. If $c_j - Z_j \leq 0$, $\forall j$ then the solution of the original problem remains optimal for the changed problem and write it down. If $c_j - Z_j > 0$ for some j, then prepare the new simplex table for the changed problem with new $c_j - Z_j$ entries and use the regular simplex method till the optimality conditions are satisfied and write the solution of the changed problem from the optimal table thus obtained.

7.2.3 Change in b_j's i.e., Availabilities (the Right-Hand Side of Constraints)

Because of any change in b_j's, solution of the changed problem may become infeasible as $X_B = B^{-1}b$. If the new basic solution is feasible, then the same simplex table will be the optimal table for the changed problem also and write the new solution. If new X_B gives infeasible solution, then use the dual simplex method to restore feasibility of the solution and write the solution of the changed problem.

7.2.4 Addition of a New Variable

Addition of a variable, adds one more column in the given optimal table of the problem which may or may not affect the optimality condition. If optimality is not affected, the solution of the given table will remain optimal for the changed problem also. If optimality conditions ($c_j - Z_j$) are affected then use the regular or simplex method to get the optimal solution.

7.2.5 Addition of a New Constraint

Addition of a constraint in the given problem means addition of one more row in the given optimal simplex table which may affect the feasibility of the solution. While adding the new constraint, first check whether the given optimal solution satisfies the constraint. If it satisfies then the same solution remains optimal for the new problem also and write it. If it does not satisfy the constraint then first express the given constraint in "≤" form if it is not in this form, add the slack variable and include it in the given optimal simplex table as a row of the added basic slack variable along with its unit column vector. This will disturb the standard format of the simplex table. Bring back the resulting simplex table in the standard format using elementary row operations and apply the dual simplex method if the basic solution thus obtained is infeasible till the optimal basic feasible solution is obtained. Note that in adding a constraint we never use the artificial variable.

7.2.6 Deletion of a Constraint

Step 1: First check whether the solution of the given optimal table satisfies the constraint to be deleted as strict equality?

(i) If not, then the same solution remains optimal for the changed problem. In other words deletion of the constraint does not affect the given optimal solution of the original problem.

(ii) If the constraint is satisfied as strict equality, then its deletion will affect the solution of the original problem.

To find the new solution for (2) use step 2.

Step 2: Introduce a new variable in the problem. Two cases arise:

Case I: If the ith constraint to be deleted is of '≤' or '=' type, then add the variable say $x_i^+ \geq 0$ to the ith constraint. In other words, add a column of new variable x_i^+ with cost '0' and column $(0, 0, 0, \ldots, 1, 0, 0, \ldots, 0)^T$ in the given problem, where entry 1 is the ith component of A_i^+.

Case II: If the ith constraint to be deleted is of '≥' type, then add the variable say $(-x_i^+)$, $x_i^+ \geq 0$ to the ith constraint. In other words add a column of new variable x_i^+ with cost '0' and column $(0, 0, 0, \ldots, -1, 0, 0, \ldots, 0)^T$ in the given problem, where entry –1 is the ith component of A_i^+.

Step 3: For both the cases in step 2, because of addition of a new variable, the optimality condition will be disturbed hence, use regular simplex method to restore it.

7.2.7 Deletion of a Variable

In deleting a variable, first check whether it is a basic or non-basic variable. If it is a non-basic variable, then the same solution will remain optimal for the changed problem also. If it is a basic variable, then multiply the row of the variable in the given table with (–1). This will make the basic solution of the simplex table infeasible. Use the dual simplex method to restore feasibility.

SOLVED EXAMPLES

Example 1: Consider the linear programming problem

Max $Z = x_1 + 2x_2$

s.t. $-x_1 + x_2 \le 1$ ...(1)

$x_1 + x_2 \le 2$ ...(2)

and $x_1, x_2 \ge 0.$

(i) Find the optimal solution of the above LPP.

(ii) From the optimal solution obtained in part (i), find the solution with the following changes in the above linear programming problem:

(a) Add variable $x_5 \ge 0$ with cost '2' and column $(1, 2)^T$.

(b) Delete variable x_1.

(c) Add constraint $x_1 - x_2 \le 1$.

(d) Add constraint $x_1 - x_2 \ge 1$.

(e) Delete constraint (1).

(f) Delete constraint (2).

(All the above parts are independent).

Solution: Given problem, written in standard form of linear programming problem is

Max $Z = x_1 + 2x_2$

s.t. $-x_1 + x_2 + x_3 = 1$

$x_1 + x_2 + x_4 = 2$

and $x_1, x_2, x_3, x_4 \ge 0$; x_3 and x_4 are slack variables.

Starting simplex table is

			Cost →	1	2	0	0	Minimum ratio
			Variable →	x_1	x_2	x_3	x_4	
Table number	C_B	Basic variable	X_B	α^1	α^2	α^3	α^4	$\frac{X_{Bi}}{\alpha_i^j}$
1	0	$\leftarrow x_3$	1	–1	[1]	1	0	$\frac{1}{1} = 1 \rightarrow$
	0	x_4	2	1	1	0	1	$\frac{2}{1} = 2$
		$c_j - Z_j$		1	2↑	0	0	
2	2	x_2	1	–1	1	1	0	
	0	$\leftarrow x_4$	1	[2]	0	–1	1	
		$c_j - Z_j$		3↑	0	–2	0	
3	2	x_2	3/2	0	1	1/2	1/2	
	1	x_1	1/2	1	0	–1/2	1/2	
		$c_j - Z_j$		0	0	–1/2	–3/2	

In the above table 3, $\Delta_j \leq 0$ for all j, therefore the solution is optimal. Hence, optimal solution is $x_1 = 1/2$, $x_2 = 3/2$ and Max. $Z = 7/2$.

(ii) (a) To add variable x_5, given $c_5 = 2$ and $A_5 = (1, 2)^T$.

$$\alpha^5 = B^{-1} A_5 = \begin{pmatrix} 1/2 & 1/2 \\ -1/2 & 1/2 \end{pmatrix} \begin{pmatrix} 1 \\ 2 \end{pmatrix} = \begin{pmatrix} 3/2 \\ 1/2 \end{pmatrix}, \; C_B{}^T = (2, 1)$$

$$c_j - Z_j = 2 - C_B{}^T B^{-1} A_5 = 2 - (2, 1) \begin{pmatrix} 3/2 \\ 1/2 \end{pmatrix} = 2 - \left(3 + \frac{1}{2}\right) = -\frac{3}{2}.$$

Simplex table for the changed problem is

		Cost →		1	2	2	0	0	Minimum
		Variable →		x_1	x_2	x_5	x_3	x_4	ratio
Table number	C_B	Basic variable	X_B	α^1	α^2	α^5	α^3	α^4	$\frac{X_{B_i}}{a_i^j}$
3	2	x_2	3/2	0	1	3/2	1/2	1/2	
	1	x_1	1/2	1	0	1/2	−1/2	1/2	
		$c_j - Z_j$		0	0	−3/2	−1/2	−3/2	

The optimal solution is $x_1 = 1/2$, $x_2 = 3/2$ and Max. $Z = 7/2$.

(ii) (b) To delete basic variable x_1, multiply x_1 row by (−1) and delete x_1 column from table in (i) to get the following simplex table

		Cost →		2	0	0	
		Variable →		x_2	x_3	x_4	Minimum ratio
Table number	C_B	Basic variable	X_B	α^2	α^3	α^4	
3	2	x_2	3/2	1	1/2	1/2	Solution is infeasible, use dual simplex method
	1	← x_1	−1/2	0	1/2	[−1/2]	
		$c_j - Z_j$		0	−1/2	−3/2 ↑	
4	2	x_2	1	1	1	0	
	0	x_4	1	0	−1	1	
		$c_j - Z_j$		0	−2	0	

The optimal solution is $x_1 = 0$, $x_2 = 1$ and Max. $Z = 2$.

(ii) (c) Optimal solution of the original problem is $x_1 = 1/2$, $x_2 = 3/2$ and the given constraint $x_1 - x_2 = (1/2) - (3/2) = -1 > 1$ is satisfied. The optimal solution of the changed problem is $x_1 = 1/2$, $x_2 = 3/2$ and Max $Z = 7/2$.

(ii) (d) The constraint $x_1 - x_2 \geq 1$ is not satisfied by the given optimal solution $x_1 = 1/2$, $x_2 = 3/2$. It can be written as $-x_1 + x_2 + x_5 = -1$, x_5 is the slack variable. Addition of this constraint in the problem, adds a row of basic variable x_5 and also a column of x_5 with cost '0' and column (0, 0, 1; 0) in the given optimal table. The resulting simplex table is

	Cost →			1	2	0	0	0	Minimum ratio $\frac{X_{B_i}}{\alpha_i^j}$
	Variable →			x_1	x_2	x_3	x_4	x_5	
Table number	$\mathbf{C_B}$	Basic variable	$\mathbf{X_B}$	α^1	α^2	α^3	α^4	α^5	
	2	x_2	3/2	0	1	1/2	1/2	0	
1	1	x_1	1/2	1	0	–1/2	1/2	0	
	0	x_5	–1	–1	1	0	0	1	
		$c_j - Z_j$		0	0	–1/2	–3/2	0	

From the above table we see that addition of x_5 row, disturbed the standard format of the simplex table as x_2 being the first basic variable its column should be $(1, 0, 0)^T$ and not $(1, 0, 1)^T$. Similarly, x_1 being the second basic variable its column should be $(0, 1, 0)^T$ and not $(0, 1, -1)^T$. To bring the above table in standard format add x_1 row to x_5 row to bring zero in place of (–1) and subtract x_2 row from x_5 row to get zero in place of 1. Applying these elementary row operations, the new table obtained is

	Cost →			1	2	0	0	0	Minimum ratio $\frac{X_{Bi}}{\alpha_i^j}$
	Variable →			x_1	x_2	x_3	x_4	x_5	
Table number	$\mathbf{C_B}$	Basic variable	$\mathbf{X_B}$	α^1	α^2	α^3	α^4	α^5	
	2	x_2	3/2	0	1	1/2	1/2	0	
1	1	x_1	1/2	1	0	–1/2	1/2	0	
	0	← x_5	–2	0	0	[–1]	0	1	
		$c_j - Z_j$		0	0	–1/2 ↑	–3/2	0	
	2	x_2	1/2	0	1	0	1/2	1/2	
2	1	x_1	3/2	1/2	0	0	1/2	–1/2	
	0	x_3	2	0	0	1	0	–1	
		$c_j - Z_j$		0	0	0	–3/2	–1/2	

The optimal solution is $x_1 = 3/2$, $x_2 = 1/2$ and Max. $Z = 5/2$.

(ii) (e) Deletion of a constraint means addition of a variable and hence can affect the optimality of the solution. To delete constraint $-x_1 + x_2 \leq 1$, we write it in the form $-x_1 + x_2 + x_3 - x_3^+ = 1$, $x_3, x_3^+ \geq 0$

$$-x_1 + x_2 + x_3 = 1 \text{ when } x_3^+ = 0$$

and $$-x_1 + x_2 - x_3^+ = 1 \text{ when } x_3 = 0$$

Thus, the constraint $-x_1 + x_2 \le 1$ is satisfied by the points lying on either side of the line $-x_1 + x_2 = 1$ and hence is deleted as its existence has no significance for the solution of the problem. Thus deletion of a constraint results into the addition of a variable. So we add variable x_3^+ with cost '0' and column $(-1, 0)^T$ in the given problem. Then we have

$$\alpha^{3+} = B^{-1} A_{3+}$$

$$= \begin{pmatrix} 1/2 & 1/2 \\ -1/2 & 1/2 \end{pmatrix} \begin{pmatrix} -1 \\ 0 \end{pmatrix} = \left(-\frac{1}{2}, \frac{1}{2}\right)$$

$$c_j - Z_j = c_j - C_B^T B^{-1} A_3^+$$

$$= 0 - (2, 1)\begin{pmatrix} -1/2 \\ 1/2 \end{pmatrix} = \frac{1}{2}$$

The simplex table for the changed problem is

Cost →				1	2	0	0	0	Minimum ratio
Variable →				x_1	x_2	x_3^+	x_3	x_4	$\frac{X_{B_i}}{\alpha_i^j}$
Table number	C_B	Basic variable	X_B	α^1	a^2	a^{3+}	a^3	a^4	
1	2	x_2	3/2	0	1	–1/2	1/2	1/2	
	1	$\leftarrow x_1$	1/2	1	0	[1/2]	–1/2	1/2	
		$c_j - Z_j$		0	0	1/2↑	–1/2	–3/2	
2	2	x_2	2	1	1	0	0	1	
	1	x_3^+	1	1	0	1	–1	1	
		$c_j - Z_j$		–1	0	0	0	–2	

The optimal solution is $x_1 = 0$, $x_2 = 2$ and Max. $Z = 4$.

(ii) (f) Again deletion of constraint (2) can affect the optimality condition if it is satisfied as strict equality by the given optimal solution.

Here the second constraint $x_1 + x_2 \le 2$ is satisfied as strict equality by the given optimal solution $x_1 = 1/2$ and $x_2 = 3/2$. It can be written as

$$x_1 + x_2 + x_4 - x_4^+ = 2,\ x_4,\ x_4^+ \ge 0$$

and x_4^+ is a variable with cost '0' in objective function and column $(0, -1)^T = A_4^+$.

$$\alpha^{4+} = B^{-1} A_4^+$$

$$= \begin{pmatrix} 1/2 & 1/2 \\ -1/2 & 1/2 \end{pmatrix} \begin{pmatrix} 0 \\ -1 \end{pmatrix} = \left(-\frac{1}{2}, -\frac{1}{2}\right)^T$$

$$c_j - Z_j = c_j - C_B^T B^{-1} A_4^+$$

$$= 0 - (2, 1)\begin{pmatrix} -1/2 \\ -1/2 \end{pmatrix} = \frac{3}{2}.$$

The simplex table for the new problem is

Cost →				1	2	0	0	0	Minimum ratio
Variable →				x_1	x_2	x_4^+	x_3	x_4	
Table number	$\mathbf{C_B}$	Basic variable	$\mathbf{X_B}$	α^1	α^2	α^{4+}	α^3	α^4	$\dfrac{X_{B_i}}{\alpha_i^j}$
	2	x_2	3/2	0	1	–1/2	1/2	1/2	
1	1	x_1	1/2	1	0	–1/2	–1/2	1/2	
	$c_j - Z_j$			0	0	3/2 ↑	–1/2	–3/2	

Here variable x_4^+ can enter but there is no variable to leave the basis as all column entries of column vector α^{4+} are negative. Hence the resulting problem has unbounded solution.

Example 2: Consider the linear programming problem

Max $$Z = 5x_1 + 6x_2$$

s.t. $$3x_1 + 4x_2 \leq 18 \quad \ldots(1)$$

$$2x_1 + x_2 \leq 7 \quad \ldots(2)$$

and $$x_1, x_2 \geq 0.$$

The optimal table of this LPP with x_3, x_4 as slack variables added to constraints (1) and (2) respectively is:

Cost →				5	6	0	0
Variable →				x_1	x_2	x_3	x_4
Table number	$\mathbf{C_B}$	Basic variable	$\mathbf{X_B}$	α^1	α^2	α^3	α^4
	6	x_2	3	0	1	2/5	–3/5
1	5	x_1	2	1	0	–1/5	4/5
	$c_j - Z_j$			0	0	–7/5	–2/5

Using sensitivity analysis, find the optimal solution with the following changes in above linear programming problem:

(a) Change b_2 (RHS of the second constraint) from (i) 7 to 11 (ii) 7 to 15.
(b) Change c_1 (the cost of x_1) from 5 to 1.
(c) Change c_1 from 5 to 1 and c_2 from 6 to 1.

(All the above parts are independent).

Solution: (a) (i) Change in b_2 from 7 to 11 can affect feasibility.

We calculate new $X_B = B^{-1}$ (new b) $= \begin{pmatrix} 2/5 & -3/5 \\ -1/5 & 4/5 \end{pmatrix} \begin{pmatrix} 18 \\ 11 \end{pmatrix} = \left(\dfrac{3}{5}, \dfrac{26}{5}\right)^T$, which is feasible.

The optimal solution is $x_1 = 26/5$, $x_2 = 3/5$ and Max. $Z = 148/5$.

(ii) New $X_B = (x_2, x_1)^T = B^{-1}$ (new b), where new $b = (18, 15)^T$.

or new $X_B = \begin{pmatrix} 2/5 & -3/5 \\ -1/5 & 4/5 \end{pmatrix}\begin{pmatrix} 18 \\ 15 \end{pmatrix} = \left(-\frac{9}{5}, \frac{42}{5}\right)^T$, which is infeasible.

Incorporating this in given optimal table, we have

Cost →				5	6	0	0	Minimum ratio
Variable →				x_1	x_2	x_3	x_4	
Table number	C_B	Basic variable	X_B	α^1	α^2	α^3	α^4	
1	6	$\leftarrow x_2$	–9/5	0	1	2/5	[–3/5]	
	5	x_1	42/5	1	0	–1/5	4/5	
	$c_j - Z_j$			0	0	–7/5	–2/5 ↑	
2	0	x_4	3	0	–5/3	–2/3	1	
	5	x_1	6	1	4/3	1/3	0	
	$c_j - Z_j$			0	–2/3	–5/3	0	

The optimal solution is $x_1 = 6$, $x_2 = 0$ and Max. $Z = 30$.

(b) c_1 is changed from 5 to 1. This can change the optimality condition.

$$C_B^T B^{-1} = (6, 1)\begin{pmatrix} 2/5 & -3/5 \\ -1/5 & 4/5 \end{pmatrix} = \left(\frac{11}{5}, -\frac{14}{5}\right)$$

For x_3 column, $c_j - Z_j = 0 - C_B^T B^{-1} A_3$

$$= 0 - \left(\frac{11}{5}, -\frac{14}{5}\right)\begin{pmatrix} 1 \\ 0 \end{pmatrix} = -\frac{11}{5}$$

For x_4 column, $c_j - Z_j = 0 - C_B^T B^{-1} A_4$

$$= 0 - \left(\frac{11}{5}, -\frac{14}{5}\right)\begin{pmatrix} 0 \\ 1 \end{pmatrix} = \frac{14}{5}$$

Simplex table for the changed problem is

Cost →				1	6	0	0	Minimum ratio
Variable →				x_1	x_2	x_3	x_4	
Table number	C_B	Basic variable	X_B	α^1	α^2	α^3	α^4	
1	6	x_2	3	0	1	2/5	–3/5	
	1	$\leftarrow x_1$	2	1	0	–1/5	[4/5]	
	$c_j - Z_j$			0	0	–11/5	14/5↑	

2	6	x_2	9/2	3/4	1	1/4	0	
	0	x_1	5/2	5/4	0	–1/4	1	
	$c_j - Z_j$			–7/2	0	–3/2	0	

Thus the optimal solution of the changed problem is $x_1 = 0$, $x_2 = 6$ and Max. $Z = 36$.

(c) For the changed problem, new $c_1 = 1$ and new $c_2 = 1$.

$$C_B^T B^{-1} = (1, 1)\begin{pmatrix} 2/5 & -3/5 \\ -1/5 & 4/5 \end{pmatrix} = \left(\frac{1}{5}, \frac{1}{5}\right).$$

For x_3 column, $c_j - Z_j = 0 - C_B^T B^{-1} A_3 = \left(\frac{1}{5}, \frac{1}{5}\right)\begin{pmatrix} 1 \\ 0 \end{pmatrix} = -\frac{1}{5}$.

For x_4 column, $c_j - Z_j = 0 - C_B^T B^{-1} A_4 = \left(\frac{1}{5}, \frac{1}{5}\right)\begin{pmatrix} 0 \\ 1 \end{pmatrix} = -\frac{1}{5}$.

Incorporating these changes in the simplex table, we get

Cost →				1	1	0	0	Minimum ratio
Variable →				x_1	x_2	x_3	x_4	
Table number	C_B	Basic variable	X_B	α^1	α^2	α^3	α^4	
1	1	x_2	3	0	1	2/5	–3/5	
	1	x_1	2	1	0	–1/5	4/5	
	$c_j - Z_j$			0	0	–1/5	–1/5	

Thus the optimal solution of the changed problem is $x_1 = 2$, $x_2 = 3$ and Max. $Z = 5$.

Example 3: The optimal table of the LPP is:

$$\text{Max } Z = -2x_1 - x_2$$

$$\text{s.t. } 3x_1 + x_2 = 3 \quad \ldots(1)$$

$$4x_1 + 3x_2 \geq 6 \quad \ldots(2)$$

$$x_1 + 2x_2 \leq 3 \quad \ldots(3)$$

and $x_1, x_2 \geq 0$.

Cost →				–2	–1	0	$-M$	$-M$	0
Variable →				x_1	x_2	x_3	x_4	x_5	x_6
Table number	C_B	Basic variable	X_B	α^1	α^2	α^3	α^4	α^5	α^6
1	–2	x_1	3/5	1	0	1/5	3/5	–1/5	0
	–1	x_2	6/5	0	1	–3/5	–4/5	3/5	0
	0	x_6	0	0	0	1	1	–1	1
	$c_j - Z_j$			0	0	–1/5	$(2/5) - M$	$(1/5) - M$	0

Where x_3 and x_6 are surplus and slack variables associated with constraints (2) and (3) respectively. x_4 and x_5 are artificial variables added to constraints (1) and (2) respectively both with cost $(-M)$, M is a large positive number. Using the optimal solution of table-1 above, find the solution of the problem with following changes in the above linear programming problem. (i) Column $(3, 4, 1)^T$ of coefficients of x_1 is changed to $(6, 0, -7)^T$.

Solution: (i) Column $(3, 4, 1)^T$ to be changed belongs to a basic variable x_1. So add a new variable x_1^+ with cost (-2) and column $(6, 0, -7)^T$. The variable x_1^+ will be the same as x_1 variable with cost (-2) and changed column $(6, 0, -7)^T$. We denote column and cost of x_1^+ by A_1^+ and c_1^+ respectively, then $c_1^+ = -2$ and $A_1^+ = (6, 0, -7)^T$.

$$C_B^T = (-2, -1, 0),\ \alpha^{1+} = B^{-1} A_1^+ = \begin{pmatrix} 3/5 & -1/5 & 0 \\ -4/5 & 3/5 & 0 \\ 1 & -1 & 1 \end{pmatrix} \begin{pmatrix} 6 \\ 0 \\ -7 \end{pmatrix} = \begin{pmatrix} 18/5 \\ -24/5 \\ -1 \end{pmatrix}$$

$$\text{New } c_j - Z_j = c_1^+ - C_B^T B^{-1} A_1^+ = -2 - (-2, -1, 0) \begin{pmatrix} 18/5 \\ -24/5 \\ -1 \end{pmatrix} = \frac{2}{5}.$$

Incorporating these changes in table (1) and omitting columns of artificial variables x_4, x_5, we get the following simplex table for the changed problem.

	Cost →			−2	−1	0	−2	0
	Variable →			x_1	x_2	x_3	x_1^+	x_6
Table number	C_B	Basic variable	X_B	α^1	α^2	α^3	α^{1+}	α^6
1	−2	← x_1	3/5	1	0	1/5	18/5	0
	−1	x_2	6/5	0	1	−3/5	−24/5	0
	0	x_6	0	0	0	1	−1	1
		$c_j - Z_j$		0	0	−1/5	2/5↑	0
2	−2	x_1^+	6		0	1/18	1	0
	−1	x_2	2		1	−6/18	0	0
	0	x_6	1/6		0	19/18	0	1
		$c_j - Z_j$			0	−2/9	0	0

Which is the optimal table of the changed problem. The optimal solution is $x_1 = x_1^+ = 1/6$, $x_2 = 2$ and Max $Z = -13/6$ (Max Z is obtained by putting the value of $x_1 = x_1^+$ and x_2 in the objective function).

EXERCISE 7.1

1. What is sensitivity analysis? Explain it. Which type of changes in sensitivity analysis affect the: (i) feasibility and (ii) optimality.

2. The linear programming problem is

$$\text{Max} \quad Z = 3x_1 + 5x_2$$
$$\text{s.t.} \quad x_1 \leq 4$$
$$x_2 \leq 6$$
$$3x_1 + 2x_2 \leq 18$$
$$\text{and} \quad x_1, x_2 \geq 0$$

Find the solution of the new problem using sensitivity analysis if c_2 is changed from 5 to 1.

3. Consider the linear programming problem

$$\text{Max} \quad Z = 2x_1 - x_2$$
$$\text{s.t.} \quad x_1 + x_2 \leq 2$$
$$x_1 - x_2 \leq 1$$
$$\text{and} \quad x_1, x_2 \geq 0$$

The optimal table of the LPP is

Cost →				2	–1	0	0
Variable →				x_1	x_2	x_3	x_4
Table number	C_B	Basic variable	X_B	α^1	α^2	α^3	α^4
	–1	x_2	1/2	0	1	1/2	–1/2
1	2	x_1	3/2	1	0	1/2	1/2
	$c_j - Z_j$			0	0	–1/2	0

Find the optimal solution of the above linear programming problem, using sensitivity analysis for the following changes:

(i) Cost c_1 (the coefficient of x_1 in objective function) is changed from 2 to 3.
(ii) Cost c_1 (the coefficient of x_1 in objective function) is changed from 2 to –1.
(iii) How much c_1 (= 2) can be changed so that the optimality is not disturbed and the above table remains the optimal table (or find the range of variation in c_1 for the same simplex table to remain optimal).

4. Solve the following linear programming problem

$$\text{Max} \quad Z = -x_1 + 2x_2 - x_3$$
$$\text{s.t.} \quad 3x_1 + x_2 - x_3 \leq 10$$
$$-x_1 + 4x_2 + x_3 \geq 6$$
$$x_2 + x_3 \leq 4$$
$$\text{and} \quad x_1, x_2, x_3 \geq 0$$

Using sensitivity analysis find the range in which b_3, the right hand side of the third constraint can be changed without affecting the feasibility.

5. Solve the following linear programming problem

$$\text{Max} \quad Z = 6x_1 - 4x_2$$
$$\text{s.t.} \quad x_1 + x_2 \geq 2 \qquad \ldots(1)$$

$$x_1 + 3x_2 \le 3 \qquad \ldots(2)$$
$$4x_2 \le 1 \qquad \ldots(3)$$
and $$x_1, x_2 \ge 0$$

Using sensitivity analysis find the solution of the new problem, for the following changes: (i) Delete constraint (3) (ii) Delete constraint (1) (iii) Delete constraint (2).

ANSWERS

2. $x_1 = 4$, $x_2 = 3$ and Max $Z = 15$.
3. (i) $x_1 = 3/2$, $x_2 = 1/2$ and Max $Z = 4$. (ii) $x_1 = 0$, $x_2 = 0$ and Max $Z = 0$. (iii) $1 \le c_1 < \infty$.
4. $3/2 \le b_3 \le 10$.
5. $x_1 = 6$, $x_2 = 0$ and Max $Z = 18$
   (i) $x_1 = 6$, $x_2 = 0$ and Max $Z = 18$. (ii) $x_1 = 3$, $x_2 = 0$ and Max $Z = 18$. (iii) Unbounded solution.

8

Chapter

Transportation Problem

8.1 INTRODUCTION

In this chapter, we shall learn, how to find the optimal solution of transportation problem starting with the initial basic feasible solution obtained by any one of the methods: (a) North West Corner rule, (b) Least-Cost Entry Method, (c) Vogel's Approximation Method (VAM). Solutions of unbalanced transportation problems and problems given in the maximization form have also been discussed.

8.2 FORMULATION OF TRANSPORTATION PROBLEM

Let us consider a transportation problem comprising three sources S_1, S_2 and S_3 with availabilities a_1, a_2 and a_3 and five destinations D_1, D_2, D_3, D_4 and D_5 with requirements b_1, b_2, b_3, b_4 and b_5. Let c_{ij} is the cost of transportation of one unit of the commodity (item) from S_i to D_j and x_{ij} are the units of the commodity be transported from S_i to D_j, $i = 1, 2,\ldots, m$; $j = 1, 2, \ldots, n$. To solve the problem first we put it in the tabular form (which is different from the simplex table) as:

Destination → *source* ↓	D_1	D_2	D_3	D_4	D_5	Availability ↓
S_1	$C_{11}(x_{11})$	$C_{12}(x_{12})$	$C_{13}(x_{21})$	$C_{14}(x_{21})$	$C_{15}(x_{21})$	a_1
S_2	$C_{21}(x_{21})$	$C_{22}(x_{22})$	$C_{23}(x_{23})$	$C_{24}(x_{24})$	$C_{25}(x_{25})$	a_2
S_3	$C_{31}(x_{31})$	$C_{32}(x_{32})$	$C_{33}(x_{33})$	$C_{34}(x_{34})$	$C_{35}(x_{35})$	a_3
Requirement →	b_1	b_2	b_3	b_4	b_5	

In the above table the rectangle obtained at the intersection of row and column may be called a cell. The cells in which allocation will be made are called basic cells and where allocation has not been made are called non-basic cells. In any balanced transportation problem with m sources and n destinations, there have to be $m + n - 1$ basic cells. The allocation made in basic cells gives us the basic feasible solution (BFS).

8.3 INITIAL BASIC FEASIBLE SOLUTION OF A TRANSPORTATION PROBLEM

The following steps are used to solve a transportation problem:

Step 1: Put the problem in tabular form.

Step 2: If the given transportation problem is unbalanced, then first balance it by introducing a additional row or column called the fictitious source or destination respectively.

For example: Consider the transportation problem

	D_1	D_2	D_3	*Availability*
S_1	4	3	2	8
S_2	2	5	0	13
S_3	3	8	6	12
Requirement	8	5	4	

Here $a_1 = 8$, $a_2 = 13$, $a_3 = 12$; $b_1 = 8$, $b_2 = 5$, $b_3 = 4$. So that

$$\sum_{i=1}^{3} \alpha_i = 33, \sum_{j=1}^{3} b_j = 17. \sum_{i=1}^{3} a_i \neq \sum_{j=1}^{3} b_j \text{ and } \sum_{i=1}^{3} a_i - \sum_{j=1}^{3} b_j = 16.$$ So it is unbalanced.

First we balance it. So we introduce an additional column of D_F (fictitious destination) and enter

zero's in this column in the position of costs and the difference 16 in the requirement row. This makes the problem a balanced transportation problem:

	D_1	D_2	D_3	D_F	*Availability*
S_1	4	3	2	0	8
S_2	2	5	0	0	13
S_3	3	8	6	0	12
Requirement	8	5	4	16	33

Step 3: If the problem is given as a maximization problem then first convert it into a minimization problem by multiplying each entry in the matrix with (–1).

Step 4: Find the initial basic feasible solution using any of the following methods:

1. North West Corner Rule (NWC)
2. Least Cost Entry Method (LCM) or Matrix Minima Method
3. Vogel's Approximation Method (VAM)

Step 5: Find the optimal solution starting from the solution obtained in step 5 by using MODI Method.

8.4 NORTH-WEST CORNER RULE

The following steps are used in this method:

Step 1: Allocate the maximum possible amount $\{x_{11} = \min(a_1, b_1)\}$ of the commodity in the extreme north-west corner of the table and neglect the row and column which is satisfied.

Step 2: Three cases may arise:

Case I: If $b_1 < a_1$ then we have $x_{11} = b_1$ and thus the first column is satisfied. Now some quantity $(a_1 - b_1)$ is left unallocated in the first row. And we move to the right hand side along the first row and make the second allocation of the amount $x_{12} = \min(a_1 - x_{11}, b_2)$ to the extreme north-west cell (1, 2). Thus, the first row is satisfied and we neglect the first row and first column for further allocation.

Case II: If $b_1 > a_1$ then we have $x_{11} = a_1$ and in this case the first row is satisfied. Now some quantity $(b_1 - a_1)$ is left unallocated in the first column. And we move vertically downward in the first column and make the second allocation of the amount $x_{21} = \min(a_2, b_1 - x_{11})$ to the extreme north-west cell (2, 1). Thus, the first row and column are satisfied and we neglect these for further allocation.

Case III: If $b_1 = a_1$ then $x_{11} = a_1 = b_1$ 0. Thus, both the first row and the first column are satisfied simultaneously. Now if we omit the satisfied row and column for further allocation then the number of basic variables will be less than $m + n - 1$. To overcome this difficulty, we imagine only one to be satisfied and put zero at other place treating as if this is not satisfied. This zero will be allocated somewhere and the solution obtained will be degenerate.

Step 3. Repeating steps 1 and 2 the allocation is completed.

Example 1: Use the north-west corner rule to determine as initial basic feasible solution to the following transportation problem:

Destination

		D_1	D_2	D_3	D_4	*Supply*
	O_1	6	4	1	5	14
Origin	O_2	8	9	2	7	16
	O_3	4	3	6	2	5
Demand		6	10	15	4	35

Solution: In the given transportation problem, the north-west corner is x_{11} position. The maximum possible allocation in x_{11} position $= \min(a_1, b_1) = \min(14, 6) = 6$. We enter 6 in x_{11} position and subtract 6 from a_1 and b_1. The row and column which is satisfied is not considered for any other allocation and is removed.

	D_1	D_2	D_3	D_4	
O_1	6(6)	4	1	5	~~14~~ 8
O_2	8	9	2	7	16
O_3	4	3	6	2	5
	~~6~~	10	15	4	

In above D_1 column is satisfied and is removed to get the resulting table:

	D_2	D_3	D_4	
O_1	4	1	5	8
O_2	9	2	7	16
O_3	3	6	2	5
	10	15	4	

Now we make maximum possible allocation in x_{13} position. The amount allocated is x_{12} = min (8, 10) = 8. We enter 8 in x_{12} position and subtract 8 from 8 and 10 to get the following table:

	D_2	D_3	D_4	
O_1	4(8)	1	5	~~8~~
O_2	9	2	7	16
O_3	3	6	2	5
	~~10~~ 2	15	4	

In above O_1 row is satisfied and is removed to get the following table:

	D_2	D_3	D_4	
O_2	9	2	7	16
O_3	3	6	2	5
	2	15	4	

Again we allocate maximum possible amount equal to min (2, 16) = 2 and enter this amount 2 in x_{22} position. Also subtract 2 from 2 and 16 to get the following table:

	D_2	D_3	D_4	
O_2	9(2)	2	7	~~16~~ 14
O_3	3	6	2	5
	~~2~~	15	4	

In the above table removing D_2 column which is satisfied we get the resulting table:

	D_3	D_4	
O_2	2	7	14
O_3	6	2	5
	15	4	

Now again we allocate maximum possible amount which is min (14, 15) = 14 and enter 14 in x_{23} position. Subtract 14 from 14 and 15 to get the table given below:

	D_3	D_4	
O_2	2(14)	7	~~14~~
O_3	6	2	5
	~~15~~ 1	4	

In above table satisfied O_2 column is removed to get the following table:

	D_3	D_4	
O_3	6	2	5
	1	4	

We allocate 1 in x_{33} and 4 in x_{34} position to complete the overall allocation so that the row totals and column totals are satisfied.

	D_3	D_4	
O_3	6(1)	2(4)	~~5~~
	~~1~~	~~4~~	

Thus, the resulting allocation made is shown in the following table:

		Destination				
		D_1	D_2	D_3	D_4	*Supply*
	O_1	6 (6)	4 (8)	1	5	14
Origin	O_2	8	9 (2)	2 (14)	7	16
	O_3	4	3	6 (1)	2 (4)	5
	Demand	6	10	15	4	35

The basic feasible solution is $x_{11} = 6$, $x_{12} = 8$, $x_{22} = 2$, $x_{23} = 14$, $x_{33} = 1$, $x_{43} = 4$.

8.5 LEAST COST ENTRY METHOD (LCM)

Here allocate the maximum possible amount to the cell with the lowest cost (c_{ij}). If the cell with the least cost is not unique, then make choice among the cells of lowest cost where more allocation can be made. In this process of allocation neglect the row or column which is satisfied for any more allocation. Repeat the process untill all a_i's and b_j's are exhausted.

Example 2: Using the least cost entry method determine the initial basic feasible solution to the following transportation problem:

	Destination				
	D_1	D_2	D_3	D_4	*Capacity*
O_1	1	2	3	4	6
O_2	4	3	2	0	8
O_3	0	2	2	1	10
Demand	4	6	8	6	24

Solution: Consider the problem of the given table. As we see in the cost matrix, the lowest cost is 0 in cells (2, 4) and (3, 1). Maximum amount that we can make in cell (2, 4) is min (8, 6) = 6. So we allocate 6 units in cell (2, 4). This satisfied the demand at D_4 and neglect D_4 column for further allocation. This gives $x_{24} = 6$.

	D_1	D_2	D_3	
O_1	1	2	3	6
O_2	4	3	2	2
O_3	0	2	2	10
	4	6	8	

In the reduced transportation table we find that there is least cost 0 in cell (3, 1). Here we allocate the maximum possible amount 4. Thus $x_{31} = 4$. This satisfied the demand at D_1. Leaving the D_1 column, we get the following table:

	D_2	D_3	
O_1	2	3	6
O_2	3	2	2
O_3	2	2	6
	6	8	

Now we see that in the above table the lowest cost is 2 in cell (1, 2), (2, 3), (3, 2) and (3, 3). Here we allocate the maximum possible amount 6 in x_{12}, x_{32} and x_{33}. Thus $x_{12} = 6$. This satisfied the requirement at O_1 and D_2. Leaving the O_1 row and D_2 column, we get the following table:

	D_3	
O_2	2	2
O_3	2	6
	8	

In the above reduced transportation table the lowest cost is 2 in cells (2, 3) and (3, 3). We allocate 2 in x_{23} and 6 in x_{33} position to complete the overall allocation. Thus the resulting allocation made is shown in the following table:

	Destination				
	D_1	D_2	D_3	D_4	*Capacity*
O_1	1	2 (6)	3	4	6
O_2	4	3	2 (2)	0 (6)	8
O_3	0 (4)	2	2 (6)	1	10
Demand	4	6	8	6	24

The basic feasible solution is $x_{12} = 6$, $x_{23} = 2$, $x_{24} = 6$, $x_{31} = 4$, $x_{41} = 7$, $x_{33} = 6$.

8.6 VOGEL'S APPROXIMATION METHOD (VAM)

Use the following steps to find a basic feasible solution by VAM:

Step 1: Find the difference between the lowest and the next higher cost for each row and column, called row and column penalties respectively and write them along side of the respective row or column.

Step 2: Now choose the row or column with the largest penalty. Allocate the maximum possible amount to the cell with lowest cost in the chosen row and column. If the largest penalty is not unique then break the tie among these by selecting the row or column having a cell with smaller cost. If still the tie persists then select the row or column with the minimum cost among the tied cells where more allocation can be made. If again the tie persists then break it arbitrarily.

Step 3: The satisfied row or column, omit from future consideration,

Step 4: Repeating the steps 1, 2 and 3 till all a_i's and b_j's are exhausted.

Example 3: Determine an initial basic feasible solution to the following transportation table using Vogel's approximation method:

		To				
		I	*II*	*III*	*IV*	*Supply*
From	A	5	1	3	3	34
	B	3	3	5	4	15
	C	6	4	4	3	12
	D	4	1	4	2	19
	Requirement	21	25	17	17	80

Solution: Consider the problem of the given table and compute the penalties for all rows and columns. Put the data in the form of the following table:

Table 1

	I	*II*	*III*	*IV*	*Available*	*Penalty*
A	5	1	3	3	34	(2)
B	3	3	5	4	15	(1)
C	6	4	4	3	12	(1)
D	4	1	4	2	19	(1)
Requirement	21	25	17	17		
penalty	(1)	(2)	(1)	(1)		

We find that maximum penalty (2) is for *A* row and II column. We choose that row or columns having a cell with smallest cost. Such are the two cells (1, 2) and (4, 2). But we can allocate more in (1, 2) and choose cell (1, 2) for allocation. Amount allocated in (1, 2) is minimum (34, 25) = 25, thus $x_{12} = 25$. As a result requirement at II is fulfilled and we omit II column for further allocation. The reduced is cost and requirement matrix is as in Table 2 given below:

Table 2

	I	*III*	*IV*	*Available*	*Penalty*
A	5	3	3	9	(2)
B	3	5	4	15	(1)
C	6	4	3	12	(1)
D	4	4	2	19	(2)
Requirement	21	17	17		
penalty	(1)	(1)	(1)		

We note that the amount still needed to requirement *A* is 9. Again we calculate the row and column penalties and entre them along sides the rows and columns as in the above table. Maximum penalty is 2 for rows *A* and *D*. We choose that rows having a cell with the smallest cost. But we can allocate more in (4, 4) and choose cell (4, 4) for allocation. Amount allocated in (4, 4) is minimum (19, 17) = 17, thus $x_{44} = 17$. As a result requirement at IV is fulfilled and we omit IV column for further allocation. The reduced is

Table 3

	I	*II*	*Available*	*Penalty*
A	5	3	9	(2)
B	3	5	15	(2)
C	6	4	12	(2)
D	4	4	2	(0)
Requirement	21	17		
penalty	(1)	(1)		

Now we again calculate the row and column penalties and enter them along sides the rows and columns as in above table. Maximum penalty is 2 for row *A*, *B* and *C*. We choose that rows having a cell with the smallest cost. Such are the two cells (1, 3), (2, 3) and (3, 3). We choose that row having a cell with smallest cost. Such is the cells (1, 3). Amount allocated in (1, 3) is minimum (9, 17) = 9, thus $x_{13} = 9$. As a result availability at *A* is fulfilled and we omit *A* row for further allocation. The reduced is

Table 4

	I	*III*	*Available*	*Penalty*
B	3	5	15	(2)
C	6	4	12	(2)
D	4	4	2	(0)
Requirement	21	8		
penalty	(1)	(1)		

Here we calculate the row and column penalties and enter them along sides the rows and columns as in the above table. Maximum penalty is 2 for row *B* and *C*. We choose that row having a cell with smallest cost. Such is the cells (2, 1). Amount allocated in (2, 1) is minimum (15, 21) = 15, thus $x_{21} = 15$. As a result requirement at *B* is fulfilled and we omit *B* row for further allocation. The reduced is

Table 5

	I	*III*	*Available*	*Penalty*
C	6	4	12	(2)
D	4	4	2	(0)
Requirement	6	8		
penalty	(2)	(0)		

Now we calculate the row and column penalties and enter them along sides the rows and columns as in the above table. Maximum penalty is 2 for row *C* and column III. We choose that row or columns having a cell with smallest cost. Such is the cells (3, 3). Amount allocated in (3, 3) is minimum (12, 8) = 8, thus $x_{33} = 8$. As a result requirement at III is fulfilled and we omit III column for further allocation. The reduced is

Table 6

	I	*Available*
C	6	4
D	4	2
Requirement	6	

We allocate 4 in x_{31} and 2 in x_{41} position to complete the overall allocation. Thus the resulting allocation made is shown in the following table:

Thus the resulting allocation made is shown in the following table:

Table 7

	I	*II*	*III*	*IV*	*Available*
A	5	1 (25)	3 (9)	3	34
B	3 (15)	3	5	4	15
C	6 (4)	4	4 (8)	3	12
D	4 (2)	1	4	2 (17)	19
Requirement	21	25	17	17	

The basic feasible solution is $x_{12} = 25$, $x_{13} = 9$, $x_{21} = 15$, $x_{31} = 4$, $x_{33} = 8$, $x_{41} = 2$, $x_{44} = 17$.

Example 4: Determine an initial basic feasible solution to the following transportation table using: (i) North-west corner method, (ii) least cost entry method, (iii) Vogel's approximation method:

	D_1	D_2	D_3	*Available*
S_1	2	7	4	5
S_2	3	3	1	8
S_3	5	4	7	7
S_4	1	6	2	14
Requirement	7	9	18	34

Solution: (i) In the given transportation problem north-west corner is x_{11} position. The maximum possible allocation in x_{11} position = min (a_1, b_1) = min (5, 7) = 5. We enter 5 in x_{11} position and subtract 5 from a_1 and b_1 to get the following table:

	D_1	D_2	D_3	
S_1	2(5)	7	4	~~5~~
S_2	3	3	1	8
S_3	5	4	7	7
S_4	1	6	2	14
	~~7~~ 2	9	18	34

In above S_1 row is satisfied and is removed to get the following table:

	D_1	D_2	D_3	
S_2	3	3	1	8
S_3	5	4	7	7
S_4	1	6	2	14
	2	9	18	

Now we make maximum possible allocation in x_{21} position. The amount allocated is x_{21} = min (2, 8) = 2. We enter 2 in x_{21} position and subtract 2 from 2 and 8 to get the following table:

	D_1	D_2	D_3	
S_2	3(2)	3	1	~~8~~ 6
S_3	5	4	7	7
S_4	1	6	2	14
	~~2~~	9	18	

In above D_1 column is satisfied and is removed to get the following table:

	D_2	D_3	
S_2	3	1	6
S_3	4	7	7
S_4	6	2	14
	9	18	

Again we allocate maximum possible amount equal to min (6, 9) = 6 and enter this amount 6 in x_{22} position. Also subtract 6 from 6 and 9 to get the following table:

	D_2	D_3	
S_2	3(6)	1	~~6~~
S_3	4	7	7
S_4	6	2	14
	~~9~~ 3	18	

In the above table removing S_2 row which is satisfied we get the resulting table:

	D_2	D_3	
S_3	4	7	7
S_4	6	2	14
	3	18	

Now again we allocate the maximum possible amount which is min (3, 7) = 3 and enter 3 in x_{32} position. Subtract 3 from 3 and 7 to get the table given below:

	D_2	D_3	
S_3	4(3)	7	~~7~~ 4
S_4	6	2	14
	~~3~~	18	

In above table satisfied D_2 column is removed to get the following table:

	D_3	
S_3	7	4
S_4	2	14
	18	

We allocate 4 in x_{33} and 14 in x_{43} position to complete the overall allocation so that the row totals and column totals are satisfied.

	D_3	
S_3	7 (4)	~~4~~
S_4	2 (14)	~~14~~
	~~18~~	

Thus, the resulting allocation made is shown in the following table:

	D_1	D_2	D_3	*Available*
S_1	2(5)	7	4	5
S_2	3(2)	3(6)	1	8
S_3	5	4(3)	7(4)	7
S_4	1	6	2(14)	14
Requirement	7	9	18	34

The basic feasible solution is $x_{11} = 5$, $x_{21} = 2$, $x_{22} = 6$, $x_{32} = 3$, $x_{33} = 4$, $x_{43} = 14$.
Total cost of transportation $= 5 \times 2 + 2 \times 3 + 6 \times 3 + 3 \times 4 + 4 \times 7 + 14 \times 2 = 102$.

(ii) As we see that in the given transportation problem, the lowest cost is 1 in cells (2, 3) and (4, 1). Maximum amount that we can make in cell (2, 3) is min (8, 18) = 8. So we allocate 8 units in cell (2, 3). This exhausts the availability from S_2 and neglect S_2 row for further allocation. This gives $x_{23} = 8$.

	D_1	D_2	D_3	
S_1	2	7	4	5
S_3	5	4	7	7
S_4	1	6	2	14
	7	9	10	

In the reduced transportation table we find that there is least cost 1 in cell (4, 1). Here we allocate the maximum possible amount 7. Thus $x_{41} = 7$. This satisfied the requirement at D_1.

Leaving the D_1 column, we get the following table:

	D_2	D_3	
S_1	7	4	5
S_3	4	7	7
S_4	6	2	7
	9	10	

Now we see that in the above table the lowest cost is 2 in cell (4, 3). Here we allocate the maximum possible amount 7. Thus $x_{43} = 7$. This satisfied the requirement at S_4. Leaving the S_4 row, we get the following table:

	D_2	D_3	
S_1	7	4	5
S_3	4	7	7
	9	3	

In the above reduced transportation table the lowest cost is 4 in cells (1, 3) and (3, 2). Maximum amount that we can make in cell (3, 2) is min (7, 9) = 7. So we allocate 7 units in cell (3, 2). This exhausts the availability from S_3 and neglect S_3 row for further allocation. This gives $x_{32} = 7$.

	D_2	D_3	
S_1	7	4	5
	2	3	

We allocate 2 in x_{12} and 3 in x_{13} position to complete the overall allocation.
Thus, the resulting allocation made is shown in the following table:

	D_1	D_2	D_3	*Available*
S_1	2	7 (2)	4 (3)	5
S_2	3	3	1 (8)	8
S_3	5	4 (7)	7	7
S_4	1 (7)	6	2 (7)	14
Requirement	7	9	18	34

The basic feasible solution is $x_{12} = 2$, $x_{13} = 3$, $x_{23} = 8$, $x_{32} = 7$, $x_{41} = 7$, $x_{43} = 7$. Total cost of transportation = $7 \times 2 + 4 \times 3 + 1 \times 8 + 4 \times 7 + 1 \times 7 + 2 \times 7 = 83$.

(iii) In the given transportation problem first we compute the penalties for all rows and column. Put the data in the form of the following table:

Table 1

	D_1	D_2	D_3	*Available*	*Penalty*
S_1	2	7	4	5	(2)
S_2	3	3	1	8	(2)
S_3	5	4	7	7	(1)
S_4	1	6	2	14	(1)
Requirement	7	9	18	34	
penalty	(1)	(1)	(1)		

Here we find that maximum penalty (2) is for S_1 and S_2 rows. But S_2 row has a cell with minimum cost 1 as compared to S_1 row with a cell with smaller cost 2. So we prefer to select S_2 row for all allocation and allocate maximum possible amount in cell (2, 3). Thus x_{23} = min (8, 18) = 8. This exhausts the availability from S_2. We omit S_2 row for further allocation. The reduced cost and requirement matrix is as table 2 given below:

Table 2

	D_1	D_2	D_3	*Available*	*Penalty*
S_1	2	7	4	5	(2)
S_3	5	4	7	7	(1)
S_4	1	6	2	14	(1)
Requirement	7	9	10		
pnalty	(1)	(2)	(2)		

We note that amount still needed to destination D_3 is 10. Again we calculate the row and column penalties and entre them along sides the rows and columns as in above table. Maximum penalty is 2 for columns D_2 and D_3 and also for row S_1. We choose that row or columns having a cell with smallest cost. Such are the two cells (1, 1) and (4, 3). But we can allocate more in (4, 3) and choose cell (4, 3) for allocation. Amount allocated in (4, 3) is minimum(10, 14) = 10, thus x_{43} = 10. As a result requirement at D_3 is fulfilled and we omit D_3 column for further allocation. There duced is

Table 3

	D_1	D_2	*Available*	*Penalty*
S_1	2	7	5	(5)
S_3	5	4	7	(1)
S_4	1	6	4	(5)
Requirement	7	9		
penalty	(1)	(2)		

Now again determine the row and column penalties and enter them along sides the rows and columns as in the above table. Maximum penalty is 5 for row S_1 and S_3. We choose that rows having a cell with smallest cost. Such are the two cells (1, 1) and (4, 1). But we can allocate more in (4, 1) and choose cell (4, 1) for allocation. Amount allocated in (4, 1) is minimum(4, 7) = 4, thus x_{41} = 4. As a result requirement at S_4 is fulfilled and we omit S_4 row for further allocation. The reduced is

Table 4

	D_1	D_2	*Available*	*Penalty*
S_1	2	7	5	(5)
S_3	5	4	7	(1)
Requirement	3	9		
penalty	(3)	(3)		

Here we find the row and column penalties and enter them along sides the rows and columns as in the above table. Maximum penalty is 5 for row S_1. We choose that column having a cell with the smallest cost. Such is the cell (1, 1). Amount allocated in (1, 1) is minimum (3, 5) = 3, thus $x_{11} = 3$. As a result requirement at D_1 is fulfilled and we omit D_1 column for further allocation. The reduced is

Table 5

	D_2	*Available*
S_3	7	2
S_1	4	7
Requirement	9	

We allocate 2 in x_{12} and 7 in x_{32} position to complete the overall allocation. Thus the resulting allocation made is shown in the following table:

Hence, the resulting allocation made is shown in the following table:

Table 6

	D_1	D_2	D_3	Available
S_1	2(3)	7(2)	4	5
S_2	3	3	1(8)	8
S_3	5	4(7)	7	7
S_4	1(4)	6	2(10)	14
Requirement	7	9	18	34

The basic feasible solution is $x_{11} = 3$, $x_{12} = 2$, $x_{23} = 8$, $x_{32} = 7$, $x_{41} = 4$, $x_{43} = 10$. Total cost of transportation $= 2 \times 3 + 7 \times 2 + 1 \times 8 + 4 \times 7 + 1 \times 4 + 2 \times 10 = 80$.

8.7 TEST OF OPTIMALITY OF TRANSPORTATION PROBLEM

The following steps are used to get the optimal solution by MODI method:

Step 1: Prepare a transportation table entering the costs c_{ij} at the upper right corner of the cells. Write the availabilities a_i's against each row on extreme right of the source row and requirements b_j's at the bottom of each destination column. Enter the initial BFS obtained by one of the three methods in appropriate cells within braces. The cells where allocation is made are called basic cells and remaining are non-basic cells.

Step 2: Associate dual variables u_i's to rows and v_j's to columns such that $u_i + v_j = c_{ij}$ holds good for basic cells. These will be $(m + n - 1)$ equations in $(m + n)$ unknowns when m is the number of rows (sources) and n is the number of columns (destinations). Solve these equations to find

u_i's and v_j's. It is convenient to take that dual variable zero which corresponds to the row or column which has maximum number of basic cells. Write down u_i's in rows on extreme right and v_j's in columns at the bottom of the table.

Step 3: After u_i's and v_j's are known, find $u_i + v_j - c_{ij} = d_{ij}$ for each non-basic cell and enter this number in the south-west corner of the cell.

(i) If at least one $d_{ij} > 0$, then the solution is not optimal. In that case go to step 4.

(ii) If all $d_{ij} \leq 0$, then the existing BFS is optimal. Write the optimal solution. If at least one $d_{ij} = 0$, then it is the indication of existence of an alternate optimal solution.

Step 4: (a) Entering variable rule: Choose the variable with the largest positive d_{ij} in south-west corner to enter the basis. To maintain the total member $m + n - 1$ of the basic variables, one of the existing basic variables has to leave the basis.

(b) Leaving variable rule: For this we construct a loop as follows. Start from the cell of entering variable, move horizontally and vertically taking turn at a basic cell only and return back to the starting cell. The closed loop thus obtained is unique and will have its edges at the basic cells.

(c) Suppose a is the maximum amount that can be allocated at the entering variable cell. Subtract and add a from the row edges and column edges respectively maintaining row and column totals. The value of a is the minimum of the allocations from which the amount a has been subtracted. This will be a new BFS. Now repeat steps 2, 3 and 4 until there is no positive entry in south-west corner of any non-basic cells. The solution thus obtained will be the optimal solution. For example,

Cost Matrix

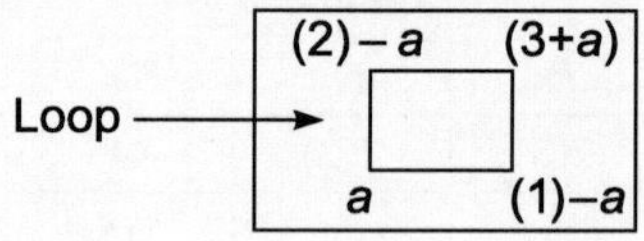

Here $a = \min(1, 2)$

Step 5: Write the solution of the problem and mention the value of the basic variables and the minimum cost.

Note: Degeneracy also crept into the solution while forming a loop to get the optimal solution. In a loop at the corners of the amount from which Q is subtracted. If the amount in a loop at the corners from which Q is subtracted are all different while applying the leaving variable rule the cell with minimum allocation will become non-basic. However, if there are more than one such basic cells, then any one of them may be considered non-basic cells and in the remaining basic cells zero amount is allocated. The BFS thus obtained will be degenerate.

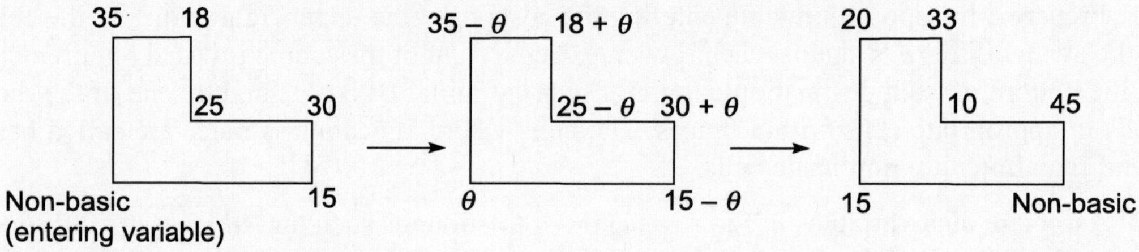

Here $\theta = \min(15, 25, 35) = 15$.

If loop is such that we get zero at more than one place then only one cell is kept empty and zero is entered at other basic cells. The BFS thus obtained will be degenerate.

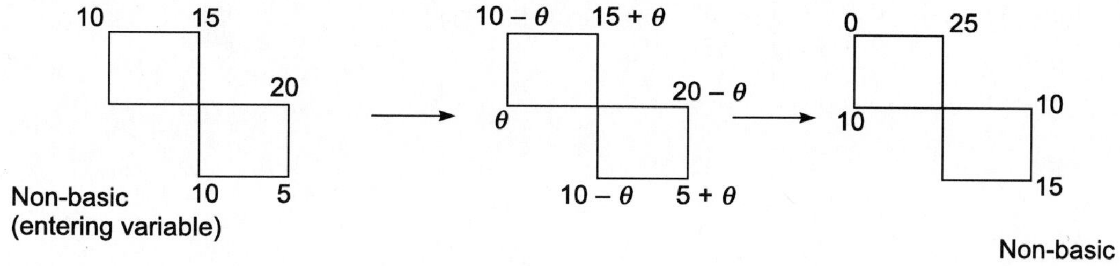

Here $\theta = \min(10, 20, 10) = 10$.

Example 5: Determine the optimum basic feasible solution to the following transportation problem:

	D_1	D_2	D_3	D_4	$a_i \downarrow$
O_1	5	3	6	2	19
O_2	4	7	9	1	37
O_3	3	4	7	5	34
$b_j \rightarrow$	16	18	31	25	

Solution: Step 1: First we find the initial basic feasible solution using VAM, obtained as follows:

	D_1	D_2	D_3	D_4	
O_1	5	3 (18)	6 (1)	2	19
O_2	4 (12)	7	9	1 (25)	37
O_3	3 (4)	4	7 (30)	5	34
	16	18	31	25	

The total transportation cost = 18 × 3 + 1 × 6 + 12 × 4 + 25 × 1 + 4 × 3 + 30 × 7 = Rs. 355.

Step 2: Calculate u_i's and v_j's taking $u_i + v_j = c_{ij}$ for basic cells

$u_1 + v_2 = 3$, $u_1 + v_3 = 6$, $u_2 + v_1 = 4$, $u_2 + v_4 = 1$, $u_3 + v_1 = 3$ and $u_3 + v_3 = 7$.

Solving these with $u_1 = 0$, we get

$u_1 = 0$, $u_2 = 2$, $u_3 = 1$, $v_1 = 2$, $v_2 = 3$, $v_3 = 6$ and $v_4 = -1$.

					$u_i \downarrow$
	5	3 (18)	6 (1)	2	0
	4 (12)	7	9	1 (25)	2
	3 (4)	4	7 (30)	5	1
$v_j \rightarrow$	2	3	6	–1	

Step 3: Now calculate $d_{ij} = u_i + v_j - c_{ij}$ for non-basic cells and enter this number in south-west corner of the cell

$$d_{11} = u_1 + v_1 - c_{11} = 0 + 2 - 5 = -3; \quad d_{14} = u_1 + v_4 - c_{14} = 0 - 1 - 2 = -3;$$
$$d_{22} = u_2 + v_2 - c_{22} = 2 + 3 - 7 = -2; \quad d_{23} = u_2 + v_3 - c_{23} = 2 + 6 - 9 = -1;$$
$$d_{32} = u_3 + v_2 - c_{32} = 1 + 3 - 4 = 0; \quad d_{34} = u_3 + v_4 - c_{34} = 1 - 1 - 5 = -5.$$

					$u_i \downarrow$
	5 −3	3 (18)	6 (1)	2 −3	0
	4 (12)	7 −2	9 −1	1 (25)	2
	3 (4)	4 0	7 (30)	5 −5	1
$v_j \rightarrow$	2	3	6	−1	

Here all d_{ij}'s are non-positive, i.e., $d_{ij} \leq 0$. Thus existing BFS is optimal. Hence the solution of the problem is $x_{12} = 18$, $x_{13} = 1$, $x_{21} = 12$, $x_{24} = 25$, $x_{31} = 4$, $x_{33} = 30$ and the minimum transportation cost $= 18 \times 3 + 1 \times 6 + 12 \times 4 + 25 \times 1 + 4 \times 3 + 30 \times 7 =$ Rs. 355.

Note: In this problem one $d_{ij} = 0$ ($d_{32} = 0$) which is the indication of existence of an alternate optimum solution.

Example 6: Solve the following transportation problem in which the cell entries represent unit cost:

	To			Available
	2	7	4	5
	3	3	1	8
From	5	4	7	7
	1	6	2	14
Required	7	9	18	

Solution: **Step 1:** First we find the initial basic feasible solution using VAM, obtained as follows:

2 (5)	7	4	5
3	3	1 (8)	7
5	4 (7)	7	8
1 (2)	6 (2)	2 (10)	14
7	9	18	

The total transportation cost = 2 × 5 + 1 × 8 + 4 × 7 + 1 × 2 + 6 × 2 + 2 × 10 = Rs. 80.

Step 2: Calculate u_i's and v_j's taking $u_i + v_j = c_{ij}$ for basic cells

$u_1 + v_1 = 2$, $u_2 + v_3 = 1$, $u_3 + v_2 = 4$, $u_4 + v_1 = 1$, $u_4 + v_2 = 6$ and $u_4 + v_3 = 2$.

Solving these with $u_4 = 0$, we get

$u_1 = 1$, $u_2 = -1$, $u_3 = -2$, $u_4 = 0$, $v_1 = 1$, $v_2 = 6$ and $v_3 = 2$.

				$u_i \downarrow$
	2 (5)	7	4	1
	3	3	1 (8)	–1
	5	4 (7)	7	–2
	1 (2)	6 (2)	2 (10)	0
$v_j \rightarrow$	1	6	2	

Step 3: Now calculate $d_{ij} = u_i + v_j - c_{ij}$ for non-basic cells and enter this number in south-west corner of the cell

$d_{12} = u_1 + v_2 - c_{12} = 1 + 6 - 7 = 0$; $d_{13} = u_1 + v_3 - c_{13} = 1 + 2 - 4 = -1$;
$d_{21} = u_2 + v_1 - c_{21} = -1 + 1 - 3 = -3$; $d_{22} = u_2 + v_2 - c_{22} = -1 + 6 - 3 = 2$;
$d_{31} = u_3 + v_1 - c_{31} = -2 + 1 - 5 = -6$; $d_{33} = u_3 + v_3 - c_{33} = -2 + 2 - 7 = -7$.

				$u_i \downarrow$
	2 (5)	7, 0	4, –1	1
	3, –3	3, 2	1 (8)	–1
	5, –6	4 (7)	7, –7	–2
	1 (2)	6 (2)	2 (10)	0
$v_j \rightarrow$	1	6	2	

Here all d_{ij}'s are not negative. Thus existing BFS is not optimal. The solution can be improved as shown in the next step.

Step 4: Here $d_{22} = 2$ is most positive, so we select the variables x_{22} to enter the basis. Now to maintain the member of basic variable some of the existing basic variable has to leave the basis. For this we construct a closed loop starting from the cell of entering variable

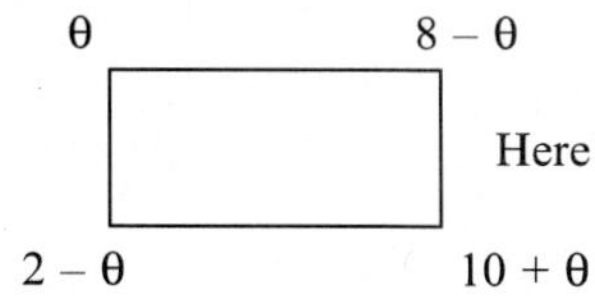

Here = min (2, 8) = 2.

and

				$u_i \downarrow$
	(5) 2	0 7	−1 4	1
	−3 3	(2) 3	(6) 1	−1
	−6 5	(7) 4	−7 7	−2
	(2) 1	6	(12) 2	0
$v_j \rightarrow$	1	6	2	

The total transportation cost

$$= 2 \times 5 + 3 \times 2 + 1 \times 6 + 4 \times 7 + 1 \times 2 + 2 \times 12 = \text{Rs. } 76.$$

Now again we shall test the optimality. Proceeding steps 2, 3 and 4, we get the following table:

				$u_i \downarrow$
	(5) 2	0 7	−1 4	1
	−3 3	(2) 3	(6) 1	−1
	−6 5	(7) 4	−7 7	0
	(2) 1	0 6	(12) 2	0
$v_j \rightarrow$	1	4	2	

Here all d_{ij}'s are non-positive, i.e., $d_{ij} \leq 0$. Thus existing BFS is optimal.

Hence, the solution of the problem is $x_{11} = 5, x_{22} = 2, x_{23} = 6, x_{32} = 7, x_{41} = 2, x_{43} = 12$

and the minimum transportation cost $= 2 \times 5 + 3 \times 2 + 1 \times 6 + 4 \times 7 + 1 \times 2 + 2 \times 12$

$= \text{Rs. } 76.$

Example 7: A manufacture wants to ship 8 loads of his product as shown in the table. The matrix gives the mileage from origin O to destination D. Shipping costs are Rs. 10 per load per mile. What shipping schedule should be used?

	D_1	D_2	D_3	*Available*
O_1	50	30	220	1
O_2	90	45	170	3
O_3	250	200	50	4
Required	4	2	2	

Solution: Step 1: First we find the initial basic feasible solution using VAM, obtained as follows:

	D_1	D_2	D_3	
O_1	50 (1)	30	220	1
O_2	90 (3)	45 (0)	170	3
O_3	250	200 (2)	50 (2)	4
	4	2	2	

or another alternate solution is

	D_1	D_2	D_3	
O_1	50 (1)	30	220	1
O_2	90 (3)	45	170	3
O_3	250 (0)	200 (2)	50 (2)	4
	4	2	2	

Here one of the starting basic variable is zero, so the solution obtained is degenerate basic solution.

The total transportation cost

$$= 1 \times 50 + 3 \times 90 + 0 \times 45 + 2 \times 200 + 2 \times 50 = 820 \text{ i.e., Rs } 8200.$$

or the total transportation cost

$$= 1 \times 50 + 3 \times 90 + 0 \times 250 + 2 \times 200 + 2 \times 50 = 820 \text{ i.e., Rs } 8200.$$

Step 2: Calculate u_i's and v_j's taking $u_i + v_j = c_{ij}$ for basic cells

$$u_1 + v_1 = 50,\ u_2 + v_1 = 90,\ u_2 + v_2 = 45,\ u_3 + v_2 = 200,\ u_3 + v_3 = 50.$$

Solving these with $u_3 = 0$, we get

$$u_1 = -195,\ u_2 = -155,\ u_3 = 0,\ v_1 = 245,\ v_2 = 200,\ v_3 = 50.$$

				$u_i \downarrow$
	50 (1)	30	220	–195
	90 (3)	45 (0)	170	–155
	250	200 (2)	50 (2)	0
$v_j \rightarrow$	245	200	50	

Step 3: Now calculate $d_{ij} = u_i + v_j - c_{ij}$ for non-basic cells and enter this number in south-west corner of the cell as follow:

				$u_i \downarrow$
	50 (1)	30 −25	220 −365	−195
	90 (3)	45 (0)	170 −275	−155
	250 −5	200 (2)	50 (2)	0
$v_j \rightarrow$	245	200	50	

Here all d_{ij}'s are non-positive, i.e., $d_{ij} \leq 0$. Thus, existing BFS is optimal.
Hence, the solution of the problem is $x_{11} = 1$, $x_{21} = 3$, $x_{22} = 0$, $x_{32} = 2$, $x_{33} = 2$
and the minimum transportation cost $= 1 \times 50 + 3 \times 90 + 0 \times 45 + 2 \times 200 + 2 \times 50 = 820$ i.e., Rs 8200.

8.8 UNBALANCED TRANSPORTATION PROBLEM

If in a transportation problem $\sum_{i=1}^{m} a_i \neq \sum_{j=1}^{n} b_j$, then it is a unbalanced transportation problem. There are two different situations of getting unbalanced transportation problem:

(i) Excess of availabilities. $\left(\text{i.e., if } \sum_{i=1}^{m} a_i > \sum_{j=1}^{n} b_j\right)$ then introduce a dummy destinations (column) denoted by D_F with zero costs and write the amount $\left(\sum_{i=1}^{m} a_i - \sum_{j=1}^{n} b_j\right)$ its requirement.

Example 8: Consider a transportation problem

	D_1	D_2	D_3	$a_i \downarrow$
S_1	4	1	0	10
S_2	3	2	4	7
S_3	7	7	2	14
S_4	9	6	4	15
$b_j \rightarrow$	5	15	10	

Solution: Here $\sum_{i=1}^{4} a_i = 10 + 7 + 14 + 15 = 46$ and $\sum_{j=1}^{3} b_j = 5 + 15 + 10 = 30$ $\sum_{i=1}^{4} a_i > \sum_{j=1}^{3} b_j$

and is an unbalanced transportation problem. To make this a balanced transportation problem, we introduce a dummy destination D_F (column) with amount $b_4 = \sum_{i=1}^{m} a_i - \sum_{j=1}^{n} b_j = 46 - 30 =$ 16 and with zero costs. The balanced problem thus obtained is:

	D_1	D_2	D_3	D_F	$a_i \downarrow$
S_1	4	1	0	0	10
S_2	3	2	4	0	7
S_3	7	7	2	0	14
S_4	9	6	4	0	15
$b_j \rightarrow$	5	15	10	16	

(ii) Shortage in availabilities $\left(\text{i.e., if } \sum_{i=1}^{m} a_i > \sum_{j=1}^{n} b_j\right)$ then introduce a dummy source (row) with zero costs and write the amount $\left(\sum_{j=1}^{n} b_j - \sum_{i=1}^{m} a_i\right)$ its availabilities. The amount transported from dummy source to a destination is the shortage at that destination. Consider a transportation problem

	D_1	D_2	D_3	D_4	$a_i \downarrow$
S_1	12	0	22	11	12
S_2	11	7	94	20	8
S_3	0	14	15	18	7
$b_j \rightarrow$	5	15	15	10	

Here $\sum_{i=1}^{3} a_i = 12 + 8 + 7 = 27$ and $\sum_{j=1}^{4} b_j = 5 + 15 + 15 + 10 = 45$. Thus, $\sum_{i=1}^{3} a_i < \sum_{j=1}^{4} b_j$, demand is more than requirement i.e., it is an unbalanced transportation problem. We add a fictitious (dummy) source S_F with availability $b_4 = \sum_{j=1}^{4} b_j - \sum_{i=1}^{3} a_i = 18$ and prices zeros in S_F row to get the following balanced transportation problem:

	D_1	D_2	D_3	D_4	$a_i \downarrow$
S_1	12	0	22	11	10
S_2	11	7	94	20	8
S_3	0	14	15	18	7
S_4	0	0	0	0	18
$b_j \rightarrow$	5	15	15	10	

Example 9: Solve the following unbalanced transportation problem (symbols have their usual meanings).

	D_1	D_2	D_3	$a_i \downarrow$
O_1	4	3	2	10
O_2	2	5	0	13
O_3	3	8	6	12
$b_j \rightarrow$	8	5	4	

Solution: In the given problem,

$$\sum_{i=1}^{3} a_i = 10 + 13 + 12 = 35 \text{ and } \sum_{j=1}^{3} b_j = 8 + 5 + 4 = 17.$$

Thus, $\sum_{i=1}^{3} a_i > \sum_{j=1}^{4} b_j$ the problem is unbalanced. To convert this problem in balanced problem

we introduce a fictitious destination D_4 with requirement $35 - 17 = 18$ having all the transportation costs equal to zero. The balanced transportation table is given below:

	D_1	D_2	D_3	D_4	$a_i \downarrow$
O_1	4	3	2	0	10
O_2	2	5	0	0	13
O_3	3	8	6	0	12
$b_j \rightarrow$	8	5	4	18	

Step 1: First we find the initial basic feasible solution using VAM, obtained as follows:

	D_1	D_2	D_3	D_4	$a_i \downarrow$
	4	3 (5)	2	0 (5)	10
	2 (8)	5	0 (4)	0 (1)	13
	3	8	6	0 (12)	12
$b_j \rightarrow$	8	5	4	18	

The total transportation cost

$$= 3 \times 5 + 0 \times 5 + 2 \times 8 + 0 \times 4 + 0 \times 1 + 0 \times 12 = \text{Rs. } 31.$$

Step 2: Calculate u_i's and v_j's taking $u_i + v_j = c_{ij}$ for basic cells

$$u_1 + v_2 = 3,\ u_1 + v_4 = 0,\ u_2 + v_1 = 2,\ u_2 + v_3 = 0,\ u_2 + v_4 = 0,\ u_3 + v_4 = 0.$$

Solving these with $u_2 = 0$, we get

$$u_1 = 0,\ u_2 = 0,\ u_3 = 0,\ v_1 = 2,\ v_2 = 3,\ v_3 = 0 \text{ and } v_4 = 0.$$

10. Obtain an optimum BFS to the following degenerate transportation problem:

	To			Available
From	7	3	4	2
	2	1	3	3
	3	4	6	5
Demand	4	1	5	

11. Solve the following transportation problem:

	To				Supply
From	21	16	25	13	11
	17	18	14	23	13
	32	27	18	41	19
Demand	6	10	12	15	

12. Solve the following cost minimizing transportation problem:

	D_1	D_2	D_3	D_4	D_5	*Available*
O_1	5	5	6	4	2	9
O_2	6	9	7	8	5	13
O_3	5	6	4	6	3	9
Requirement	3	7	8	5	8	27

13. Solve the following transportation problem:

	D_1	D_2	D_3	D_4	D_5	D_6	*Available*
O_1	9	12	9	6	9	10	5
O_2	7	3	7	7	5	5	6
O_3	6	5	9	11	3	11	2
O_4	6	8	11	2	2	10	9
Required	4	4	6	2	4	4	22

14. Solve the following transportation problem:

	To			
	D_1	D_2	D_3	$a_i \downarrow$
O_1	7	4	0	5
O_2	6	8	0	15
O_3	3	9	0	9
$b_j \rightarrow$	15	6	8	

15. Solve the following cost minimizing transportation problem:

	D_1	D_2	D_3	D_4	*Available*
O_1	5	3	1	5	15
O_2	10	7	12	4	11
O_3	7	5	8	4	13
Demand	8	12	13	6	

16. Solve the following transportation problem:

							Available
A	5	3	7	3	8	5	3
B	5	6	12	5	7	11	4
C	2	1	2	4	8	2	2
D	9	6	10	5	10	9	8
Required	3	3	6	2	1	2	

17. Determine an optimum BFS to the following transportation problem:

	D_1	D_2	D_3	D_4	*Capacity*
O_1	1	2	3	4	6
O_2	4	3	2	0	8
O_3	0	2	2	1	10
Demand	4	6	8	6	24

Where O_i and D_j denote i^{th} origin and j^{th} destination respectively.

18. A company has three plants P_1, P_2, P_3 and three warehouses W_1, W_2, W_3. Number of units available at the plants is 60, 70 and 80 respectively. Demand at W_1, W_2, W_3 are 50, 80 and 80 respectively. Unit costs of transportation are as follows:

	W_1	W_2	W_3
P_1	8	7	3
P_2	3	8	9
P_3	11	3	5

What should be your transportation plan? Give minimum distribution cost.

19. Given below the unit cost array with supplies a_i; $i = 1, 2, 3$ and demand b_j; $j = 1, 2, 3$ and 4.

					Available units
	6	1	9	3	70
	11	5	2	8	55
	10	12	4	7	90
Required units	85	35	50	45	

Find the optimum solution to the above problem.

20. The cost-transportation table for the transportation problem is given below:

	W_1	W_2	W_3	W_4	W_5	*Available*
F_1	4	3	1	2	6	40
F_2	5	2	3	4	5	30
F_3	3	5	6	3	2	20
F_4	2	4	4	5	3	10
Required	30	30	15	20	5	

Obtain the optimal solution of the problem.

21. A steel company has three open hearth furnaces and five rolling mills. Transportation costs (rupees per quintal) for shipping steel from furnaces to rolling mills are shown in the following table:

		Mills					
		M_1	M_2	M_3	M_4	M_5	*Available*
	F_1	4	2	3	2	6	8
Furnaces	F_2	5	4	5	2	1	12
	F_3	6	5	4	7	3	14
Required		4	4	6	8	8	

What is the optimal shipping schedule?

ANSWERS

2. $x_{11} = 2, x_{21} = 1, x_{22} = 3, x_{23} = 2, x_{33} = 2, x_{43} = 5.$
3. $x_{11} = 3, x_{12} = 1, x_{22} = 2, x_{23} = 4, x_{24} = 2, x_{34} = 3, x_{35} = 6.$
4. $x_{11} = 9, x_{12} = 8, x_{14} = 17, x_{23} = 15, x_{31} = 12, x_{42} = 17, x_{43} = 2.$
5. (i) $x_{11} = 20, x_{13} = 10, x_{22} = 20, x_{23} = 20, x_{24} = 10, x_{32} = 20.$
   (ii) $x_{11} = 20, x_{13}=10, x_{22} = 20, x_{23} = 20, x_{24} = 10, x_{32} = 20.$
6. (i) $x_{11}=5, x_{12} = 2, x_{22} = 6, x_{23} = 3, x_{33} = 4, x_{34} = 14$. Minimum transportation cost = Rs. 1015.
   (ii) $x_{14} = 7, x_{21} = 2, x_{23} = 7, x_{31} = 3, x_{32} = 8, x_{34} = 7$. Minimum transportation cost = Rs. 814.
   (iii) $x_{11} = 5, x_{14} = 2, x_{23} = 7, x_{24} = 2, x_{32} = 8, x_{34} = 10$. Minimum transportation cost = Rs. 779.
7. $x_{13} = 8, x_{24} = 4, x_{25} = 3, x_{31} = 5, x_{32} = 4, x_{41} = 1, x_{43} = 2, x_{52} = 4, x_{55} = 1$. Minimum transportation cost is Rs. 1102.
8. $x_{11} = 20, x_{13} = 10, x_{22} = 20, x_{23} = 20, x_{24} = 10, x_{32} = 20$. Minimum transportation cost is Rs. 180.
9. $x_{11} = 6, x_{13} = 3, x_{15} = 4, x_{24} = 4, x_{25} = 2, x_{32} = 7, x_{34} = 1$. Minimum transportation cost is Rs. 163.
10. $x_{13} = 2, x_{22} = 1, x_{23} = 2, x_{31} = 42, x_{33} = 1$. Minimum transportation cost is Rs. 33.
11. $x_{14}= 11, x_{21} = 6, x_{22} = 3, x_{24} = 4, x_{32} = 7, x_{33} = 12$. Minimum transportation cost is Rs. 796.

12. $x_{12} = 4$, $x_{14} = 5$, $x_{21} = 3$, $x_{22} = 2$, $x_{25} = 8$, $x_{32} = 1$, $x_{33} = 8$. Minimum transportation cost is Rs. 154.
13. $x_{13} = 5$, $x_{22} = 3$, $x_{23} = 1$, $x_{26} = 2$, $x_{31} = 1$, $x_{32} = 1$, $x_{41} = 3$, $x_{44} = 2$, $x_{45} = 4$. Minimum transportation cost is Rs. 112.
14. $x_{12} = 5$, $x_{21} = 6$, $x_{22} = 1$, $x_{23} = 8$, $x_{31} = 9$. Minimum transportation cost is Rs. 91.
15. $x_{11} = 8$, $x_{12} = 7$, $x_{22} = 5$, $x_{24} = 6$, $x_{33} = 13$. Minimum transportation cost is Rs. 224.
16. $x_{12} = 1$, $x_{16} = 2$, $x_{21} = 3$, $x_{25} = 1$, $x_{32} = 2$, $x_{42} = 2$, $x_{43} = 4$, $x_{44} = 2$. Minimum transportation cost is Rs. 101.
17. $x_{12} = 6$, $x_{23} = 2$, $x_{24} = 6$, $x_{31} = 4$, $x_{33} = 6$. Minimum transportation cost is Rs. 28.
18. $x_{13} = 60$, $x_{21} = 50$, $x_{23} = 20$, $x_{32} = 80$. Minimum transportation cost is Rs. 750.
19. $x_{11} = 25$, $x_{12} = 2$, $x_{14} = 23$, $x_{23} = 40$, $x_{32} = 30$. Minimum transportation cost is Rs. 848.
20. $x_{11} = 5$, $x_{13} = 15$, $x_{14} = 20$, $x_{22} = 30$, $x_{31} = 15$, $x_{35} = 5$, $x_{41} = 10$. Minimum transportation cost is Rs. 210.
21. $x_{11} = 4$, $x_{12} = 4$, $x_{24} = 8$, $x_{25} = 4$, $x_{33} = 6$, $x_{35} = 4$. Minimum transportation cost is Rs. 80.

9 Chapter

Assignment Problems

9.1 INTRODUCTION

Every business faces the problem of assigning duties to various persons for performing different jobs. Each person can perform every job but they differ in their efficiency for doing different jobs. In this chapter, we shall discuss to tackle minimal, maximal, unbalanced and assignment problems under certain restrictions using Hungarian Algorithm. Travelling salesman problem using Hungarian Algorithm has also been discussed. Let us define

$$x_{ij} = \begin{cases} 1, & \text{if } i^{th} \text{ person is assigned } j^{th} \text{ job} \\ 0, & \text{if } i^{th} \text{ person is not assigned } j^{th} \text{ job} \end{cases}$$

$$c_{ij} = \text{inefficiency of } i^{th} \text{ person in doing } j^{th} \text{ job.}$$

Now question arises as how to assign jobs to persons such that overall inefficiency of doing jobs is minimized. Minimize $Z = \sum_{j=1}^{n} \sum_{i=1}^{n} c_{ij}\, x_{ij}$

Subject to

$\sum_{i=1}^{n} x_{ij} = 1, j = 1, 2, \ldots, n$ i.e., i^{th} person is assigned only one job. $\sum_{j=1}^{n} x_{ij} = 1, i = 1, 2, \ldots, n$ i.e., j^{th} job is assigned to only one person. Assignment problem can be represented in the form of following matrix called effectiveness (cost) matrix:

Jobs

		J_1	J_2	J_3	…	…	…	J_n
	P_1	c_{11}	c_{12}	c_{13}	…	…	…	c_{1n}
	P_2	c_{21}	c_{22}	c_{23}	…	…	…	c_{2n}
Persons	P_3	c_{31}	c_{32}	c_{33}	…	…	…	c_{3n}
	…	…	…	…	…	…	…	…
	…	…	…	…	…	…	…	…
	…	…	…	…	…	…	…	…
	P_n	C_{n1}	c_{n2}	c_{n3}	…	…	…	c_{nn}

9.2 FORMULATION OF ASSIGNMENT PROBLEM USING HUNGARIAN ALGORITHM

To get the optimum assignment using Hungarian Algorithm, the following steps are used:

Step 1: First we find the minimum entry in each row of the cost matrix and subtract it from all other entries of that row to get the reduced matrix having at least one zero entry in each row.

Step 2: Repeat step 1 for each column to get the second reduced matrix containing at least one zero in every column.

Step 3: Now go through the first row of the matrix obtained in step 2. If there is only one zero, then surround this zero by a square as $\boxed{0}$ and cross (X) all other zeros in the column of this surrounded zero. Repeat it for all the rows. If there is more than one zero in any row then skip it and pass onto the next row. Repeat this procedure for columns also.

Step 4: If we get exactly one surrounded zero in every row and column, then the complete optimal assignment is done. The mark $\boxed{0}$ represents the assignment of the job to respective person and we stop at this stage. If the total surrounded zeros are not exactly n, then we go to next step 5.

Step 5: Draw the minimum member of vertical or horizontal lines to cover all zeros. Select the smallest element, say x among all elements uncovered by these lines. Subtract this value x from all elements not covered by the lines and add x to all those elements that lie at the intersection of the lines drawn. Repeat steps 3 and 4 untill we get the optimum assignment.

Remark: To draw the minimum number of lines the following procedure can be followed. If a zero is surrounded while going around a row (column), then draw a vertical (horizontal) line through this zero. When this is done for all surrounded zeroes, it will give minimum number of lines covering all (surrounded or crossed) zeroes.

9.3 MAXIMAL ASSIGNMENT PROBLEM

If the given assignment problem is a maximization problem, then we convert it into the minimization assignment problem by applying any of the following methods:

(i) Multiply every entry of the assignment matrix by (–1) and proceed as usual.

(ii) Choose the largest entry in the assignment matrix and replace each entry after subtracting it from this largest entry and then proceed as usual.

9.4 UNBALANCED ASSIGNMENT PROBLEM

When number of jobs (tasks) is not equal to the number of persons (facilities) in an assignment problem, then it is called an unbalanced problem. To solve an unbalanced assignment problem, we first balance the problem by adding a dummy (fictitious) row or column with zero costs in new created cells. Now solve the problem as usual using Hungarian algorithm.

9.5 ASSIGNMENT PROBLEM WITH RESTRICTIONS

Due to some unavoidable reasons in the assignment problem a particular job is not assigned to a particular person. To tackle such type of problem, assign a very high cost ∞ in the corresponding cell and solve the problem as usual. For example,

Persons

Jobs	P_1	…	…	P_j	…	…	P_n
J_1	…	…	…	c_{1j}	…	…	…
	…	…	…	c_{2j}	…	…	…
	…	…	…	…	…	…	…
J_i	c_{i1}	c_{i2}	…	c_{ij}	…	…	c_{in}
…	…	…	…	…	…	…	…
…	…	…	…	…	…	…	…
J_n	…	…	…	c_{nj}	…	…	…

Given cost matrix

→

Persons

Jobs	P_1	…	…	P_j	…	…	P_n
J_1	…	…	…	c_{1j}	…	…	…
	…	…	…	c_{2j}	…	…	…
	⋮	⋮	⋮	⋮	⋮	⋮	⋮
J_i	c_{i1}	c_{i2}	…	∞	…	…	c_{in}
…	…	…	…	…	…	…	…
…	…	…	…	…		…	…
J_n	…	…	…	c_{nj}	…	…	…

Modified cost matrix

In particular,

Jobs	1	2	3
J_1	2	3	5
J_2	6	4	Not Suitable
J_3	3	2	8

Given cost matrix

→

Jobs	1	2	3
J_1	2	3	5
J_2	6	4	∞
J_3	3	2	8

Modified cost matrix

But in some situations when a particular person is assigned only a specific job then assign a very high cost ∞ in all the column cells except the corresponding cell where we keep the entry unchanged.

For example,

Given cost matrix (Persons across the top, Jobs down the side)

	P_1	…	…	P_j	…	…	P_n
J_1	…	…	…	c_{1j}	…	…	…
	…	…	…	c_{2j}	…	…	…
	…	…	…	…	…	…	…
Jobs J_i	c_{i1}	c_{i2}	…	c_{ij}	…	…	c_{in}
…	…	…	…	…	…	…	…
…	…	…	…	…	…	…	…
J_n	…	…	…	c_{nj}	…	…	…

→

Modified cost matrix (Persons across the top, Jobs down the side)

	P_1	…	…	P_j	…	…	P_n
J_1	…	…	…	∞	…	…	…
	…	…	…	∞	…	…	…
	…	…	…	…	…	…	…
Jobs J_i	c_{i1}	c_{i2}	…	c_{ij}	…	…	c_{in}
…	…	…	…	…	…	…	…
…	…	…	…	…	…	…	…
J_n	…	…	…	∞	…	…	…

In particular,

Jobs	1	2	3
J_1	2	3	Not Suitable
J_2	6	4	4
J_3	3	2	Not Suitable

Given cost matrix

→

Jobs	1	2	3
J_1	2	3	∞
J_2	6	4	4
J_3	3	2	∞

Modified cost matrix

9.6 TRAVELLING SALESMAN PROBLEM

A travelling salesman is to visit certain number of cities. The costs (distances or time) involved in going from one city to another city for all city pairs are known. Now the problem is, how to choose the route so as to visit all the cities once and return back to the city from where he started with minimum cost (distance or time) involved. This problem can be solved using the Hungarian algorithm.

Note. Sometimes when the assignment is not possible we use the method of numeration by assigning the next minimum element of the matrix in place of zero.

SOLVED EXAMPLES

Example 1: A department head has four subordinates, and four tasks have to be performed. Subordinates differ in efficiency and tasks differ in their intrinsic difficulty. Time each man would

Tasks \ Subordinates	I	II	III	IV
A	8	26	17	11
B	13	28	4	26
C	38	19	18	15
D	19	26	24	10

take to perform each task is given in the effectiveness matrix below. How should the tasks be allocated, one to a man, so as to minimize the total man hours?

Solution:

Step 1: First we find the minimum entry in each row of the cost matrix and subtract it from all other entries of that row to get the reduced matrix having at least one zero entry in each row.

	I	II	III	IV
A	0	18	9	3
B	9	24	0	22
C	23	4	3	0
D	9	16	14	0

Step 2: Repeat step 1 for each column to get the second reduced matrix containing at least one zero in every column.

	I	II	III	IV
A	0	14	9	3
B	9	20	0	22
C	23	0	3	0
D	9	12	14	0

Step 3: Now go through the first row of the matrix obtained in step 2. If there is only one zero, then surround this zero by a square as □ and cross all these zeros in the column of this surrounded zero. Repeat it for all the rows. If there is more than one zero in any row then skip it and pass onto the next row. Repeat this procedure for columns also. We get the following table:

	I	II	III	IV
A	[0]	14	9	3
B	9	20	[0]	22
C	23	[0]	3	~~0~~
D	9	12	14	[0]

In the above table we see that all zeros are either marked by □ or cancelled and there is an assignment □ in every row and every column. Hence we get the solution of the given problem. The optimum solution is $A \to$ I, $B \to$ III, $C \to$ II, $D \to$ IV.

Example 2: Find the optimal assignment for the problem with the given cost matrix

	1	2	3	4
A	10	12	19	11
B	5	10	7	8
C	12	14	13	11
D	8	15	11	9

Solution:

Step 1: First we find the minimum entry in each row of cost matrix and subtract it from all other entries of that row to get the reduced matrix having at least one zero entry in each row.

	1	2	3	4
A	0	2	9	1
B	0	5	2	3
C	1	3	2	0
D	0	7	3	1

Step 2: Repeat step 1 for each column to get the second reduced matrix containing at least one zero in every column.

	1	2	3	4
A	0	0	7	1
B	0	3	0	3
C	1	1	0	0
D	0	5	1	1

Step 3: Now go through the first row of the matrix obtained in step 2. There are two zeros in this row and hence skip it and pass onto the next row. In the fourth row only one zero is there, so surround this zero by a square as □. Repeat this procedure for columns also. We get the following table:

~~0~~	[0]	7	1
~~0~~	3	[0]	3
1	1	~~0~~	[0]
[0]	5	1	1

In the above table we see that all zeros are either marked by □ or cancelled and there is an assignment □ in every row and every column. Hence, we get the solution of the given problem.

The optimum solution is $A \rightarrow 2$, $B \rightarrow 3$, $C \rightarrow 4$, $D \rightarrow 1$.

Example 3: An airlines that operates seven days a week has a time table shown below. Crews must have minimum layover of 6 hours between flights. Obtain the pairing of flights that minimizes layover time away from home. For any given pairing the crew will be based at the city that results in the smaller layover.

Delhi-Dehradun			*Dehradun-Delhi*		
Flight No.	*Departure*	*Arrival*	*Flight No.*	*Departure*	*Arrival*
1	7.00 A.M.	9.00 A.M.	101	9.00 A.M.	11.00 A.M.
2	9.00 A.M.	11.00 A.M.	102	10.00 A.M.	12.00 Noon
3	1.30 P.M.	3.30 P.M.	103	3.30 P.M.	5.30 P.M.
4	7.30 P.M.	9.30 P.M.	104	8.00 P.M.	10.00 P.M.

For each pair also mention the town where the crew should be based.

Solution: First we prepare the tables for layover times between the flights. Consider we pair the flight no. 101 when the crew is based at Delhi. The time of stay at Dehradun will be the layover time away from home. Now a plane of flight no. 1 which reaches Dehradun at 9.00 A.M., cannot fly at 9.00 A.M. on the same day from Dehradun as the minimum layover time is 6 hours. So it will depart from Dehradun on the next day at 9.00 A.M. which will result in a layover time of 24 hours. Similarly, other layover time can be calculated.

Tables for layover times in hours

Crew based at Delhi

	101	102	103	104
1	24	25	6.5	11
2	22	23	28.5	9
3	17.5	18.5	24	28.5
4	11.5	12.5	18	22.5

Crew based at Dehradun

	1	2	3	4
101	20	22	26.5	8.5
102	19	21	25.5	7.5
103	13.5	15.5	20	26
104	9	11	27.5	21.5

Here we are assuming either the layover times in terms of quarter hour (0.25 hours or 15 minutes) as one unit of time or the layover times for weeks, to avoid the fractions. Thus multiplying by 4 in each table.

Crew based at Delhi

	101	102	103	104
1	96	100	26	44
2	88	92	114	36
3	70	74	96	114
4	46	50	72	90

Crew based at Dehradun

	1	2	3	4
101	80	88	106	34
102	76	84	102	30
103	54	62	80	104
104	36	44	110	86

For combining the above two tables, we choose that base which gives a lesser layover time for each pairing. The layover times marked with'*' denote that crew is based at Dehradun; otherwise the crew is based at Delhi.

	101	102	103	104
1	80*	76*	26	36*
2	88*	84*	62*	36
3	70	74	80*	110*
4	34*	30*	72	86*

Step 1: First we find the minimum entry in each row of the cost matrix and subtract it from all other entries of that row to get the reduced matrix having at least one zero entry in each row.

	101	102	103	104
1	54*	50*	0	10*
2	52*	58*	26*	0
3	0	4	10*	40*
4	4*	0*	42	56*

Step 2: Repeat step 1 for each column to get the second reduced matrix containing at least one zero in every column.

	101	102	103	104
1	54*	50*	0	10*
2	52*	58*	26*	0
3	0	4	10*	40*
4	4*	0*	42	56*

Step 3: Now go through the first row of the matrix obtained in step 2. If there is only one zero, then surround this zero by a square as □ and cross all these zeros in the column of this surrounded zero. Repeat it for all the rows. If there is more than one zero in any row then skip it and pass onto the next row. Repeat this procedure for columns also. We get the following table:

	101	102	103	104
1	54*	50*	[0]	10*
2	52*	58*	26*	[0]
3	[0]	4	10*	40*
4	4*	[0]*	42	56*

In the above table we see that all zeros are either marked by □ and there is an assignment □ in every row and every column. Hence, we get the solution of the given problem. The optimum solution is 1 → 103, 2 → 104, 3 → 101, 4 → 103, i.e., crew at Dehradun-Delhi-Delhi-Dehradun.

Example 4: A marketing manager has 5 salesmen and 5 sales districts. Considering the capabilities of the salesmen and the nature of districts, the marketing manager estimates that sales per month (in hundred Rs.) for each salesman in each district would be as follows:

		Districts				
		A	*B*	*C*	*D*	*E*
Person	1	32	38	40	28	40
	2	40	24	28	21	36
	3	41	27	33	30	37
	4	22	38	41	36	36
	5	29	33	40	35	39

Find the assignment of salesmen to districts that will result in maximization sales.

Solution: First we convert the given maximization problem into minimization problem by multiplying each element by (– 1). The table thus obtained is

–32	–38	–40	–28	–40
–40	–24	–28	–21	–36
–41	–27	–33	–30	–37
–22	–38	–41	–36	–36
–29	–33	–40	–35	–39

Step 1: First we find the minimum entry in each row of cost matrix and subtract it from all other entries of that row to get the reduced matrix having at least one zero entry in each row.

8	2	0	12	0
0	16	12	19	4
0	14	8	11	4
19	3	0	5	5
11	7	0	5	1

Step 2: Repeat step 1 for each column to get the second reduced matrix containing at least one zero in every column.

8	0	0	7	0
0	14	12	14	4
0	12	8	6	4
19	1	0	0	5
11	5	0	0	1

Step 3: Now go through the first row of the matrix obtained in step 2. There are three zeros in this row and hence skip it and pass onto the next row. In the second row only one zero is there, so surround this zero by a square as □ and cross all these zeros in the column of this surrounded zero. Repeat it for all the rows. If there is more than one zero in any row then skip it and pass onto the next row. Repeat this procedure for columns also. We get the following table:

8	[0]	~~0~~	7	~~0~~
[0]	14	12	14	4
~~0~~	12	8	6	4
19	1	[0]	~~0~~	5
11	5	~~0~~	[0]	1

Step 4: Here we see that in the above table row 3 and columns 5 have no assignment, then we go to next step.

Step 5: Draw the minimum number of vertical or horizontal lines to cover all zeros.

8	[0]	~~0~~	7	~~0~~
[0]	14	12	14	4
~~0~~	12	8	6	4
19	1	[0]	~~0~~	5
11	5	~~0~~	[0]	1

Select the smallest element, among all elements uncovered by lines, i.e., 4. Subtract this value 4 from all elements not covered by the lines and add 4 to all those elements that lie at the intersection of the lines.

12	0	0	7	0
0	10	8	10	0
0	8	4	2	0
23	1	0	0	5
15	5	0	0	1

Repeat steps 3 and 4 to get the optimum assignment.

12	[0]	~~0~~	7	~~0~~
[0]	10	8	10	~~0~~
~~0~~	8	4	2	[0]
23	1	[0]	~~0~~	5
15	5	~~0~~	[0]	1

In the above table we see that all zeros are either marked by □ or cancelled and there is an assignment □ in every row and every column. Hence, we get the solution of the given problem. The optimum solution is $1 \rightarrow B$, $2 \rightarrow A$, $3 \rightarrow E$, $4 \rightarrow C$, $5 \rightarrow D$. The maximum sales = 38 + 40 + 37 + 41 + 35 = 191, i.e., 19100.

Example 5: A department head has four tasks to be performed with three subordinates. Subordinates differ in efficiency. The estimates of the time, each subordinate would take to perform, are given below in the matrix. How should he allocate the tasks, one to each man, so as to minimize the total man hours?

		Subordinates		
		1	2	3
Tasks	I	9	26	15
	II	13	27	6
	III	35	20	15
	IV	18	30	20

Solution: The given problem is unbalanced. To solve the problem, we first balance the problem by adding a dummy (fictitious) column with zero costs in new created cells.

	1	2	3	4
I	9	26	15	0
II	13	27	6	0
III	35	20	15	0
IV	18	30	20	0

Step 1: First we find the minimum entry in each row of cost matrix and subtract it from all other entries of that row to get the reduced matrix having at least one zero entry in each row.

	1	2	3	4
I	9	26	15	0
II	13	27	6	0
III	35	20	15	0
IV	18	30	20	0

Step 2: Repeat step 1 for each column to get the second reduced matrix containing at least one zero in every column.

	1	2	3	4
I	0	6	9	0
II	4	7	0	0
III	26	0	9	0
IV	9	10	14	0

Step 3: Now go through the first row of the matrix obtained in step 2. There are two zeros in this row and hence skip it and pass onto the next row. In the fourth row only one zero is there, so surround this zero by a square as □. Repeat this procedure for columns also. We get the following table:

	1	2	3	4
I	[0]	6	9	~~0~~
II	4	7	[0]	~~0~~
III	26	[0]	9	~~0~~
IV	9	10	14	0

In the above table we see that all zeros are either marked by □ or cancelled and there is an assignment □ in every row and every column. Hence we get the solution of the given problem. The optimum solution is I → 1, II → 3, III → 2, task IV remains unsigned. The total time (man hours) = 9 + 6 + 20 = 35.

Example 6: Four engineers are available to design four projects. Engineer 2 is not competent to design the project B. Given the following time estimates needed to each engineer to design a given project, find how the engineers should be assigned to projects so as to minimize the total design time of four projects.

		Projects			
		A	*B*	*C*	*D*
	1	12	10	10	8
Engineers	2	14	Not suitable	15	11
	3	6	10	16	4
	4	8	10	9	7

Solution: To solve the problem, we assign a very high cost say ∞ in c_{22} cell. Thus the given problem of the above table is reduced to

12	10	10	8
14	∞	15	11
6	10	16	4
8	10	9	7

Step 1: First we find the minimum entry in each row of the cost matrix and subtract it from all other entries of that row to get the reduced matrix having at least one zero entry in each row.

4	2	2	0
3	∞	4	0
2	6	12	0
1	3	2	0

Step 2: Repeat step 1 for each column to get the second reduced matrix containing at least one zero in every column.

3	0	0	0
2	∞	2	0
1	4	10	0
0	1	0	0

Step 3: Now go through the first row of the matrix obtained in step 2. There are two zeros in this row and hence skip it and pass onto the next row. In the second row only one zero is there, so surround this zero by a square as □ and cross all these zeros in the column of this surrounded zero. Repeat it for all the rows. If there is more than one zero in any row then skip it and pass onto the next row. Repeat this procedure for columns also. We get the following table:

3	[0]	~~0~~	~~0~~
2	∞	2	[0]
1	4	10	~~0~~
[0]	1	~~0~~	~~0~~

Step 4: Here we see that in the above table row 3 and columns 3 have no assignment, then we go to the next step.

Step 5: Draw (by trail) minimum number of vertical or horizontal lines to cover all zeros.

3	[0]	~~0~~	~~0~~
2	∞	2	[0]
1	4	10	~~0~~
[0]		~~0~~	~~0~~

Select the smallest element, among all the elements uncovered by lines, i.e., 1. Subtract this value 1 from all elements not covered by the lines and add 1 to all those elements that lie at the intersection of the lines.

3	0	0	1
1	∞	1	0
0	3	9	0
0	1	0	1

Repeat steps 3 and 4 to get the optimum assignment.

3	[0]	~~0~~	1
1	∞	1	[0]
[0]	3	9	~~0~~
~~0~~	1	[0]	1

In the above table we see that all zeros are either marked by □ or cancelled and there is an assignment □ in every row and every column. Hence we get the solution of the given problem. The optimum solution is $1 \to B$, $2 \to D$, $3 \to A$, $4 \to C$. The total minimum time = 10 + 11 + 6 + 9 = 36.

Example 7: Solve the 'travelling salesman problem' given by the following data: $c_{12} = 4$, $c_{13} = 7$, $c_{14} = 3$, $c_{23} = 6$, $c_{24} = 3$ and $c_{34} = 7$, where $c_{ij} = c_{ji}$

Solution: From the given data, the cost matrix is as follows:

	1	2	3	4
1	∞	4	7	3
2	4	∞	6	3
3	7	6	∞	7
4	3	6	7	∞

Step 1: First we find the minimum entry in each row of cost matrix and subtract it from all other entries of that row to get the reduced matrix having at least one zero entry in each row.

	1	2	3	4
1	∞	1	4	0
2	1	∞	3	0
3	1	0	∞	1
4	0	3	4	∞

Step 2: Repeat step 1 for each column to get the second reduced matrix containing at least one zero in every column.

	1	2	3	4
1	∞	1	1	0
2	1	∞	0	0
3	1	0	∞	1
4	0	3	1	∞

Step 3: Now go through the first row of the matrix obtained in step 2. If there is only one zero, then surround this zero by a square as □ and cross all these zeros in the column of this surrounded zero. Repeat it for all the rows. If there is more than one zero in any row then skip it and pass onto the next row. Repeat this procedure for columns also. We get the following table:

	1	2	3	4
1	∞	1	1	[0]
2	1	∞	[0]	~~0~~
3	1	[0]	∞	1
4	[0]	3	1	∞

In the above table we see that all zeros are either marked by □ or cancelled and there is an assignment □ in every row and every column. Hence we get the solution of the given problem. The optimum solution is $1 \to 4$, $2 \to 3$, $3 \to 2$, $4 \to 1$, i.e, $1 \to 4 \to 1$, $2 \to 3 \to 2$.

But it is not the solution of the travelling salesman problem. Now we try to get the next best solution which satisfies the extra condition of visiting all the cities. The smallest element in the table other than zero is 1. So we try to bring 1 into the solution. Since 1 occurs at 6 places, so we consider all cases and keep the acceptable solution.

Case I: Shifting the assignment from cell (1, 4) to cell (1, 3) then using step 3, we get

	1	2	3	4
1	∞	1	[0]	~~0~~
2	1	∞	~~0~~	[0]
3	1	[0]	∞	1
4	[0]	3	1	∞

In the above table we see that all zeros are either marked by □ or cancelled and there is an assignment □ in every row and every column. Hence, we get the solution of the given problem. The optimum solution is $1 \to 3 \to 2 \to 4 \to 1$ and the total minimum cost = 7 + 6 + 3 + 3 = 19.

Case II: If we shift the assignment from cell (1, 4) to cell (1, 2) then using step 3, we get

	1	2	3	4
1	∞	$\boxed{0}$	1	~~0~~
2	1	∞	$\boxed{0}$	~~0~~
3	1	~~0~~	∞	$\boxed{1}$
4	$\boxed{0}$	3	1	∞

In the above table we see that all zeros are either marked by □ or cancelled and there is an assignment □ in every row and every column. Hence we get the solution of the given problem. The optimum solution is $1 \to 2 \to 3 \to 4 \to 1$ and the total minimum cost = 4 + 6 + 7 + 3 = 20.

In all other cases the cost in the reduced matrix will be more than as obtained in case I. Hence the optimum solution is $1 \to 3 \to 2 \to 4 \to 1$ and the total optimum cost = 7 + 6 + 3 + 3 = 19.

EXERCISE 9.1

1. Write a short note on assignment algorithm (Hungarian assignment method).
2. Write a short note on assignment problem under certain restriction and travelling salesman problem.
3. A computer centre has got three expert programmers. The centre needs three application programmes to be developed. The head of the computer centre, after studying carefully the programmes to be developed, estimates the computer time in minutes required by the experts to the application programmes as follows:

		Programme		
		A	*B*	*C*
Programmer	1	120	100	80
	2	70	90	110
	3	110	140	120

   Assign the programmers to the programmes in such a way that the total computer time is least.

4. Solve the following minimal assignment problems:

		Man				
		1	2	3	4	5
Job	I	12	8	7	15	4
	II	7	9	17	14	10
	III	9	6	12	6	7
	IV	7	6	14	6	10
	V	9	6	12	10	6

5. Solve the following minimal assignment problems:

	I	II	III	IV
A	2	3	4	5
B	4	5	6	7
C	7	8	9	8
D	3	5	8	4

6. A car hire company has one car at each of five depots *a*, *b*, *c*, *d* and *e*. A customer requires a car in each town namely, *A*, *B*, *C*, *D* and *E*. Distance (in km) between depots (origins) and towns (destinations) are given in the following distance matrix:

	a	*b*	*c*	*d*	*e*
A	160	130	175	190	200
B	135	120	130	160	175
C	140	110	155	170	185
D	50	50	80	80	110
E	55	35	70	80	105

How should cars be assigned to the customer so as to minimize the distance travelled?

7. Solve the following minimal assignment problem:

		Job				
		I	II	III	IV	V
Machine	*A*	5	11	10	12	4
	B	2	4	6	3	5
	C	3	12	5	14	6
	D	6	14	4	11	7
	E	7	9	8	12	5

8. Solve the following minimal assignment problems:

		Man				
		I	II	III	IV	V
Job	*A*	1	3	2	3	6
	B	2	4	3	1	5
	C	5	6	3	4	6
	D	3	1	4	2	2
	E	1	5	6	5	4

9. One car is available at each of the stations 1, 2, 3, 4, 5, 6 and one car is required at each of the stations 7, 8, 9, 10, 11, 12. The distances between the various stations are given in the matrix below. How should the cars be despatched so as to minimize the total mileage covered?

	7	8	9	10	11	12
1	41	72	39	52	25	51
2	22	29	49	65	81	50
3	27	39	60	51	32	32
4	45	50	48	52	37	43
5	29	40	39	26	30	33
6	82	40	40	60	51	30

10. An airlines that operates seven days a week, has the time table shown below. Crews must have a minimum layover of 5 hours between the flights. Obtain the pair of flights that minimizes layover time away from home. For any given pair the crew will be based at the city that results in the smallest layover.

Delhi-Dubai			*Dubai- Delhi*		
Flight No.	*Departure*	*Arrival*	*Flight No.*	*Departure*	*Arrival*
1	7.00 A.M	8.00 A.M.	101	8.00 A.M.	9.15 A.M.
2	8.00 A.M.	9.00 A.M.	102	8.30 A.M.	9.45 A.M.
3	1.30 P.M.	2.30 P.M.	103	12.00 Noon	1.15 P.M.
4	6.30 P.M.	7.30 P.M.	104	5.30 P.M.	6.45 P.M.

For each pair, mention the town where the crew should be based.

11. Alpha Corporation has four plants each of which can manufacture any of the four products. Production costs differ from plant to plant as do the sales revenue. From the following data, obtain which product each plant should produce to maximize profit?

Sales revenue (Rs. 1000)

Product

Plant ↓	1	2	3	4
A	50	68	49	62
B	60	70	51	74
C	55	67	53	70
D	58	65	54	69

Product cost (Rs. 1000)

Product

Plant ↓	1	2	3	4
A	49	60	45	61
B	55	63	45	69
C	52	62	49	68
D	55	64	48	66

12. Four salesmen are to be assigned to four districts. Estimates of the sales revenue in hundred of Rs. are as below:

Salesmen/Districts	*A*	*B*	*C*	*D*
1	320	350	400	280
2	400	250	300	220
3	420	270	340	300
4	250	390	410	350

13. A company has four territories open and four salesmen available for assignment. The territories are not equally rich in their sales potential; it is estimated that a typical salesman operating in each territory would bring in the following annual sales:

Territory	I	II	III	IV
Annual sales (Rs)	60000	50000	40000	30000

Four salesmen are also considered to differ in their ability, it is estimated that, working under the same conditions, their yearly sales would be proportionately as follows:

Salesman	A	B	C	D
Proportion	7	5	5	4

If the criterion is maximum expected total sales, then intuitive answer is to assign the best salesman to the richest territory, the next best salesman to the second richest, and so on. Verify this answer by the assignment technique.

14. A company has four machines to do four jobs. Each job can be assigned to one and only one machine. The cost of each job on each machine is given in the following table:

		Machine			
		W	X	Y	Z
Job	A	18	24	28	32
	B	8	13	17	19
	C	10	15	19	22

What are the job assignments which will minimize the cost?

15. There are three persons p_1, p_2, p_3, p_4 and five jobs j_1, j_2, j_3, j_4. Each person can do only one job and a job is to be done by one person only. Using Hungarian method, find which two jobs should be left undone in the following cost minimizing cost minimizing assignments problem.

	J_1	J_2	J_3	J_4	J_5
P_1	7	8	6	5	9
P_2	9	6	7	6	10
P_3	8	7	9	5	6

16. A company is faced with the problem of assigning six different machines to five different jobs. The costs are estimated as follows (in hundreds of Rs.):

		Jobs				
		1	2	3	4	5
Machine	1	2.5	5	1	6	1
	2	2	5	1.5	7	5
	3	3	6.5	2	8	3
	4	3.5	7	2	9	4.5
	5	4	7	3	9	6
	6	6	9	5	10	6

Solve the problem assuming that the objective is to minimize the total cost.

17. Five operators have to be assigned to five machines. The assignment costs are given in the table below:

Machines →

Operator		I	II	III	IV	V
	A	5	5	...	2	6
	B	7	4	2	3	4
	C	9	3	5	...	3
	D	7	2	6	7	2
	E	6	5	7	9	1

Operator *A* cannot operate machine III and *C* cannot operate machine IV. Find the optimal assignment schedule.

18. Solve the travelling salesmen problem in the matrix shown below:

	1	2	3	4	5
1	∞	6	12	6	4
2	6	∞	10	5	4
3	8	7	∞	11	3
4	5	4	11	∞	5
5	5	2	7	8	∞

19. Solve the travelling salesmen problem given by the following data:
$c_{12} = 20$, $c_{13} = 4$, $c_{14} = 10$, $c_{23} = 5$, $c_{34} = 6$, $c_{25} = 10$, $c_{35} = 6$, $c_{45} = 20$, where $c_{ij} = c_{ji}$, and there is no route between i and j if the value for c_{ij} is not given above.

20. A salesman has to visit five cities *A*, *B*, *C*, *D* and *E*. The distances (in hundred miles) between the five cities are as follows:

To

From		*A*	*B*	*C*	*D*	*E*
	A	–	7	6	8	4
	B	7	–	8	5	6
	C	6	8	–	9	7
	D	8	5	9	–	8
	E	4	6	7	8	–

If the salesman starts from city *A* and has to come back to city *A*, which route should he select so that the total distance travelled is minimum.

ANSWERS

3. 1 → *C*, 2 → *B*, 3 → *A*, total computer minutes are 280.
4. I → 3, II → 1, III → 2, IV → 4, V → 5 or I → 3, II → 1, III → 4, IV → 2, V → 5.
5. *A* → II, *B* → III, *C* → IV, *D* → 1, minimum cost is 20.
6. *A* → *e*, *B* → *c*, *C* → *b*, *D* → *a*, *E* → *d*, the minimum cost (distance travelled) is 570 km.
7. *A* → V, *B* → IV, *C* → I, *D* → III, *E* → II, in both cases the minimum cost is 23.

8. $A \to$ I, $B \to$ IV, $C \to$ III, $D \to$ II, $E \to$ V or $A \to$ II, $B \to$ IV, $C \to$ III, $D \to$ V, $E \to$ I, in both cases the minimum cost is 10.
9. $1 \to 11$, $2 \to 8$, $3 \to 7$, $4 \to 9$, $5 \to 10$, $6 \to 12$, minimum total mileage covered is 125 km.
10. (i) (1 – 102)*, (2 – 104), (3 – 101), (4 – 103)* (ii) (1 – 103)*, (2 – 104), (3 – 101), (4 – 102)*; minimum layover time is 210 quarter hours, i.e., 52 hours 30 minutes.
11. $A \to 2$, $B \to 4$, $C \to 1$, $D \to 3$.
12. $1 \to C$, $2 \to A$, $3 \to D$, $4 \to B$.
13. $A \to$ I, $B \to$ II, $C \to$ III, $D \to$ IV or $A \to$ I, $B \to$ III, $C \to$ II, $D \to$ IV.
14. $A \to W$, $B \to X$, $C \to Y$ and $A \to W$, $B \to Y$, $C \to X$.
15. $P_1 \to J_3$, $P_2 \to J_2$, $P_3 \to J_4$ or $P_1 \to J_4$, $P_2 \to J_2$, $P_3 \to J_5$ and minimum cost is 17.
16. $1 \to 4$, $2 \to 2$, $3 \to 5$, $4 \to 3$, $5 \to 1$ or $1 \to 4$, $2 \to 1$, $3 \to 5$, $4 \to 3$, $5 \to 2$ and total minimum cost is 20, i.e., Rs. 2000.
17. $A \to$ IV, $B \to$ III, $C \to$ II, $D \to$ I, $E \to$ V or $A \to$ IV, $B \to$ III, $C \to$ II, $D \to$ V, $E \to$ I or $A \to$ IV, $B \to$ III, $C \to$ V, $D \to$ II, $E \to$ I.
18. $1 \to 3 \to 5 \to 2 \to 4 \to 1$ and total minimum cost is 27.
19. $1 \to 3 \to 2 \to 5 \to 4 \to 1$ and total minimum cost is 49.
20. $A \to C \to D \to B \to E \to A$ and $A \to E \to B \to D \to C \to A$ and both cases total distance travelled = 30, i.e., 3000 miles.

10 Chapter

Sequencing

10.1 INTRODUCTION

Sequencing problems have an important application in finding the minimum possible time duration, in which some specified jobs can be completed on a given number of machines. In the present chapter, we shall discuss, three types of such problems mentioned below whose solution will be based on Johnson's algorithm.

(i) Problem of sequencing n jobs on 2 machines.
(ii) Problem of sequencing n jobs on k machines.
(iii) Problem of sequencing 2 jobs on k machines.

10.2 ASSUMPTIONS, TERMINOLOGIES AND NOTATIONS

10.2.1 Assumptions

1. Each machine involved must be of different type.
2. All the jobs must be ready when their processing starts.
3. A job, started processing on a machine must be completed before it takes up the next job for processing and the time of transfer of a job from one machine to another.

10.2.2 Terminologies

1. *Processing Order:* The order in which machines are required to complete the job (task).
2. *Processing Time:* Time required by each job on a machine.
3. *Idle Time on a Machine:* The time during which there is no job to process on the machine.
4. *Total Elapsed Time:* The time between the start of first job and finish the last job.
5. *No Passing Rule:* No change in the given order of machines for performing different jobs is permissible.
6. *Number of Machine:* The number of machines means the service facilities through which a job passes before it is completed.

7. *Optimal Job Sequence:* The order in which jobs are sequenced (put) to find the total elapsed time (optimum time to finish all the jobs).

10.2.3 Notations

M_{1j} = Processing time for job j on machine M_1
M_{2j} = Processing time for job j on machine M_2
M_{3j} = Processing time for job j on machine M_3
...
...
...
M_{ij} = Processing time for job j on machine $M_i (i = 1, 2, 3,\ldots, k)$
T = The total elapsed time (including idle time) for processing all the jobs.
I_{ij} = The idle time of machine i between the finish of $(j - 1)^{\text{th}}$ job and the start of jth job.

10.3 PROCESSING *n* JOBS THROUGH 2 MACHINES

The sequencing problem can be described as:

1. Here n specific jobs to be performed are given.
2. Two different machines M_1 and M_2 with order M_1M_2 are given through which n jobs are to be processed.
3. The performing times of each job on each machine are known as shown in the following table:

Job→ / *Machine* ↓	1	2	3	...	j	...	n
M_1	M_{11}	M_{12}	M_{13}	...	M_{1j}	...	M_{1n}
M_2	M_{21}	M_{22}	M_{23}	...	M_{2j}	...	M_{2n}

Now the problem is to sequence the job so as to minimize the total elapsed time T? To answer this, follow the steps (Johnson's algorithm):

Step 1: Least processing time among M_{ij}'s is chosen. Let it be for some job k.

Step 2: If the least processing time is for M_1, the kth job is processed first and is kept at the first place of sequencing. If it is for M_2 place it at the end of sequence (processed in the last).

Step 3: In case of tie there are three cases:

Case I: If equal minimum values are only for M_1, the job with large processing time in M_2 is considered for placing first in the job sequence.

Case II: If equal minimum values are only for M_2, the job with large processing time in M_1 is considered for placing the last in the job sequence.

Case III: If equal minimum values are for both, the job is first placed on M_1, then in the last for M_2.

Step 4: The jobs already sequenced are deleted. If all jobs are sequenced go to step 4, otherwise follow steps 1 to 3.

Step 5. Put the data in the form of the following table:

Optimum job sequence → / Machines ↓	Machine M_1		Machine M_2		Idle time for M_1	Idle time for M_2
	In	Out	In	Out		
–	0	…	…	…	…	…
–	…	…	…	…	…	…
–	…	…	…	…	…	…

Overall total elapsed time is determined. Further, idle times on machines M_1 and M_2 is computed.

(i) Idle time for machine M_1
= Total elapsed time – The time when the last job is completed on machine M_1

(ii) Idle time for machine M_2

= (The time duration in which the first job is completed on machine M_1)

$$+\sum_{j=2}^{n} \{(\text{The time when the } j^{th} \text{ job starts on machine } M_2)$$

– (The time when the $(j-1)^{th}$ job is completed on machine M_2)}

The total elapsed time

= Total time from the starting of first job on machine M_1 to the completion of last job on machine $M_2 = \sum_{j=2}^{n} m_{2j} + \sum_{j=1}^{n} I_{2j}$

where M_{2j} = the time required for processing j^{th} job on machine M_2
and I_{2j} = The time in which machine M_2 remains idle after processing $(j-1)^{th}$ job and before starting work in the jth job.

SOLVED EXAMPLES

Example 1: A company received an order to complete J_1, J_2 and J_3 three jobs. Each of these must be processed through two machines M_1 and M_2, of the company in the order M_1M_2. The table lists the number of days required by each job in each machine:

Jobs→ / Machines ↓	J_1	J_2	J_3
M_1	8	6	5
M_2	8	3	4

Find the sequence in which the 3 jobs should be processed so as to take minimum time to finish all the 3 jobs.

Solution: The minimum processing time in the given problem is 3 for machine M_2. So put job J_2 in the last in job sequence as shown below:

		J_2

Deleting job J_2, the remaining processing times are

Job	J_1	J_2
Machine M_1	8	5
Machine M_2	8	4

The minimum processing time in the reduced table is 4 for machine M_2 on job J_3. So we put the job J_3 the last J_2 and job J_1 in the beginning of empty cells to have the resulting optimal job sequencing as:

J_1	J_3	J_2

For total elapsed time, we have

Job	*Machine M_1*		*Machine M_2*		*Idle time of M_1*	*Idle time of M_2*
	In	*Out*	*In*	*Out*		
J_1	0	8	8	16	0	8
J_2	8	13	16	20	0	–
J_3	13	19	20	23	4	–

The total minimum elapsed time (T) is 23 hours. The idle time of machine M_1 is (23 – 19) = 4 hours. The idle time of machine M_2 is 8 hours.

Example 2: We have five jobs, each of which must go through the two machines M_1 and M_2 in the order M_1M_2. The processing times in hours are given in the table below:

Jobs → / *Machines* ↓	J_1	J_2	J_3	J_4	J_5
M_1	5	1	9	3	10
M_2	2	6	7	8	4

Find the optimum sequence for the five jobs that will minimize the total elapsed time. Also find the idle times on machines M_1 and M_2.

Solution: The minimum processing time in the given problem is on machine M_1 of job J_2. So put job J_2 in the beginning of the job sequence as shown below:

J_2				

Deleting J_2, the remaining processing times are:

Job	J_1	J_3	J_4	J_5
Machine M_1	5	9	3	10
Machine M_2	2	7	8	4

The minimum processing time in the reduced table is 2 on machine M_2 of J_1. So put J_1 in the end of the job sequence as shown below:

J_2				J_1

Deleting J_1, the remaining processing times are:

Job	J_3	J_4	J_5
Machine M_1	9	3	10
Machine M_2	7	8	4

The minimum processing time in the reduced table is 3 on machine M_1 of J_4. So put J_4 perform next to J_2 as shown below:

J_2	J_4			J_1

Deleting J_4, the remaining processing times are:

Job	J_3	J_5
Machine M_1	9	10
Machine M_2	7	4

Now, the minimum processing time in the reduced table is 4 on machine M_2 of J_5. So put J_5 perform just before J_1 and J_3 next to J_4 to get the final optimal sequence as given below:

J_2	J_4	J_3	J_5	J_1

For total elapsed time, we have

Job	*Machine M_1*		*Machine M_2*		*Idle time of M_1*	*Idle time of M_2*
	In	*Out*	*In*	*Out*		
J2	0	1	1	7	0	1
J4	1	4	7	15	0	–
J3	4	13	15	22	0	–
J5	13	23	23	27	0	1
J1	23	28	28	30	2	1

The total minimum elapsed time (T) is 30 hours. The idle time of machine M_1 is (30 – 28) = 2 hours. The idle time of machine M_2 is 3 hours.

10.4 PROCESSING *n* JOBS THROUGH *k* MACHINES

Solving the problem of n jobs through k machines is reduced to n jobs through 2 machines if at least one of the following conditions is satisfied:

(C_1) Minimum $M_{1j} \geq$ Maximum M_{ij} $(i = 2, 3, \ldots, k - 1; j = 1, 2, \ldots, n)$.

or (C_2) Minimum $M_{kj} \geq$ Maximum M_{ij} $(i = 2, 3, \ldots, k - 1; j = 1, 2, \ldots, n)$

where M_{ij} is the processing time of job j on machine i. In that case the following procedural method can be applied:

Given n jobs to be processed on k machines in the order $M_1\ M_2\ M_3\ \ldots\ M_k$ with processing times given as per the following table:

Jobs → / Machines ↓	*1*	*2*	*3*	...	*j*	...	*n*
M_1	M_{11}	M_{12}	M_{13}	...	M_{1j}	...	M_{1n}
M_2	M_{21}	M_{22}	M_{23}	...	M_{2j}	...	M_{2n}
M_3	M_{31}	M_{32}	M_{33}	...	M_{3j}	...	M_{3n}
...	...	...	...	...	...	...	...
M_i	M_{i1}	M_{i2}	M_{i3}	...	M_{ij}	...	M_{in}
...	...	...	...	...	...	...	...
M_k	M_{k1}	M_{k2}	M_{k3}	...	M_{kj}	...	M_{kn}

The solution procedure for solving the sequencing problem is given as follow:

Step 1: For the check of conditions C_1 and C_2, find the minimum M_{1j} minimum M_{kj} and maximum M_{ij} for all $j = 1, 2, 3, \ldots, n$; $i = 1, 2, \ldots, k - 1$. If C_1 or C_2 or both conditions are satisfied, then go to step 2, otherwise the method fails.

Step 2: Now the problem is converted in n jobs and 2 machines G and H with processing times G_{gj} and H_{hj} $(j = 1, 2, 3, \ldots, n)$ are imagined, such that

$$G_{gj} = M_{1j} + M_{2j} + M_{3j} + \ldots + M_{(k-1)j} \text{ (for } g = 1, 2, 3, \ldots, k - 1; j = 1, 2, 3, \ldots, n)$$

and $$H_{hj} = M_{2j} + M_{3j} + M_{4j} + \ldots + M_{kj} \text{ (for } h = 2, 3, 4, \ldots, k; j = 1, 2, 3, \ldots, n)$$

For computational purpose put the data as per the following table:

Jobs → / Machines ↓	*1*	*2*	*3*	...	*j*	...	*n*
G	M_{11} $+M_{21}$ $+ M_{31}$ $+\ldots$ $+M_{(k-1)1}$	M_{12} $+M_{22}$ $+M_{23}$ $+\ldots$ $+M_{(k-1)2}$	M_{13} $+M_{23}$ $+M_{33}$ $+\ldots$ $+M_{(k-1)3}$	... ...	M_{1j} $+M_{2j}$ $+M_{3j}$ $+\ldots$ $+M_{(k-1)j}$	... ...	M_{1n} $+M_{2n}$ $+M_{3n}$ $+\ldots$ $+M_{(k-1)n}$
H	M_{21} $+ M_{31}$ $+ M_{41}$ $+\ldots$ $+M_{k1}$	M_{22} $+M_{23}$ $+M_{42}$ $+\ldots$ $+M_{k2}$	M_{23} $+M_{33}$ $+M_{43}$ $+\ldots$ $+M_{k3}$	... ...	M_{2j} $+M_{3j}$ $+M_{4j}$ $+\ldots$ $+M_{kj}$	... ...	M_{2n} $+M_{3n}$ $+M_{4n}$ $+\ldots$ $+M_{kn}$

Now solve the problem as n jobs on 2 machines G and H with prescribed order GH as done in the previous article.

SOLVED EXAMPLES

Example 3: We have 4 jobs each of which has to go through the machines M_i, $i = 1, 2, 3, \ldots, 6$ in the order of $M_1\ M_2\ M_3\ M_4\ M_5\ M_6$. The processing time (in hours) is given below:

Jobs → / Machines ↓	A	B	C	D
M_1	18	17	11	20
M_2	8	6	5	4
M_3	7	9	8	3
M_4	2	6	5	4
M_5	10	8	7	8
M_6	25	19	15	12

Determine a sequence of these four jobs that minimizes the total elapsed time.

Solution: Here, minimum $M_{1j} = 11$, minimum $M_{6j} = 12$ and maximum of M_{2j}, M_{3j}, M_{4j} and M_{5j} 8, 9, 6 and 10 respectively. Since the conditions

$$\text{Minimum } M_{1j} \geq \text{Maximum } M_{ij} \quad (i = 2, 3, 4, 5)$$

and

$$\text{Minimum } M_{6j} \geq \text{Maximum } M_{ij} \quad (i = 2, 3, 4, 5)$$

are satisfied. So, the given sequencing problem can be converted into a 4 jobs and 2 machines. The two factitious machines denoted by G and H and the corresponding processing times are shown in the following table:

Jobs → / Machines ↓	A	B	C	D
G	45	46	36	39
H	52	48	40	31

where $G_j = \sum_{i=1}^{5} M_{ij}$ and $H_j = \sum_{i=2}^{6} M_{ij}$ for $j = 1, 2, 3, 4$.

Now solving this new problem using the n jobs and 2 machines method, we get the following optimal job sequence:

C	A	B	D

For total elapsed time, we have

Job	Machine					
	M_1	M_2	M_3	M_4	M_5	M_6
	In - Out	In - Out	In - Out	In - Out	In - Out	In - Out
C	0-11	11-16	16-24	24-29	29-36	36-51
A	11-29	29-37	37-44	44-46	46-56	56-81
B	29-46	46-52	52-61	61-67	67-75	81-100
D	46-66	66-70	70-73	73-77	77-85	100-112

The total minimum elapsed time (T) is 112 hours.

Example 4: Given the following data: (i)

Jobs→ / Machines↓	J_1	J_2	J_3	J_4	J_5	J_6
M_1	12	10	9	14	7	9
M_2	7	6	6	5	4	4
M_3	6	5	6	4	2	4

(ii) The order of processing jobs on machines is $M_1\ M_3\ M_2$.
(iii) Sequence suggested: Jobs $J_5, J_3, J_6, J_2, J_1, J_4$.

(a) Determine the total elapsed time for the sequence suggested.
(b) Is the given sequence optimal?
(c) If your answer to (b) is 'No', determine the optimal sequence and the total elapsed time associated with it.

Solution: Here, minimum M_{1j} = 7, minimum M_{2j} = 4 and maximum M_{3j} = 6. Since the condition

$$\text{Minimum } M_{1j} \geq \text{Maximum } M_{3j} (j = 2, 3, 4, 5)$$

is satisfied. So, the given sequencing problem can be converted into 6 jobs and 2 machines. The two factitious machines denoted by *G* and *H* and the corresponding processing times are shown in the following table:

Jobs → / Machines ↓	J_1	J_2	J_3	J_4	J_5	J_6
G	18	15	15	18	9	13
H	13	11	12	9	6	8

where $G_j = M_{1j} + M_{3j}$ and $H_j = M_{3j} + M_{2j}$ for $j = 1, 2, \ldots, 6$.

Now using the *n* jobs and 2 machines method solved this new problem. We get the following optimal sequence

J_1	J_3	J_2	J_4	J_6	J_5

For total elapsed time, we have

Job	*Machines*		
	M1	*M2*	*M3*
	In - Out	In - Out	In - Out
J1	0-12	12-18	18-25
J3	12-21	21-27	27-33
J2	21-31	31-36	36-42
J4	31-45	45-49	49-54
J6	45-54	54-58	58-62
J5	54-61	61-63	63-67

The total minimum elapsed time (T) is 67 hours.

(iii) Given sequence of jobs are J_5, J_3, J_6, J_2, J_1, J_4.

Now for total elapsed time according to given sequence J_5, J_3, J_6, J_2, J_1, J_4, we have

Job	*Machines*		
	M1	*M2*	*M3*
	In - Out	In - Out	In - Out
J_5	0-7	7-9	9-13
J_3	7-16	16-22	22-28
J_6	16-25	25-29	29-31
J_2	25-35	35-40	40-46
J_1	35-47	47-53	53-60
J_4	47-61	61-65	65-70

The total minimum elapsed time (T) is 70 hours.

(b) Given sequence is not optimal.

(c) Hence, the total minimum elapsed time is 67 hours.

10.5 PROCESSING 2 JOBS THROUGH *k* MACHINES

The sequencing problem of 2 jobs and k machines can be described as:

(i) Two jobs J_1 and J_2 are to be processed through k machines say M_1, M_2, M_3, …, M_k.

(ii) The ordering of k machines for each job is fixed. However, the ordering may not be the same for jobs J_1 and J_2.

(iii) The processing time M_{ij} ($i = 1, 2, 3,\ldots, k$; $j = 1, 2$) is known and shown as in the following table:

Job→ / *Machines* ↓	*Job* J_1	J_2
Machines	$M_{\alpha 1}$	$M_{\beta 1}$
	$M_{\alpha 2}$	$M_{\beta 2}$
	$M_{\alpha 3}$	$M_{\beta 3}$
	…	…
	…	…
	$M_{\alpha k}$	$M_{\beta k}$

where all α_i's and β_i's are any of the numbers $i = 1, 2, 3, \ldots, k$. And

Job → / *Machine* ↓	*Job* J_1	J_2
M_1	M_{11}	M_{12}
M_2	M_{21}	M_{22}
M_3	M_{31}	M_{32}
…	…	…
…	…	…
M_k	M_{k1}	M_{k2}

The solution procedure of the problem is given as follows:

Step 1: Draw two perpendicular lines, horizontal and vertical. The horizontal line called x-axis represents the processing times for job J_1 on different k machines and another the vertical line called y-axis represents the processing times for job J_2 on different k machines.

Step 2: Mark the processing times for job J_1 on x-axis and job J_2 on y-axis according to the given order of k machines.

Step 3: Now construct the various block diagrams starting from the origin by pairing the same machine until the end point.

Step 4: After constructing blocks, we draw a line from the origin included at an angle of 45° with the horizontal, will represent the work on the both jobs simultaneously. The horizontal segment of this line means that job J_1 is processing while job J_2 is idle and vertically segment (moving upwards) means that job J_2 is processing while job J_1 is idle and diagonal segment (moving diagonally) means that both the jobs J_1 and J_2 are being processed simultaneously.

Step 5: An optimum path is that which minimizes the idle time for job J_1 and J_2, i.e., a path in which diagonal movement is maximum.

Step 6: The total elapsed time is obtained by adding the idle time for either job to the processing time for that job, i.e.,

Total elapsed time = Processing time of job J_1 + Idle time for job J_1.

or

Total elapsed time = Processing time of job J_2 + Idle time for job J_2.

SOLVED EXAMPLES

Example 5: A machine shop has five machines *A*, *B*, *C*, *D* and *E*. Two jobs must be processed through each of these machines. The time (in hours) taken on each of the machines and the necessary sequence of jobs through the shop are given below:

			Job 1		
Sequence:	*A*	*B*	*C*	*D*	*E*
Time:	2	4	5	1	2
			Job 2		
Sequence:	*D*	*E*	*A*	*C*	*B*
Time:	6	4	2	3	6

Use graphic method to determine the total minimum elapsed time.

Solution:

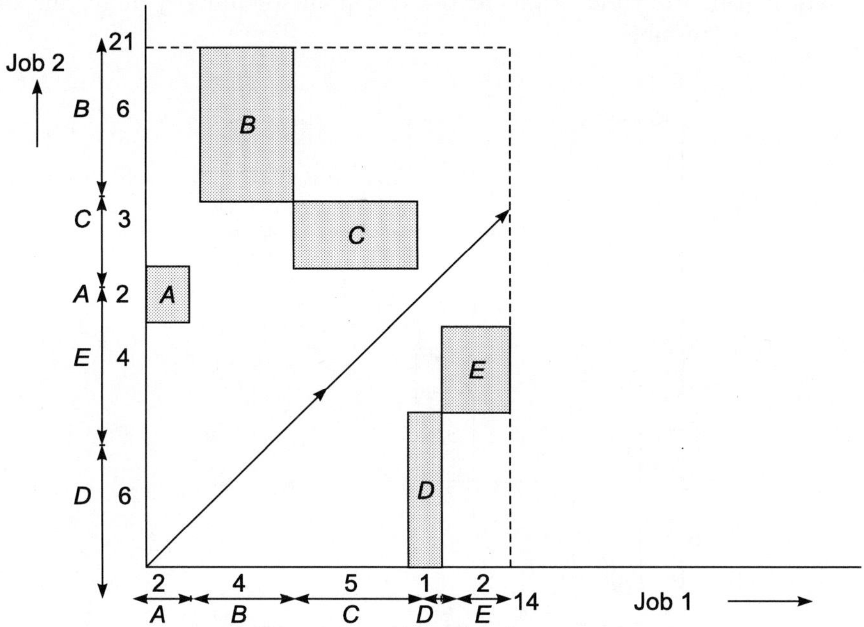

(i) Mark the processing times for job 1 and job 2 on the *x*-axis and *y*-axis respectively according to the technological order of five machines.

(ii) Construct the rectangular blocks by pairing the same machines.

(iii) Mark a path from origin to the end point of moving along the 45° line at much as possible.

Hence the total elapsed time = Processing time of job J_1 + Idle time for job J_1.

$= 14 + 7 = 21$ hours.

or Total elapsed time $=$ Processing time of job J_2 + Idle time for job J_2.

$= 21$ hours.

Example 6: Use graphical method to minimize the time added to process the following jobs on the machines shown, i.e., for each machine find the job which should be done first. Also calculate the total time elapsed to complete both the jobs:

Job 1

Machines					
Sequence:	A	B	C	D	E
Time:	3	4	2	6	2

Job 2

Machines					
Sequence:	B	C	A	D	E
Time:	5	4	3	2	6

Solution: (i) Mark the processing times for job 1 and job 2 on the x-axis and y-axis respectively according to the technological order of five machines.

(ii) Construct the rectangular blocks by pairing the same machines.

(i) Mark a path from the origin to the end point of moving along the 45° line at much as possible.

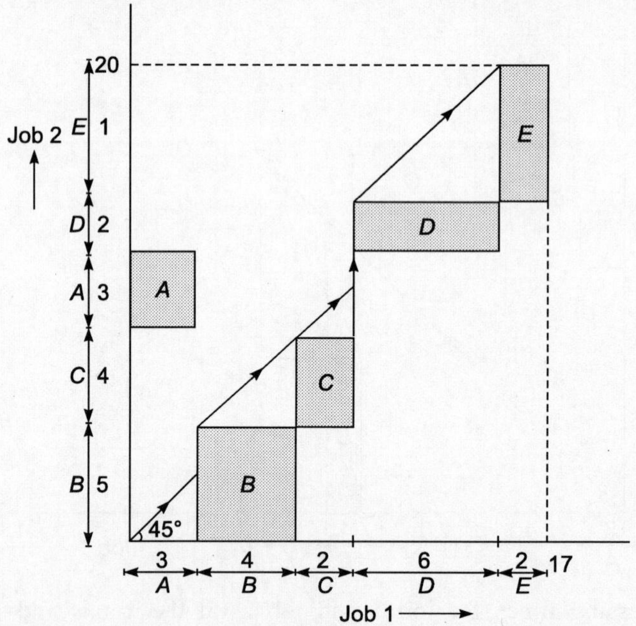

Hence, the total elapsed time $=$ Processing time of job J_1 + Idle time for job J_1.

$= 17 + (2 + 3) = 22$ hours.

or Total elapsed time = Processing time of job J_2 + Idle time for job J_2.

= 20 + 2 = 20 hours.

EXERCISE 10.1

1. Explain the optimal sequencing procedure for solving the sequencing problem of processing n jobs through two machines.
2. In a factory there are six jobs to be performed, each of which should go through two machines M_1 and M_2 in the order M_1M_2. The processing timing (in hours) for the jobs are given in the table given below:

Jobs → / Machines ↓	J_1	J_2	J_3	J_4	J_5	J_6
M_1	1	3	8	5	6	3
M_2	5	6	3	2	2	10

Find the optimal job sequence to minimize the total elapsed time and also find the total elapsed time.

3. Six jobs go first over machine M_1 and then on machine M_2. The order of the completion of jobs has no significance. The following table gives the machine times in hours for six jobs on two machines:

Jobs → / Machines ↓	J_1	J_2	J_3	J_4	J_5	J_6
M_1	5	9	4	7	8	6
M_2	7	4	8	3	9	5

Find the sequence of the jobs that minimizes the total elapsed time to complete the jobs.

4. A book binder has one printing press machine, one binding machine and the manuscripts of six different books. The time required to perform the printing and binding operations for each book is shown below. Determine the order in which the books should be processed, so that the total time required to process all the books is minimized.

Books → / Machines ↓	B_1	B_2	B_3	B_4	B_5	B_6
Printing Machine Time (hours)	30	120	50	20	90	100
Binding Machine Time (hours)	80	100	90	60	30	10

5. A company has six jobs J_1, J_2, J_3, J_4, J_5, and J_6 to perform. All the jobs have to go through two machines M_1 and M_2 in the order M_1M_2. The time required for each job on each machine, in hours, is given below:

Machines ↓ \ Jobs →	J_1	J_2	J_3	J_4	J_5	J_6
M_1	3	12	18	9	15	6
M_2	9	18	24	24	3	15

Determine the optimum sequence of jobs that minimizes the total elapsed time to complete the jobs. Also compute the minimum complete time.

6. Seven jobs are to be processed on two machines M_1 and M_2 in the order M_1M_2. Each machine can process only one job at a time. The processing times (in hours) are as follows:

Machines ↓ \ Jobs →	J_1	J_2	J_3	J_4	J_5	J_6	J_7
M_1	10	12	13	7	14	5	16
M_2	15	11	8	9	6	7	16

Determine the optimum sequence of processing the jobs and the total elapsed time.

7. We have seven jobs each of which has to go through the machines M_1 and M_2 in the order M_1M_2. The processing times (in hours) are given as:

Machines ↓ \ Jobs →	J_1	J_2	J_3	J_4	J_5	J_6	J_7
M_1	3	12	15	6	10	11	9
M_2	8	10	10	6	12	1	3

Determine a sequence of these jobs that will minimize the total elapsed time T.

8. In a machine shop, 8 different products are being manufactured, each requiring time on two machines M_1 and M_2 in the order M_1M_2 as given below:

Machines ↓ \ Jobs →	P_1	P_2	P_3	P_4	P_5	P_6	P_7	P_8
M_1	30	45	15	20	80	120	65	10
M_2	20	30	50	35	36	40	50	20

Find the optimum sequence of processing of different products in order to minimize the total manufacturing time for all the products.

9. Explain the optimal sequencing procedure for solving the sequencing problem of processing n jobs through k machines.

10. Solve the following sequencing problem, giving an optimal solution when passing is not allowed:

Machines ↓ \ Jobs →	J_1	J_2	J_3	J_4	J_5
M_1	10	12	8	15	16
M_2	3	2	4	1	5
M_3	5	6	4	7	3
M_4	14	7	12	8	10

11. There are 5 jobs each of which must go through three machines M_1, M_2, M_3 in the order of M_1 M_2 M_3. The processing times are given in the following table:

Machines ↓ \ Jobs →	J_1	J_2	J_3	J_4	J_5
M_1	7	9	5	6	10
M_2	4	5	1	2	3
M_3	3	8	7	5	4

Determine a sequence for five jobs that will minimize the total elapsed time.

12. Solve the following sequencing problem when passing out is not allowed:

Machines ↓ \ Jobs →	I	II	III	IV
M_1	15	12	16	17
M_2	5	2	3	3
M_3	4	10	5	4
M_4	15	12	16	17

13. We have 5 jobs, each of which must be processed on machines M_1, M_2, M_3 in the order of $M_1 M_2 M_3$. The processing time (in hours) for each job on each machine is given below. Determine a sequence which minimizes the total elapsed time. Also find idle time on each machine.

Machines ↓ \ Jobs →	J_1	J_2	J_3	J_4	J_5
M_1	8	5	4	6	5
M_2	6	2	9	7	4
M_3	10	13	11	10	12

14. A book binder has one printing press, one binding machine and the manuscripts of a number of different books. The times required to perform the printing and binding operation for each book are known. Determine the order in which the books should be processed in order to minimize the total time required to process all the books. Find also the total time required.

Processing time (in minutes)

Books	1	2	3	4	5
Printing time	40	90	80	60	50
Binding time	50	60	20	30	40

(ii) Suppose that an additional operation is added to the process described in (i), viz, finishing. The time required for operations are given below:

Finishing time (in minutes)

Books	1	2	3	4	5
Finishing time	80	100	60	70	100

What is the order in which the books should be processed? Find also the minimal total elapsed time.

15. When passing is not allowed; solve the sequencing problem giving an optimal sequence:

Jobs → / Machines ↓	J_1	J_2	J_3	J_4
M_1	11	18	12	13
M_2	4	1	2	2
M_3	2	8	3	2
M_4	11	8	12	13

16. A readymade garments manufacturer has to process 7 times through two stages to production, viz, cutting and sewing. The times taken for each of these at the different stages are given below in appropriate units:

Processing time

Items	1	2	3	4	5	6	7
Cutting	5	7	3	4	6	7	12
Sewing	2	6	7	5	9	5	8

(ii) Find an order in which these items are to be processed through these stages so as to minimize the total processing time.

(iii) Suppose a third stage of production is added, viz., pressing and packing, with processing time for these items as follows:

Items	*1*	*2*	*3*	*4*	*5*	*6*	*7*
Processing time (Pressing & Packing)	10	12	11	13	12	10	11

Find an order in which these seven items are to be processed so as to minimize the time taken to process all the items through all the three stages.

17. Explain the optimal sequencing procedure for solving the sequencing problem of processing 2 jobs through k machines.

18. Two jobs are to be processed on four machines A, B, C and D. The technological order for these jobs on machines is as follows:

Job 1	*A*	*B*	*C*	*D*
Job 2	*D*	*B*	*A*	*C*

The processing times (in hours) are given in the following table:

Jobs	*Machines*			
	A	*B*	*C*	*D*
1	4	6	7	3
2	4	7	5	8

Find the optimal sequence of jobs on each of the machines.

19. Use the graphical method to find the minimum total elapsed time sequence of 2 jobs and 5 machines, when we are given the following information:

Job 1

	Machines				
Sequence:	A	B	C	D	E
Time (hours):	2	3	4	6	2

Job 2

	Machines				
Sequence:	*C*	*A*	*D*	*E*	*B*
Time (hours):	4	5	3	2	6

20. Using the graphical method, calculate the minimum time needed to process jobs 1 and 2 on five machines A, B, C, D and E, i.e., for each machine find the job which should be done first. Also calculate the total time needed to calculate both jobs.

Job 1

	Machines				
Sequence:	A	B	C	D	E
Time (hours):	6	8	4	12	4

	Job 2				
	Machines				
Sequence:	B	C	A	D	E
Time (hours):	10	8	6	4	12

21. Use the graphical method to minimize the time needed to process the following jobs on the machines shown below:

	Job 1					
	Machines					
Sequence:	*A*	*B*	*C*	*D*	*E*	*F*
Time (hours):	8	10	2	6	12	10

	Job 2					
	Machines					
Sequence:	*B*	*A*	*C*	*F*	*D*	*E*
Time (hours):	12	6	4	8	6	10

22. A machine shop has six machines *A*, *B*, *C*, *D*, *E* and *F*. Two jobs must be processed through each of the machines. The time on machines and the necessary sequence of the jobs through the shop are given below:

	Job 1					
	Machines					
Sequence:	A	C	D	B	E	F
Time (hours):	20	10	10	30	25	16

	Job 2					
	Machines					
Sequence:	A	C	B	D	F	E
Time (hours):	10	30	15	10	15	20

Determine the optimum sequence for the job in order to minimize the total time necessary to finish the jobs.

ANSWERS

2. Optimal sequence 2-4-3-5-1; Elapsed time = 29 hours. Idle time for machine M_1 = 3 hours and M_2 = 1 hour.
3. Optimal sequence 3-1-5-6-2-4; Elapsed time = 42 hours.
4. Optimal sequence 4-1-3-2-5-6; Elapsed time = 430 hours. Idle time for printing machine is 10 hours and for binding machine is 60 hours.
5. Optimal sequence 1-6-4-2-3-5; Elapsed time = 96 hours.
6. Optimal sequence 6-4-1-7-2-3-5; Elapsed time = 83 hours.

7. Optimal sequence 1-4-5-3-2-7-6 or 1-4-5-2-3-7-6; Elapsed time = 67 hours.
8. Optimal sequence 8-3-4-7-6-5-2-1; Elapsed time = 6 hours and 45 minutes.
10. Optimal sequence 3-1-5-4-2; Elapsed time = 76 hours.
11. Optimal sequence 3-2-1-4-5 or 3-2-1-5-4 or 3-2-5-4-1 or 3-2-4-5-1 or 3-2-4-1-5- or 3-2-5-1-4; Elapsed time = 44 hours.
12. Optimal sequence 1-2-3-4 or 1-2-4-3 or 1-3-4-2 or 1-4-2-3 etc.; Elapsed time = 84 hours.
13. Optimal sequence 2-5-4-3-1; Elapsed time = 63 hours. Idle time for machine M_1, M_2 and M_3 are 35, 35 and 7 hours respectively.
14. (i) Optimal sequence 1-2-5-4-3; Elapsed time = 5 hours and 40 minutes.
    (ii) Optimal sequence 5-1-4-2-3; Elapsed time = 8 hours and 30 minutes.
15. Optimal sequence 1-2-3-4 or 1-2-4-3 etc.; Elapsed time = 61 hours.
16. (i) Optimal sequence 3-4-5-7-2-6-1; Elapsed time = 46 hours.
    (ii) Optimal sequence 1-4-3-6-2-5-7; Elapsed time = 86 hours.
18. Total elapsed time = 24 hours.
19. Total elapsed time = 20 hours.
20. Total elapsed time = 44 hours.
21. Total elapsed time = 26 hours.
22. Total elapsed time = 150 hours.

11 Chapter

Project Scheduling

11.1 INTRODUCTION

For any organization, there is a great need of careful study and analysis of the entire project before getting the financial approval and the signal to go ahead with the project. It is essential that the planning and control on the project are precise and provide ample warning of the difficulties that may arise later on. A project comprises several activities and it has a definite beginning and ending. For example, take a project of publishing a book, which consists of the following activities:

1. First writing the manuscript of the book
2. Editing the material
3. Printing of the book
4. Binding of the book.

11.2 PROJECT MANAGEMENT

In managing a project three basic steps are involved:

Planning: In planning various activities and their sequences are recognized to calculate the project.

Scheduling: In scheduling, we prepare the network diagram, starting and completion time for each activity, recognition of critical activities.

Controlling: To complete the project in time proper monitoring is done.

There are various techniques to manage a project. Some of them are:

(i) Critical Path Method (CPM)
(ii) Programming Evaluation and Review Technique (PERT)
(iii) Review Analysis of Multiple Project (RAMP)
(iv) Graphical Evaluation and Review Techniques (GERT)
(v) Least Cost Scheduling (LCS)

(vi) Man Power Allocation Procedure (MPAP)
(vii) Multi-Operation Scheduling System (MOSS)
(viii) Resource Planning and Scheduling Method (RPSM)

Here we shall discuss only critical path method and programming evaluation and review techniques.

11.3 NETWORK DIAGRAMS

A network diagram is a graphical representation of a project comprising activities and nodes. An activity is denoted by an arrow with time duration over it, as

$$\xrightarrow{4}$$

The start and end point of an activity is denoted by a circle and is called a node. Thus graphically we represent an activity of time duration 4 starting from node 1 and ending at node 2 as (1, 2), shown below. The number alone arrow indicates its duration.

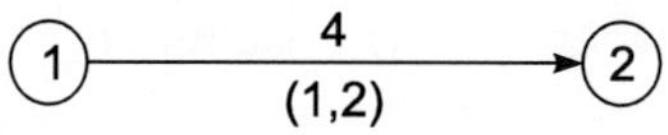

The following conditions must be satisfied by a network diagram:

(i) Each activity is represented by an arrow.
(ii) The starting and ending points of an activity is called nodes and are denoted by some integral figure.
(iii) One or more than one activity can start from a node or end to a node.
(iv) A situation like

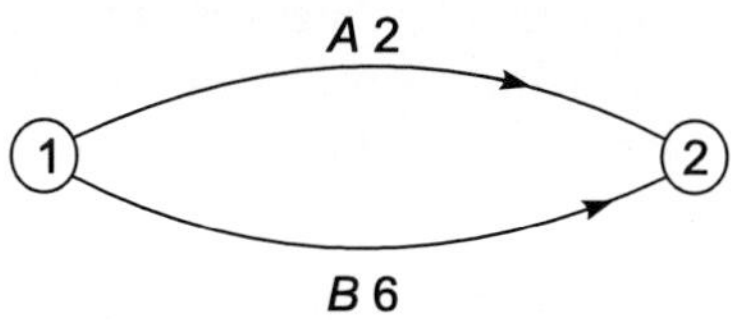

is not permissible in a network diagram i.e., no two distinct activities should have same starting and end nodes. In the above situation, we introduce a dummy activity of '0' duration either at the starting node 1 or at the end node 2, i.e., we correct the above situation as:

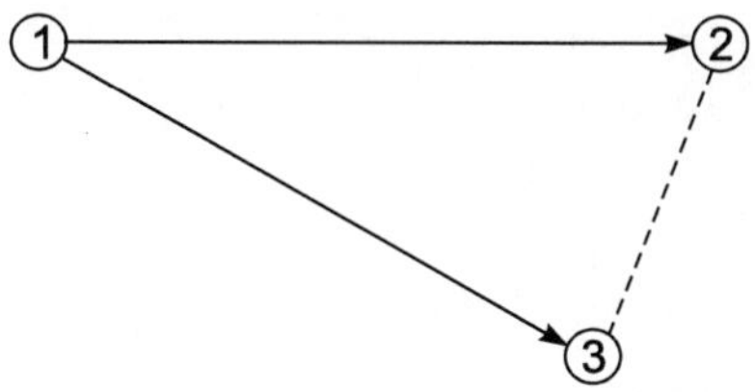

Note that the dummy activity is denoted by a broken line.

(v) A closed loop such as in the following figure is not allowed.

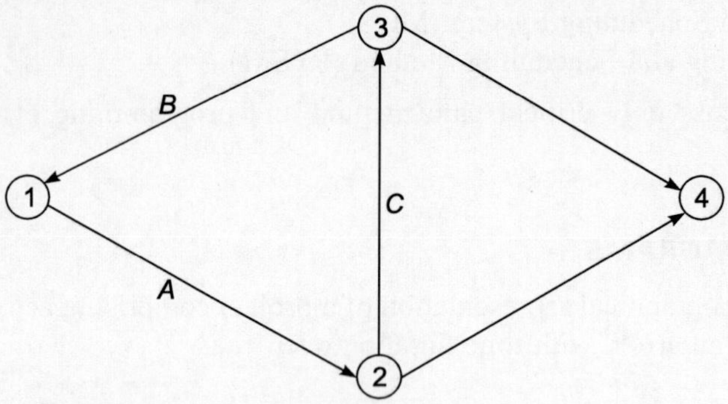

Here ACB is a closed loop.

(vi) Starting and ending nodes should be unique.

Notations

(i) The numeral inside the circle denotes the node number.
(ii) We shall use capital alphabets for activities.
(iii) We shall use ordered pair (i, j) to represent an activity from node 'i' to node 'j'.

11.4 CRITICAL PATH METHOD (CPM)

Critical path method is a deterministic model for managing a project. During planning and scheduling of the project the duration and the sequential order of various activities in which they are taken up to complete the project is decided. On the basis of all this the project is put in graphical form and we call it the network diagram of the project. Once the network diagram is completed, planning is done to finish the project in minimum possible duration. To carry out necessary computations find the duration of the project, we define certain terms and introduce notations for them as below.

(i) Event: We define an event as a point in time at which activities are started or terminated. In terms of network an event can be treated as a node. Thus if we consider an activity (1, 2)

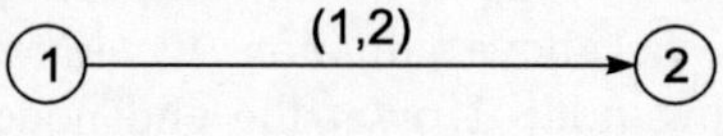

then 1 is an event and so is 2.

(ii) E_i = Earliest occurrence of event i.
(iii) L_i = Latest occurrence of event i.
(iv) t_{ij} = Duration of the activity (i, j)

Now to find the duration of the project we make use of two types of computation (calculation).

11.4.1. Forward Pass Calculations

Initial Step: Set $E_1 = 0$, to indicate that the project starts at $t = 0$.

General Step: Given that the nodes $i_1, i_2, \ldots, i_n$ are connected directly to node j by incoming activities $(i_1, j), (i_2, j), \ldots, (i_n, j)$ entering into node j and the earliest occurrence times of events (nodes) $i_1, i_2, \ldots, i_n$ have already been calculated, then

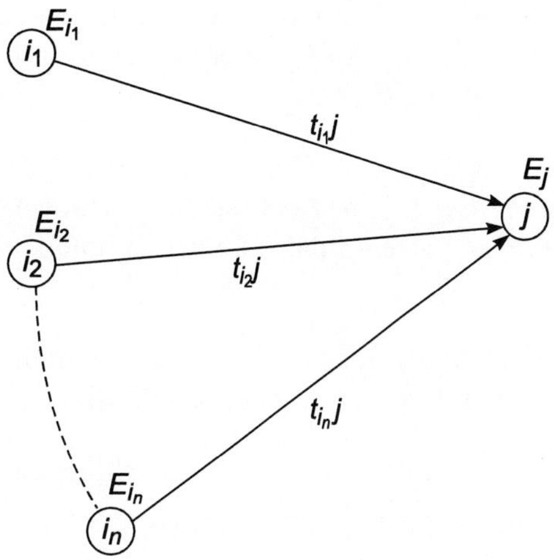

$$E_j = \max_i \{E_i + t_{ij}\}, \text{ where } i = i_1, i_2, \ldots, i_n.$$

Forward pass is completed if final node n is reached.

$$ES_{ij} = \text{Earliest start time of activity } (i, j)$$
$$= E_i \text{ for all activity } (i, j)$$
$$EF_{ij} = \text{Earliest finish time of activity } (i, j)$$
$$= ES_{ij} + t_{ij} = E_i + t_{ij} \text{ for all activity } (i, j)$$

11.4.2. Backward Pass Calculations

Here we calculate the latest occurrence times L_j. After the completion of forward pass backward computations start from node n and proceed back to node 1.

Initial Step: Let $L_n = E_n$ to indicate that the earliest and latest occurrences of the last node are the same.

General Step: For the node j, let $(j, k_1), (j, k_2), \ldots, (j, k_n)$ be the activities going out of node j to nodes $k_1, k_2, \ldots, k_n$ and the latest occurrence times of the events (nodes) $k_1, k_2, \ldots, k_n$ have already been calculated, then

$$L_j = \max_k \{L_k - t_{jk}\}, k = k_1, k_2, \ldots, k_n$$

Backward pass is completed when L_1 is completed.

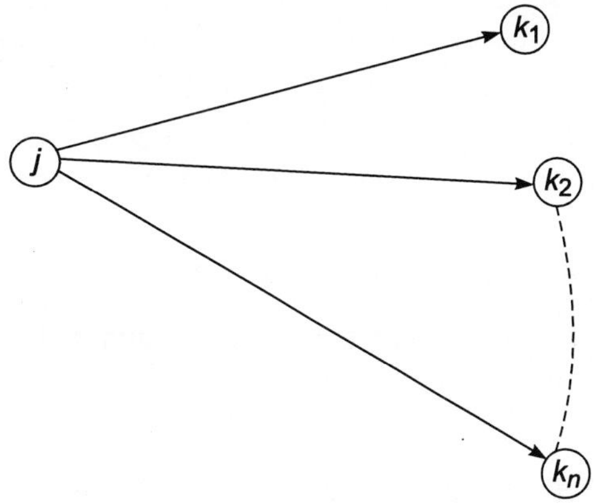

Critical activity

Activity (i, j) is called critical if

(i) $E_i = L_i$
(ii) $E_j = L_j$
(iii) $E_j - E_i = L_j - L_i = t_{ij}$

An activity (i, j) not satisfying all the above conditions is called a non-critical activity.

Float of activities

Some common floats are:

(i) Total float (ii) Free float (iii) Event float
(iv) Independent float (v) Interference float

Total float: The total float of any activity is the excess of the time span from the earliest occurrence of event i to latest occurrence of event j over the duration t_{ij} of the activity (i, j) i.e.,

$$TF_{ij} = (L_j - E_i) - t_{ij}$$

Free float: The excess of the time span from the earliest occurrence of event I to earliest occurrence of event j over the duration t_{ij} of the activity (i, j) is called the free float of (i, j) and denoted as FF_{ij}, i.e.,

$$FF_{ij} = E_j - E_i - t_{ij}$$

Event float: The float (slack) of an event is the difference between its latest time and its earliest time, i.e., Event float = Latest completion time – Earliest starting time

$$EF_i = L_i - E_i$$

Independent float: The time in which any activity can be rescheduled without affecting the preceding or the succeeding activities is known as independent float.

$$\text{Independent float} = E_j - (L_i + t_{ij}).$$

Interference float: The difference between the total float and free float is known as interference float. The basic difference between slack and float is that the slack is used for events only, whereas float is applied for activities.

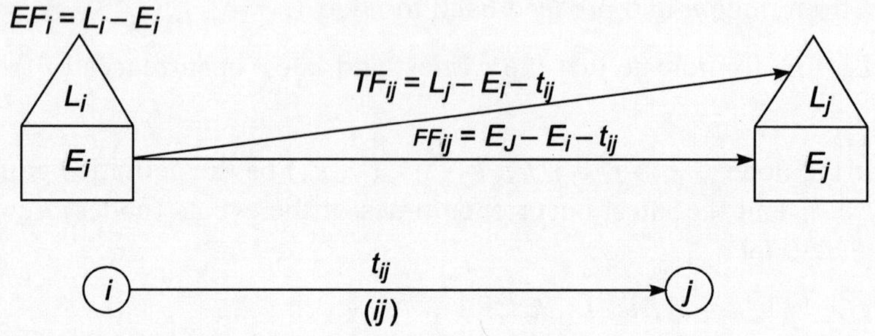

In the network diagram, we enter the earliest time inside the rectangle and the latest time inside the triangle surmounted on the rectangle.

Total Duration of the Project

Total duration of the project is denoted by T and is the sum of the durations of the activities along the critical path.

11.5 PROGRAMME EVALUATION AND REVIEW TECHNIQUE (PERT)

PERT is a probabilistic model for managing a project when the exact durations of the activities are not known. The manager of the project decides on the basis of his/her experience the three estimated durations of each activity involved in the project.

(i) t_o = Optimistic duration of the activity
(ii) t_p = Pessimistic time of the activity
(iii) t_m = Most likely time of the activity

Note that most likely time t_m lies between t_o and t_p, i.e., $t_o < t_m < t_p$.

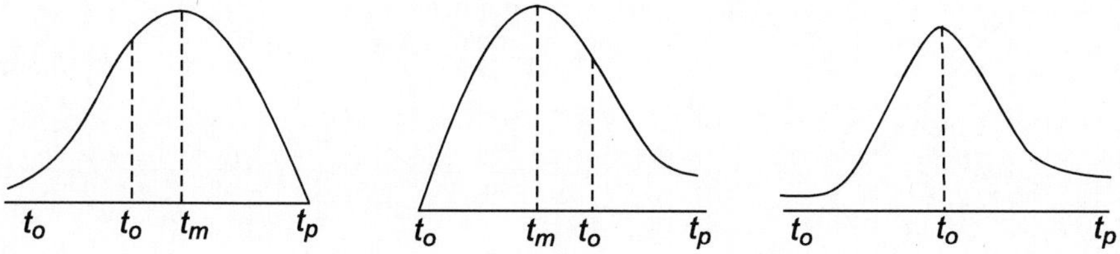

Generally, normal distribution function is used to calculate the approximate probability density function. Here end points are non-negative and also the shapes are not symmetrical, we cannot use the normal distribution. So beta distribution may follow and fulfill the requirements.

The expected time $t_e = \dfrac{t_o - 4t_m + t_p}{6}$

Standard deviation $\sigma_t = \dfrac{t_p - t_o}{6}$

and variance $V_{ij} = \sigma_t^2 = \left(\dfrac{t_p - t_o}{6}\right)^2$

Now to determine the probability that the project will be completed in specific period of time. If T is the total duration of the project, then probability of completing the project within the scheduled time T_s is given as

$$P\,(T \le T_s) = P\left(\frac{T - T_e}{\sigma_t} \le \frac{T_s - T_e}{\sigma_t}\right) = P\left(Z \le \frac{T_s - T_e}{\sigma_t}\right)$$

where Z is the standard normal variate and T_e is the expected duration calculated along critical path.

11.6 OPTIMUM SCHEDULING BY CPM

The optimum scheduling by critical path method means the studies of the time-cost relationship. An optimum project schedule implies the lowest possible cost and the associated time for the project. Every project involves two types of cost-: direct and indirect cost. Direct costs are manpower, material, loading and equipment, utilized, etc., and indirect costs are bonus, insurance, administrative, supervision, etc. The direct costs are associated with individual activities and indirect costs are those expenditures which cannot be allocated for individual activities of the project. If direct cost of a project is increased and indirect cost is decreased then the time of the project is reduced. If time is reduced then the cost would be increased but limit for this reduction. The time which cannot be further reduced is called crash time and otherwise it is normal time. The cost of completing an activity in normal time is called normal cost and in crash time is called crash cost. The crash cost slope is

$$\text{Crash cost slope} = \frac{\text{Crash cost} - \text{Normal cost}}{\text{Normal time} - \text{Crash time}}$$

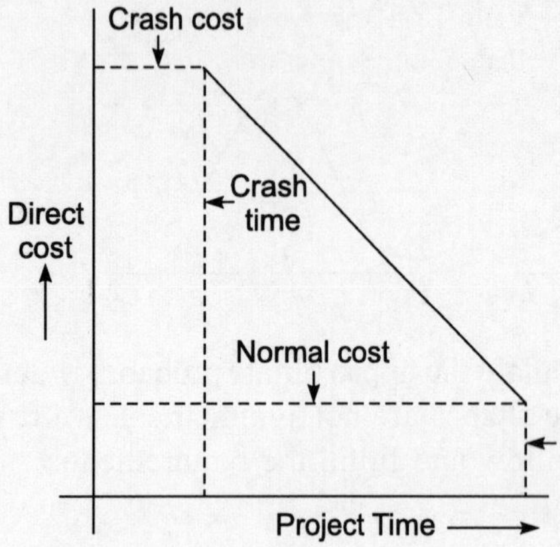

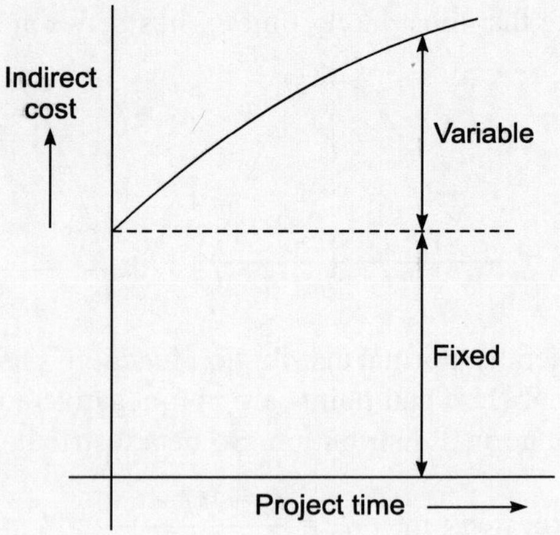

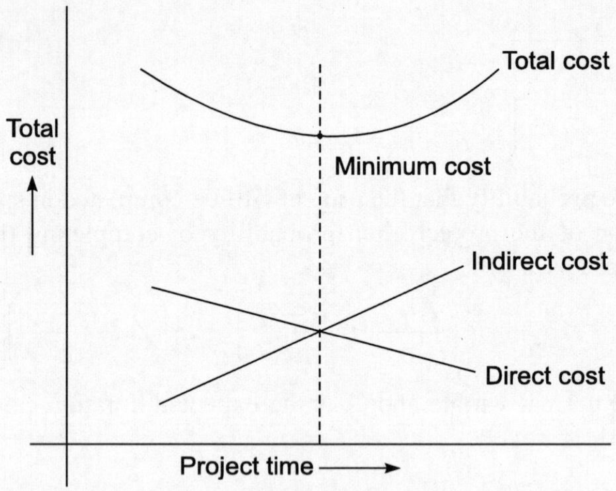

11.7 TIME-COST OPTIMIZATION ALGORITHM

The process of time-cost optimization algorithm contains the following steps:

Step 1: First we find the expected duration, standard deviation and variance for all the activities.

Step 2: Draw the network diagram and the earliest and latest occurrence times for each node.

Step 3: Find the critical path and total expected duration of the project.

Step 4: Find the cost slope for the different activities and rank the activities in the ascending order of cost slopes.

Step 5: Crashing the activities on the critical path as per their ranking. When the critical path time (duration) is reduced by crashing then additional critical paths may be developed. Now crash the activities on the critical paths. The process of crashing the activities on the critical paths is continues until when further crashing is not possible.

Step 6: Find the crash duration of the project and the total project cost by adding direct cost and indirect costs.

SOLVED EXAMPLES

Example 1: A project consists of activities *A*, *B*, *C*, *D*, *E*, *F*, *G*, and *H*. The details are as follows:

(i) *A*, *B*, *C* start immediately.
(ii) *D*, *E*, *F* can start when *A* is completed.
(iii) *G* can start after *B* and *D* are completed.
(iv) *H* can start after *C* and *E* are completed.
(v) *F*, *G*, *H* are terminal activities.

If the duration of activities are respectively 5, 3, 4, 8, 7, 5, 8, 6 then construct a network.

Solution: The network diagram for the project is

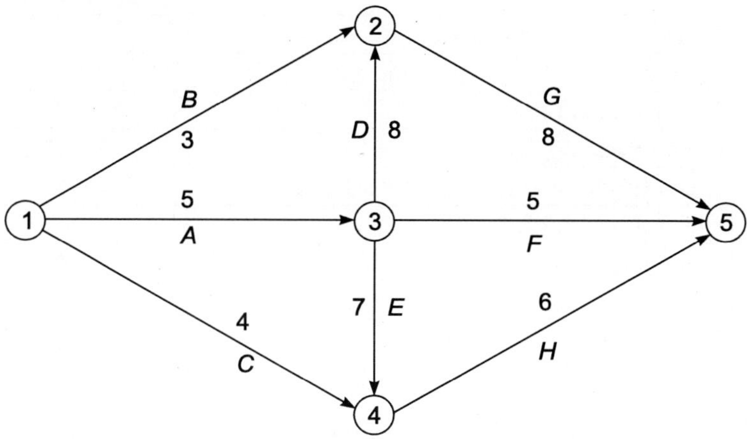

Example 2: Construct a network diagram for a project consisting of the following activities:

Activity	*Immediate predecessor(s)*
A	–
B	–
C	*A*, *B*
D	*B*
E	*D*
F	*C*, *E*
G	*D*

F and *G* are the terminal activities of the project.

Solution: The network diagram for the project is

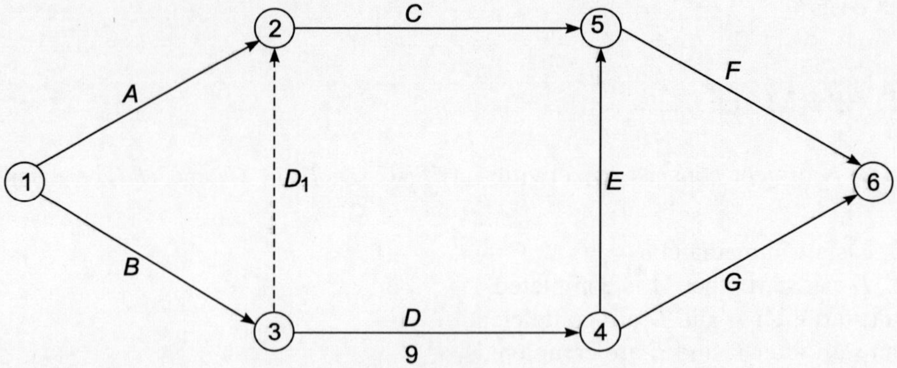

Example 3: Listed in the table are the activities and sequencing necessary for a maintenance job on the heat exchangers in a refinery.

Activity	Description	Predecessor activity
A	Dismantle pipe connections	–
B	Dismantle header, closure and floating front	*A*
C	Remove tube bundle	*B*
D	Clean bolts	*B*
E	Clean header and floating head front	*B*
F	Clean tube bundle	*C*
G	Clean shell	C
H	Replace tube bundle	*F*, *G*
I	Prepare shell pressure test	*D*, *E*, *H*
J	Prepare tube pressure test and reassemble	I

Draw a network diagram of activities for the project.

Solution: The network diagram for the project is

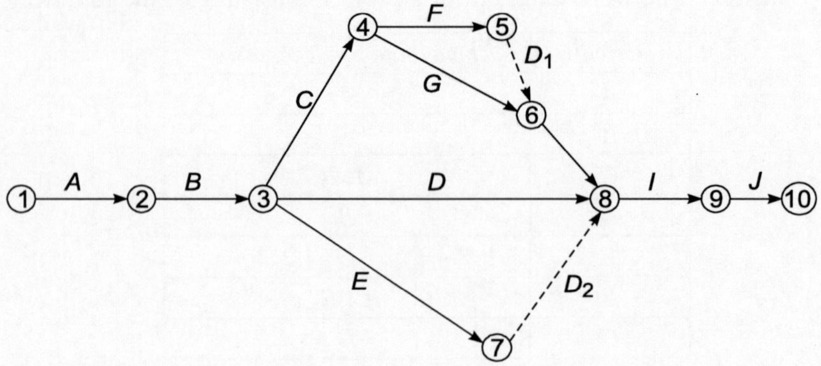

Example 4: Construct the network diagram comprising activities B, C…, Q and N such that the following constraints are satisfied:

$$B < E, F;\ C < G, L;\ E, G < H;\ L, H < I;\ L < M;\ H < N;\ H < J;\ I, J < P;\ P < Q.$$

The notation $X < Y$ means that the activity X must be finished before Y can begin.

Solution: The network diagram for the project is

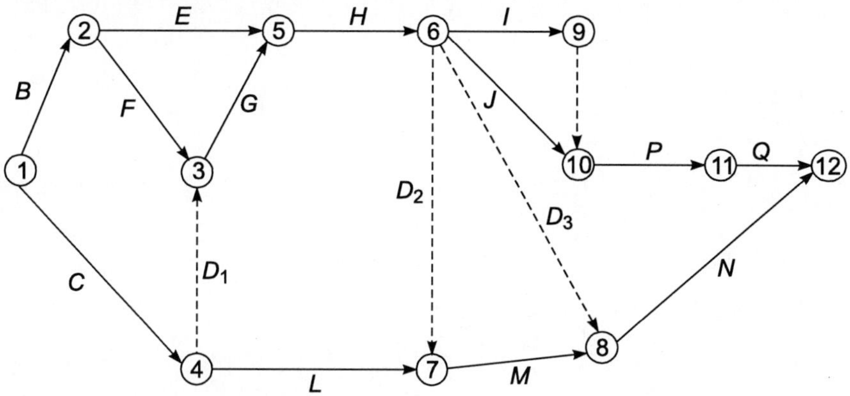

Example 5: Find the critical path for the project network given below:

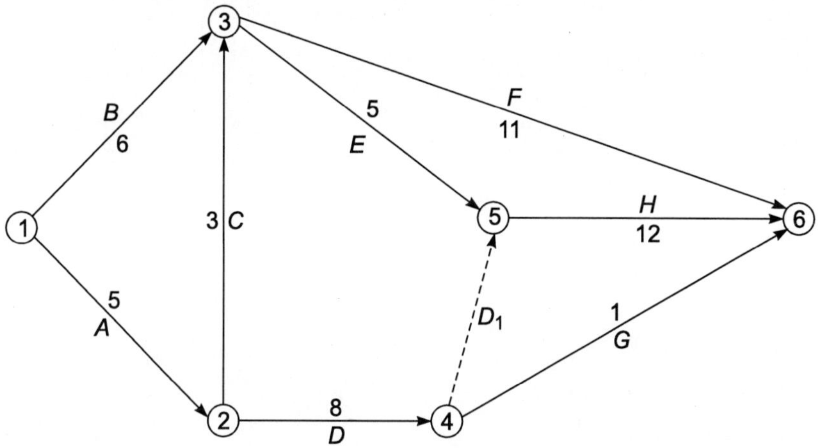

where all the durations are in days.

Solution: First we determine the forward and backward calculations.

Forward Calculations

Node 1: $E_1 = 0$

Node 2: $E_2 = E_1 + t_{12}$
$= 0 + 5$
$= 5$

Node 3: $E_3 = \max\limits_{i=1,2} (E_i + t_{i3})$
$= \max(E_1 + t_{13}, E_2 + t_{23})$
$= \max(0 + 6, 5 + 3)$
$= 8$

Node 4: $E_4 = E_2 + t_{24}$
$= 5 + 8$
$= 13$

Node 5: $E_5 = \max\limits_{i=3,4} (E_i + t_{i5})$
$= \max(E_3 + t_{35}, E_4 + t_{45})$
$= \max(8 + 5, 13 + 0)$
$= 13$

Node 6: $E_6 = \max\limits_{i=3,4,5} (E_i + t_{i6})$
$= \max(E_3 + t_{36}, E_4 + t_{46}, E_5 + t_{56})$
$= \max(8 + 11, 13 + 1, 13 + 12)$
$= 25.$

Backward Calculations

Node 6: $L_6 = E_6$
$= 25$

Node 5: $L_5 = L_6 - t_{56}$
$= 25 - 12$
$= 13$

Node 4: $L_4 = \min\limits_{j=5,6} (L_j - t_{4j})$
$= \min(L_5 - t_{45}, L_6 - t_{46})$
$= \min(13 - 0, 25 - 1)$
$= 13$

Node 3: $L_3 = \min\limits_{j=5,6} (L_j - t_{3j})$
$= \min(L_5 - t_{35}, L_6 - t_{36})$
$= \min(13 - 5, 25 - 11)$
$= 8$

Node 2: $L_2 = \min\limits_{j=3,4} (L_j - t_{2j})$
$= \min(L_3 - t_{23}, L_4 - t_{24})$
$= \min(11 - 3, 13 - 8)$
$= 5$

Node 1: $L_1 = \min\limits_{j=2,3} (L_j - t_{1j}) = \min(L_2 - t_{12}, L_3 - t_{13})$
$= \min(5 - 5, 8 - 6)$
$= 0.$

Using the above calculations, we have

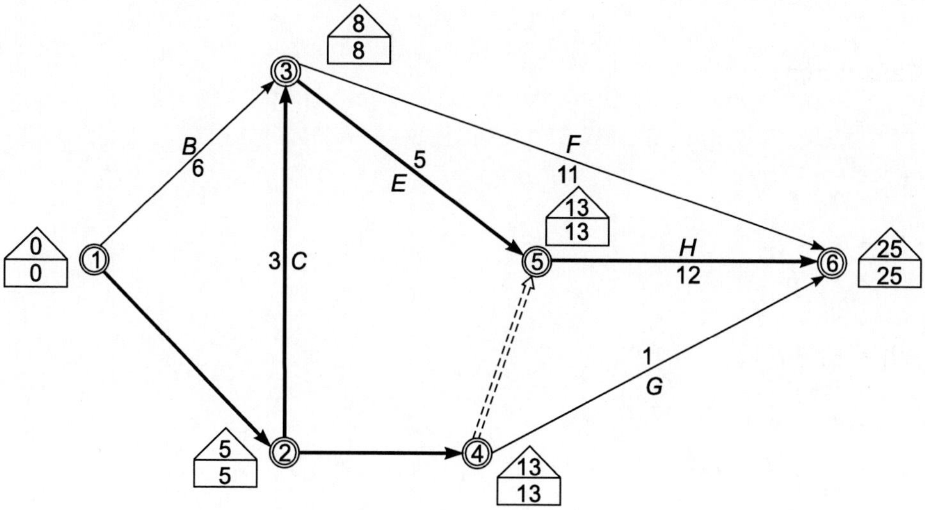

The critical path is 1 → 2 → 4 → 5 → 6 and 1 → 2 → 3 → 5 → 6, and the project durations is 25 days.

Example 6: The following data are available for a project

Activity	Duration (days)	Proceeding activity
A	5	–
B	6	*A*
C	5	*B*
D	4	*A*
E	3	*D*
F	4	*C*, *E*

(i) Prepare the network diagram for the project.
(ii) Compute the earliest event time and latest event time.
(iii) Determine the critical path and total project duration.
(iv) Compute the total float for each activity.

Solution:

(i) The network diagram for the project is

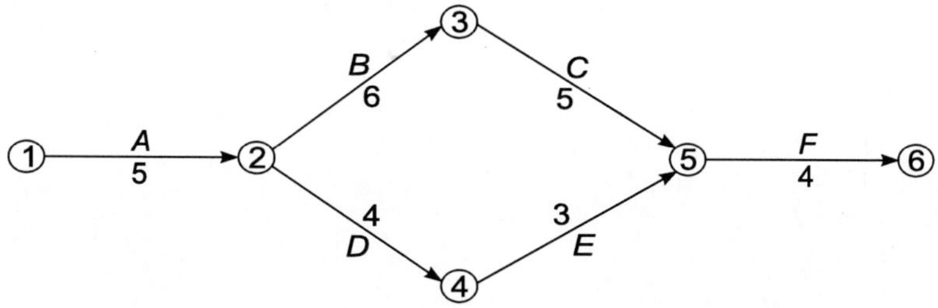

(ii) Now we determine the earliest event time and latest event time for each activity of the project.

Forward Calculations:

Node 1: $E_1 = 0$

Node 2: $E_2 = E_1 + t_{12}$
$= 0 + 5$

Node 3: $E_3 = E_2 + t_{23}$
$= 5 + 6$
$= 11$

Node 4: $E_4 = E_2 + t_{24}$
$= 5 + 4$
$= 9$

Node 5: $E_5 = \max_{i=3,4} (E_i + t_{i5})$
$= \max(E_3 + t_{35}, E_4 + t_{45})$
$= \max(11 + 5, 9 + 3)$
$= 16$

Node 6: $E_6 = E_5 + t_{56}$
$= 16 + 4$
$= 20$

Backward Calculations:

Node 6: $L_6 = E_6$
$= 20$

Node 5: $L_5 = L_6 - t_{56}$
$= 20 - 4$
$= 16$

Node 4: $L_4 = L_5 - t_{45}$
$= 16 - 3$
$= 13$

Node 3: $L_3 = L_5 - t_{35}$
$= 16 - 5$
$= 11$

Node 2: $L_2 = \min_{j=3,4} (L_j - t_{2j})$
$= \min(L_3 - t_{23}, L_4 - t_{24})$
$= \min(11 - 6, 13 - 4)$
$= 5$

Node 1: $L_1 = L_2 - t_{12}$
$= 5 - 5$
$= 0$

To find the critical nodes and total float for each of the activities, all these calculations are shown in the following table:

Activity	*Duration (Days)*	*Earliest time*		*Latest time*		*Total float*
		Start Ei	*Finish Ei + tij*	*Start Li = Lj |– tij*	*Finish Lj*	*Lj = (Ei + tij)*
A	5	0	5	0	5	0
B	6	5	11	5	11	0
C	5	11	16	11	16	0
D	4	5	9	9	13	4
E	3	9	12	13	16	4
F	4	16	20	16	20	0

The above table, we find that the critical nodes are (1, 2), (2, 3), (3, 5) and (5, 6).

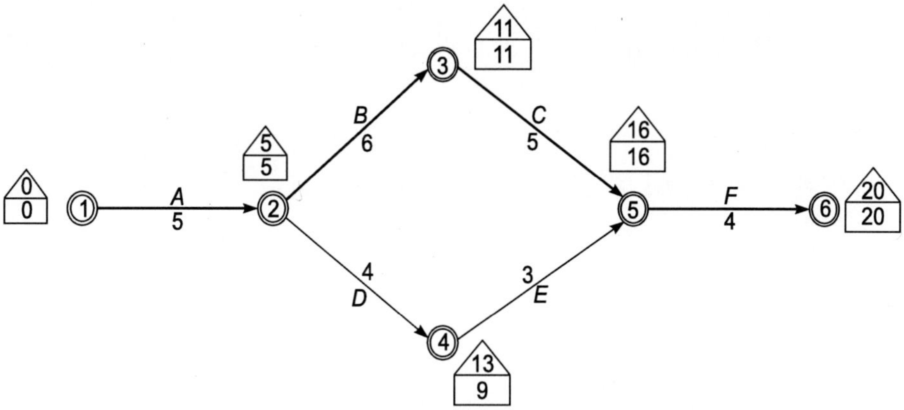

The critical path is $1 \to 2 \to 3 \to 5 \to 6$ and the total project duration time is 20 days.

Example 7: Determine the critical path for the following project network:

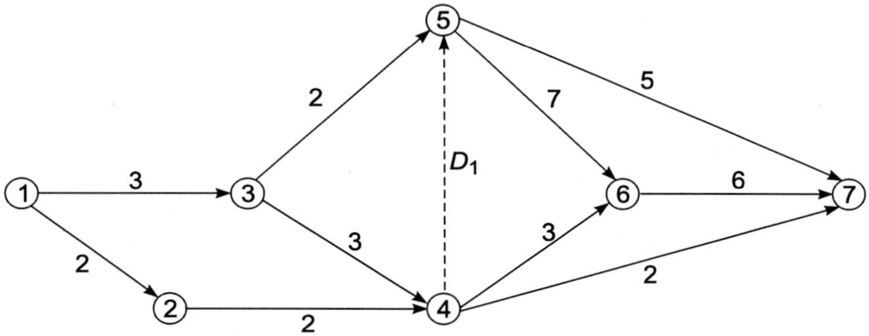

Solution: First we determine the forward and backward calculations

Forward Calculations

Node 1: $E_1 = 0$

Node 2: $E_2 = E_1 + t_{12}$
$= 0 + 2$
$= 2$

Node 3: $E_3 = E_1 + t_{13}$
$= 0 + 3$
$= 3$

Node 4: $E_4 = \max\limits_{i=2,3} (E_i + t_{i4})$
$= \max(E_2 + t_{24}, E_3 + t_{34})$
$= \max(2 + 2, 3 + 3)$
$= 6$

Node 5: $E_5 = \max\limits_{i=3,4} (E_i + t_{i5})$
$= \max(E_3 + t_{35}, E_4 + t_{45})$
$= \max(3 + 2, 6 + 0)$
$= 6$

Node 6: $E_6 = \max\limits_{i=4,5} (E_i + t_{i6})$
$= \max(E_4 + t_{46}, E_5 + t_{56})$
$= \max(6 + 3, 6 + 7)$
$= 13$

Node 7: $E_7 = \max\limits_{i=4,5,6} (E_i + t_{i7})$
$= \max(E_4 + t_{47}, E_5 + t_{57}, E_6 + t_{67})$
$= \max(6 + 2, 6 + 5, 13 + 5)$
$= 19.$

Backward Calculations:

Node 7: $L_7 = E_7$
$= 19$

Node 6: $L_6 = L_7 - t_{67}$
$= 19 - 6$
$= 13$

Node 5: $L_5 = \min\limits_{j=6,7} (L_j - t_{5j})$
$= \min(L_6 - t_{56}, L_7 - t_{57})$
$= \min(13 - 7, 19 - 5)$
$= 6$

Node 4: $L_4 = \min_{j=5,6,7} (L_j - t_{4j})$
$= \min(L_5 - t_{45}, L_6 - t_{46}, L_7 - t_{47})$
$= \min(13 - 3, 19 - 2, 6 - 0)$
$= 6$

Node 3: $L_3 = \min_{j=4,5} (L_j - t_{3j})$
$= \min(L_4 - t_{34}, L_5 - t_{35})$
$= \min(6 - 3, 6 - 2)$
$= 3$

Node 2: $L_2 = L_3 - t_{23}$
$= 6 - 2$
$= 4$

Node 1: $L_1 = \min_{j=2,3} (L_j - t_{1j})$
$= \min(L_2 - t_{12}, L_3 - t_{13})$
$= \min(4 - 2, 3 - 3)$
$= 0$

Using the above calculations, we have

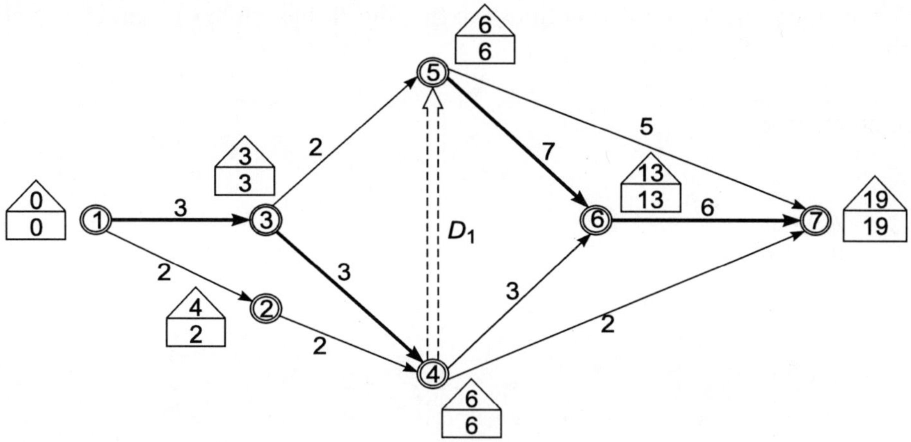

Here (1, 3), (3, 4), (4, 5), (5, 6) and (6, 7) are critical activities. The critical path is 1 → 3 → 4 → 5 → 6 → 7 and the project durations is 19.

Example 8: A project consists of a series of tasks labelled *A*, *B*, *C* ..., *G*, *H*, *I* with the following relationships ($W < X, Y$ means *X* and *Y* cannot start until *W* is completed: $X, Y < W$ means *W* cannot start until *X* and *Y* are completed) with this notation construct the network diagram having the following constraints:

$A < D, E;$ $\quad B, D < F;$ $\quad C < G;$ $\quad B, G < H;$ $\quad F, G < I$

Find also the minimum time of completion of the project, when the time (in days) of completion of each task is as follows:

Task	A	B	C	D	E	F	G	H	I
Time	23	8	20	16	24	18	19	4	10

Solution: The network diagram for the project is

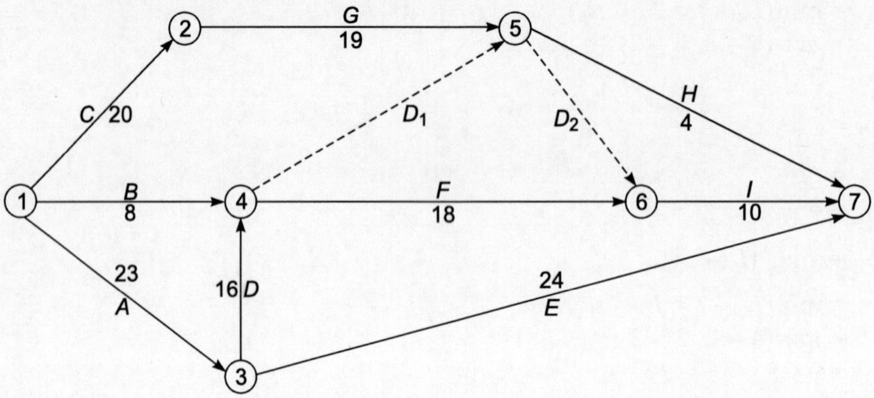

Now calculate the project completion time, we find the earliest time (E_i) and late finish time (L_j) for each activity of the project.

Forward Calculations:

Node 1: $E_1 = 0$

Node 2: $E_2 = E_1 + t_{12}$
$= 0 + 20$
$= 20$

Node 3: $E_3 = E_1 + t_{13}$
$= 0 + 23$
$= 23$

Node 4: $E_4 = \max_{i=1,3} (E_i + t_{i4})$
$= \max(E_1 + t_{14}, E_3 + t_{34})$
$= \max(0 + 8, 23 + 16)$
$= 39$

Node 5: $E_5 = \max_{i=2,4} (E_i + t_{i5})$
$= \max(E_2 + t_{25}, E_4 + t_{45})$
$= \max(20 + 19, 39 + 0)$
$= 34$

Node 6: $E_6 = \max\limits_{i=4,5} (E_i + t_{i6})$
$= \max(E_4 + t_{46}, E_5 + t_{56})$
$= \max(39 + 18, 39 + 0)$
$= 57$

Node 7: $E_7 = \max\limits_{i=3,5,6} (E_i + t_{i7})$
$= \max(E_3 + t_{37}, E_5 + t_{57}, E_6 + t_{67})$
$= \max(23 + 24, 39 + 4, 57 + 10)$
$= 67$

Backward Calculations:

Node 7: $L_7 = E_7 = 67$

Node 6: $L_6 = L_7 - t_{67}$
$= 57$

Node 5: $L_5 = \min\limits_{j=6,7} (L_j - t_{5j})$
$= \min((L_6 - t_{56}, L_7 - t_{57})$
$= \min(57 - 0, 67 - 4)$
$= 57$

Node 4: $L_4 = \min\limits_{j=5,6} (L_j - t_{4j})$
$= \min(L_5 - t_{45}, L_6 - t_{46})$
$= \min(57 - 0, 57 - 18)$
$= 39$

Node 3: $L_3 = \min\limits_{j=4,7} (L_j - t_{3j})$
$= \min(L_4 - t_{34}, L_7 - t_{37})$
$= \min(39–16, 67 - 24)$
$= 23$

Node 2: $L_2 = L_5 - t_{25}$
$= 57 - 19$
$= 38$

Node 1: $L_1 = \min\limits_{j=2,3,4} (L_j - t_{1j})$
$= \min(L_2 - t_{12}, L_3 - t_{13}, (t_4 - t_{14})$
$= \min(38 - 20, 23 - 23, 39 - 8)$
$= 0$

To find the critical nodes and total float for each of the activities, all these calculations are shown in the following table:

Activity (task) (i, j)	Normal time days	Earliest time		Latest time		Float		
		Start E_i	Finish $E_i + t_{ij}$	Start $L_i = L_j - t_{ij}$	Finish L_j	Total float $L_j = (E_i + t_{ij})$	Free $E_j = (E_i + t_{ij})$	Independent $E_j = (L_i + t_{ij})$
(1,2)	20	0	20	18	38	18	0	0
(1,3)	23	0	23	0	23	0	0	0
(1,4)	8	0	8	31	39	31	31	31
(2,5)	19	20	39	38	57	18	0	–18
(3,4)	16	23	39	23	39	0	0	0
(3,7)	24	23	47	43	67	20	20	20
(4,5)	0	39	39	57	57	18	0	0
(4,6)	18	39	57	57	57	0	0	0
(5,6)	0	39	39	57	57	18	18	0
(5,7)	4	39	43	67	67	24	24	6
(6,7)	10	57	67	67	67	0	0	0

In the above table, we find that the critical nodes are (1, 3), (3, 4), (4, 6) and (6, 7).

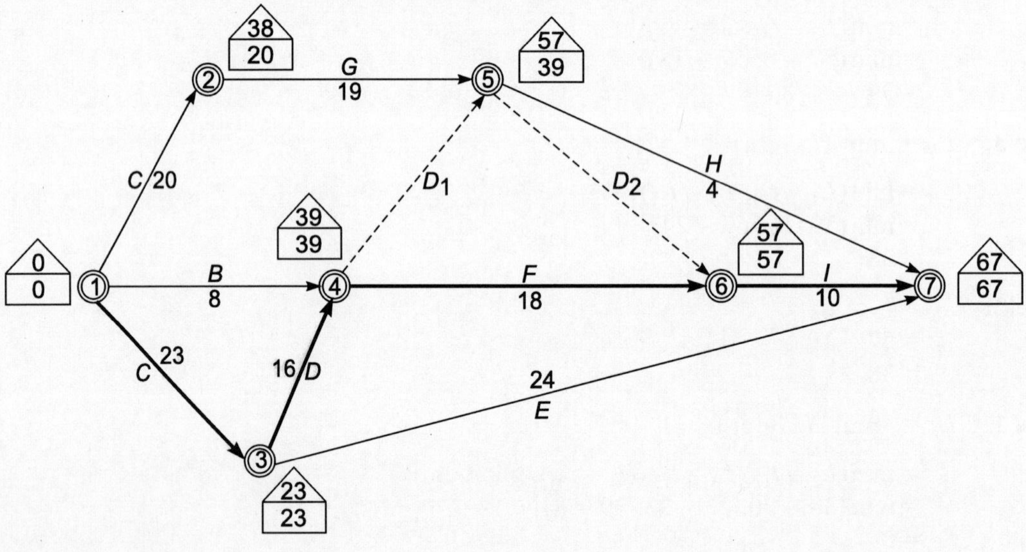

The critical path is $1 \to 3 \to 4 \to 6 \to 7$ and the project completion time is 67 days.

Example 9: A small engineering project consists of six activities. The three time estimates in number of days for each activity are given below:

Activity (task) (i, j)	Normal time days	Earliest time		Latest time		Float		
		Start E_i	Finish $E_i + t_{ij}$	Start $L_i = L_j - t_{ij}$	Finish L_j	Total float $L_j = (E_i + t_{ij})$	Free $E_j = (E_i + t_{ij})$	Independent $E_j = (L_i + t_{ij})$
(1,2)	20	0	20	18	38	18	0	0
(1,3)	23	0	23	0	23	0	0	0
(1,4)	8	0	8	31	39	31	31	31
(2,5)	19	20	39	38	57	18	0	–18
(3,4)	16	23	39	23	39	0	0	0
(3,7)	24	23	47	43	67	20	20	20
(4,5)	0	39	39	57	57	18	0	0
(4,6)	18	39	57	57	57	0	0	0
(5,6)	0	39	39	57	57	18	18	0
(5,7)	4	39	43	67	67	24	24	6
(6,7)	10	57	67	67	67	0	0	0

In the above table, we find that the critical nodes are (1, 3), (3, 4), (4, 6) and (6, 7).

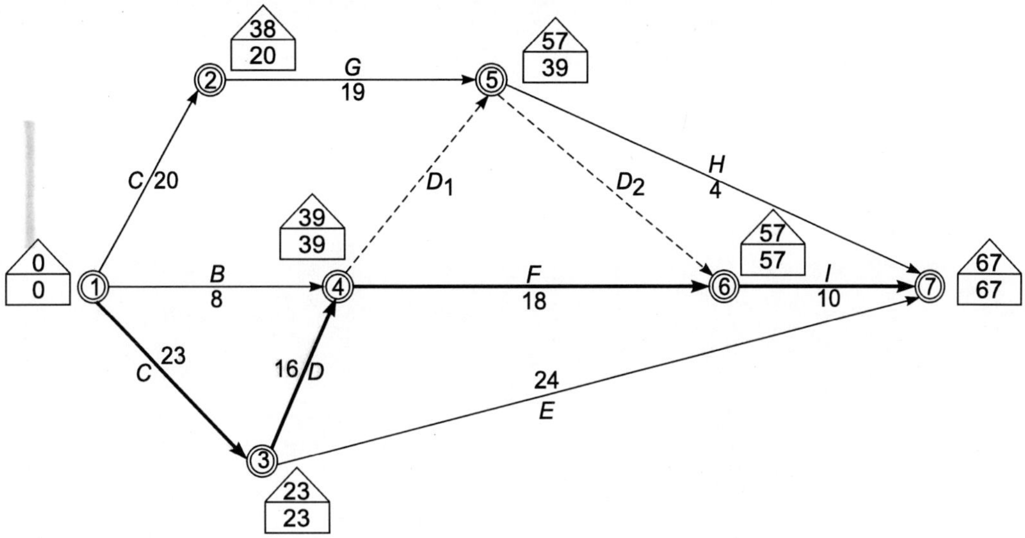

The critical path is $1 \rightarrow 3 \rightarrow 4 \rightarrow 6 \rightarrow 7$ and the project completion time is 67 days.

Example 9: A small engineering project consists of six activities. The three time estimates in number of days for each activity are given below:

Node 6: $E_6 = \max_{i=4,5} (E_i + t_{i6})$
$= \max(E_4 + t_{46}, E_5 + t_{56})$
$= \max(39 + 18, 39 + 0)$
$= 57$

Node 7: $E_7 = \max_{i=3,5,6} (E_i + t_{i7})$
$= \max(E_3 + t_{37}, E_5 + t_{57}, E_6 + t_{67})$
$= \max(23 + 24, 39 + 4, 57 + 10)$
$= 67$

Backward Calculations:

Node 7: $L_7 = E_7 = 67$

Node 6: $L_6 = L_7 - t_{67}$
$= 57$

Node 5: $L_5 = \min_{j=6,7} (L_j - t_{5j})$
$= \min((L_6 - t_{56}, L_7 - t_{57})$
$= \min(57 - 0, 67 - 4)$
$= 57$

Node 4: $L_4 = \min_{j=5,6} (L_j - t_{4j})$
$= \min(L_5 - t_{45}, L_6 - t_{46})$
$= \min(57 - 0, 57 - 18)$
$= 39$

Node 3: $L_3 = \min_{j=4,7} (L_j - t_{3j})$
$= \min(L_4 - t_{34}, L_7 - t_{37})$
$= \min(39–16, 67 - 24)$
$= 23$

Node 2: $L_2 = L_5 - t_{25}$
$= 57 - 19$
$= 38$

Node 1: $L_1 = \min_{j=2,3,4} (L_j - t_{1j})$
$= \min(L_2 - t_{12}, L_3 - t_{13}, (t_4 - t_{14})$
$= \min(38 - 20, 23 - 23, 39 - 8)$
$= 0$

To find the critical nodes and total float for each of the activities, all these calculations are shown in the following table:

Activity	t_o	t_m	t_p
1 - 2	2	5	8
2 - 3	1	1	1
3 - 5	0	6	18
5 - 6	7	7	7
1 - 4	3	3	3
4 - 5	2	8	14

(i) Find the values of expected time (t_e), standard deviation (σ_t) and variance (V_{ij}) for each activity.
(ii) Draw the network diagram.
(iii) Determine the earliest start, latest finish time and total float for each activity.
(iv) Find the critical path and project duration.

Solution: We know that the expected time $t_e = \dfrac{t_o - 4t_m + t_p}{6}$

$$\text{Standard deviation } \sigma_t = \frac{t_p - t_o}{6}$$

and

$$\text{variance } V_{ij} = \sigma_t^2 \left(\frac{t_p - t_o}{6}\right)^2$$

Activity	t_o	t_m	t_p	$t_e = \dfrac{t_o - 4t_m + t_p}{6}$	$\sigma_t = \dfrac{t_p - t_o}{6}$	$V_{ij} = \sigma_t^2 = \left(\dfrac{t_p - t_o}{6}\right)^2$
1-2	2	5	8	$\dfrac{2+4\times5+8}{6} = 5$	1	1
2-3	1	1	1	$\dfrac{1+4\times1+1}{6} = 1$	0	0
3-5	0	6	18	$\dfrac{0+4\times6+18}{6} = 7$	3	9
5-6	7	7	7	$\dfrac{7+4\times7+7}{6} = 7$	0	0
1-4	3	3	3	$\dfrac{3+4\times3+3}{6} = 3$	0	0
4-5	2	8	14	$\dfrac{2+4\times8+14}{6} = 8$	2	4

The network diagram for the project is

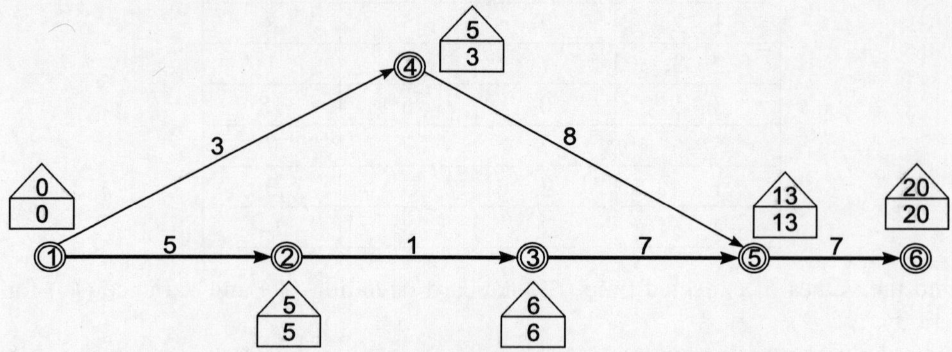

Activity	Earliest time		Latest time		Total float (Slack)
	Start	Finish	Start	Finish	$L_j - (E_i + t_{ij})$
1-2	0	5	0	5	0
2-3	5	6	5	6	0
3-5	6	13	6	13	0
5-6	13	20	13	20	0
1-4	0	3	2	5	2
4-5	3	11	5	13	2

The critical path is $1 \to 2 \to 3 \to 5 \to 6$. The project duration is 20 days.

Example 10: A project consists of the following activities and different time estimates:

Activity	t_o	t_m	t_p
1-2	3	5	8
1-3	2	4	8
1-4	6	8	12
2-5	5	9	12
3-5	3	5	9
4-6	3	6	10
5-6	2	4	8

(i) Draw the network for this project.
(ii) Determine the expected time and variance for each activity.
(iii) Find the critical path and the project variance.
(iv) What is the probability that the project will be completed in 22 days?

Solution: We know that the expected time

$$t_e = \frac{t_o - 4t_m + t_p}{6} \text{ and variance } V_{ij} = \sigma_t^2 = \left(\frac{t_p - t_o}{6}\right)^2$$

Activity	t_o	t_m	t_p	$t_e = \dfrac{t_o - 4t_m + t_p}{6}$	$V_{ij} = \sigma_t^2 \left(\dfrac{t_p - t_o}{6}\right)^2$
1-2	3	5	8	$\dfrac{3+4\times5+8}{6} = 5.167$	$\left(\dfrac{8-3}{6}\right)^2 = \dfrac{25}{36}$
1-3	2	4	8	$\dfrac{2+4\times4+8}{6} = 4.33$	$\left(\dfrac{8-2}{6}\right)^2 = 1$
1-4	6	8	12	$\dfrac{6+4\times8+12}{6} = 8.33$	$\left(\dfrac{12-6}{6}\right)^2 = 1$
2-5	5	9	12	$\dfrac{5+4\times9+12}{6} = 8.83$	$\left(\dfrac{12-5}{6}\right)^2 = \dfrac{49}{36}$
3-5	3	5	9	$\dfrac{3+4\times5+9}{6} = 5.33$	$\left(\dfrac{9-3}{6}\right)^2 = 1$
4-6	3	6	10	$\dfrac{3+4\times6+10}{6} = 6.16$	$\left(\dfrac{10-3}{6}\right)^2 = \dfrac{49}{36}$
5-6	2	4	8	$\dfrac{2+4\times4+8}{6} = 4.33$	$\left(\dfrac{8-2}{6}\right)^2 = 1$

(i) The network diagram for the project is

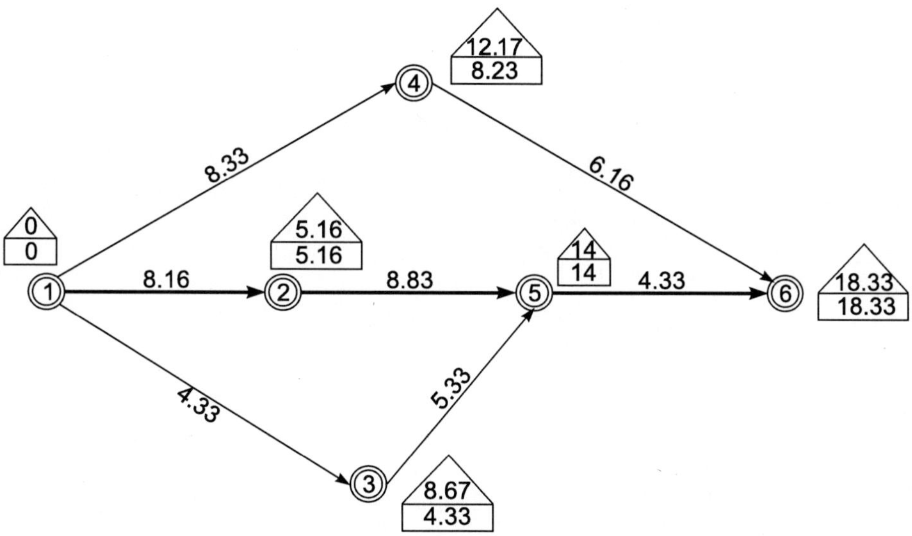

(ii) The expected time of the project is 18.33 days.

(iii) The critical path is $1 \rightarrow 2 \rightarrow 5 \rightarrow 6$.

Variance along the project length is

$$\sigma_t = \frac{25}{36} + \frac{49}{36} + 1$$

$$= \frac{75}{36}$$

$$= 2.08333$$

or $$\sigma_t = 1.4433$$

Now we have $$Z = \frac{T_s - T_e}{\sigma_t}$$

$$Z = \frac{22 - 18.33}{1.4433} = \frac{3.67}{1.4433}$$

$$= 2.540$$

Using the normal distribution table (for $z = 2.54$), we have

Probability = 98.38%.

Example 11: A project is expected to take 12 months along the critical having a standard deviation of 4 months. What is the probability of completing the project within: (i) 10 months? (ii) 16 months? (iii) 8 months?

Solution:

(i) The probability factor is given by

$$Z = \frac{T_s - T_e}{\sigma_t}$$

$$= \frac{10 - 12}{4}$$

$$= -0.5$$

By normal distribution table (for $Z = -0.5$), we have

Probability (P) = 30.85%

(ii) For $T_s = 16$ months,

We have $$Z = \frac{16 - 12}{4} = 1$$

For $Z = 1$, we get

Probability (P) = 84.13%

(iii) For $T_s = 8$ months,

We have $$Z = \frac{8 - 12}{4} = -1$$

For $Z = -1$, we get

Probability (P) = 15.87%.

Example 12: The following table gives the activities in a construction project and other related information

Activity	Optimistic time t_o (days)	Normal time (Most likely time) t_m	Pessimistic time t_p
1 - 2	20	30	46
1 - 3	9	12	21
2 - 3	3	5	7
2 - 4	2	3	4
3 - 5	1	2	3
4 - 5	12	18	24

(i) Draw a network diagram and calculate the total project duration.
(ii) Find the critical path and marked.
(iii) Find the probability that the project will be completed in 50 days.

Solution: We know that

the expected time $$t_e = \frac{t_o + 4t_m + t_p}{6}$$

Standard deviation $$\sigma_t = \frac{t_p - t_o}{6}$$

and variance $$V_{ij} = \sigma_t^2 = \left(\frac{t_p - t_o}{6}\right)^2$$

Activity	t_o	t_m	t_p	$t_e = \frac{t_o + 4t_m + t_p}{6}$	$V_{ij} = \left(\frac{t_p - t_o}{6}\right)^2$
1-2	20	30	46	$\frac{20 + 4 \times 30 + 46}{6} = 31$	$\left(\frac{46 - 20}{6}\right)^2 = 18.77$
1-3	9	12	21	$\frac{9 + 4 \times 12 + 21}{6} = 13$	$\left(\frac{21 - 9}{6}\right)^2 = 4$
2-3	3	5	7	$\frac{3 + 4 \times 5 + 7}{6} = 5$	$\left(\frac{7 - 3}{6}\right)^2 = 0.44$
2-4	2	3	4	$\frac{2 + 4 \times 3 + 4}{6} = 3$	$\left(\frac{4 - 2}{6}\right)^2 = 0.11$
3-4	1	2	3	$\frac{1 + 4 \times 2 + 3}{6} = 2$	$\left(\frac{3 - 1}{6}\right)^2 = 0.11$
4-5	12	18	24	$\frac{12 + 4 \times 18 + 24}{6} = 18$	$\left(\frac{24 - 12}{6}\right)^2 = 4$

The network diagram for the project is

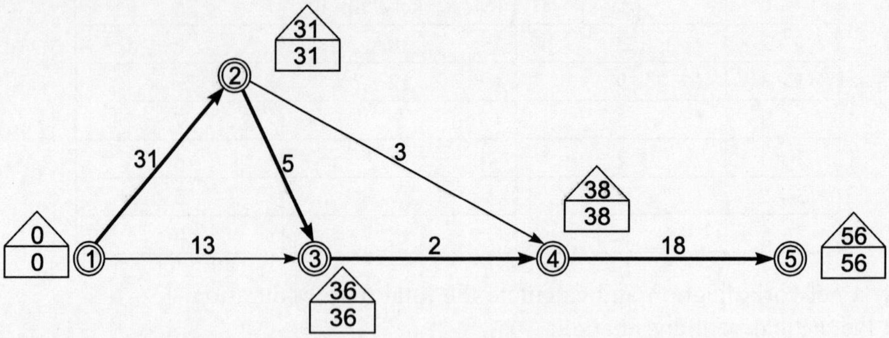

The critical path is 1→2→3→4→5.
Variance along the critical path

$$\sum V_{ij} = 18.77 + 4 + 0.49 + 0.11 + 0.11 + 4$$
$$= 27.43$$

The project duration time = 56 days.

Hence,
$$Z = \frac{T_s - T_e}{\sigma_t}$$
$$Z = \frac{50 - 56}{\sqrt{27.43}}$$
$$= -1.1456$$

Using normal distribution table (for $Z = -1.1456$) we get

$$\text{Probability } (P) = 12.62\%$$

Example 13: A small project is composed of seven activities, whose time estimates are as given below:

Activity	Estimated duration (in weeks)		Pessimistic
	Optimistic (t_o)	Most likely (t_m)	(t_p)
1 - 2	1	1	7
1 - 3	1	4	7
1 - 4	2	2	8
2 - 5	1	1	1
3 - 5	2	5	14
4 - 6	2	5	8
5 - 6	3	6	15

(i) Draw the network diagram for the project.

(ii) Find the expected duration and variance of each activity.

Also find the variance and standard deviation of project length. What is the probability that the project will be completed:

(a) At least 4 weeks later than expected?
(b) Not more than 4 weeks later than expected?

(iii) If the project due date is 19 weeks, what is the probability of meeting the due date?

Z	0.50	0.67	1.00	1.33	2.00
P	0.3085	0.2513	0.1587	0.0418	0.0228

Solution: We know that the expected time $t_e = \dfrac{t_o - 4t_m + t_p}{6}$

Standard deviation $\sigma_t = \dfrac{t_p - t_o}{6}$

and variance $V_{ij} = \sigma_t^2 = \left(\dfrac{t_p - t_o}{6}\right)^2$

Activity	t_o	t_m	t_p	$t_e = \dfrac{t_o + 4t_m + t_p}{6}$	$\sigma_t = \left(\dfrac{t_p - t_o}{6}\right)^2$	$V_{ij} = \sigma_t^2$
1-2	1	1	7	$\dfrac{1+4\times4+7}{6} = 2$	1	1
1-3	1	4	7	$\dfrac{1+4\times4+7}{6} = 4$	1	1
1-4	2	2	8	$\dfrac{2+4\times2+8}{6} = 3$	1	1
2-5	1	1	1	$\dfrac{1+4\times1+1}{6} = 1$	0	0
3-5	2	5	14	$\dfrac{2+4\times5+14}{6} = 6$	2	4
4-6	2	5	8	$\dfrac{2+4\times5+8}{6} = 5$	1	1
5-6	3	6	15	$\dfrac{3+4\times6+15}{6} = 7$	2	4

The network diagram for the project is

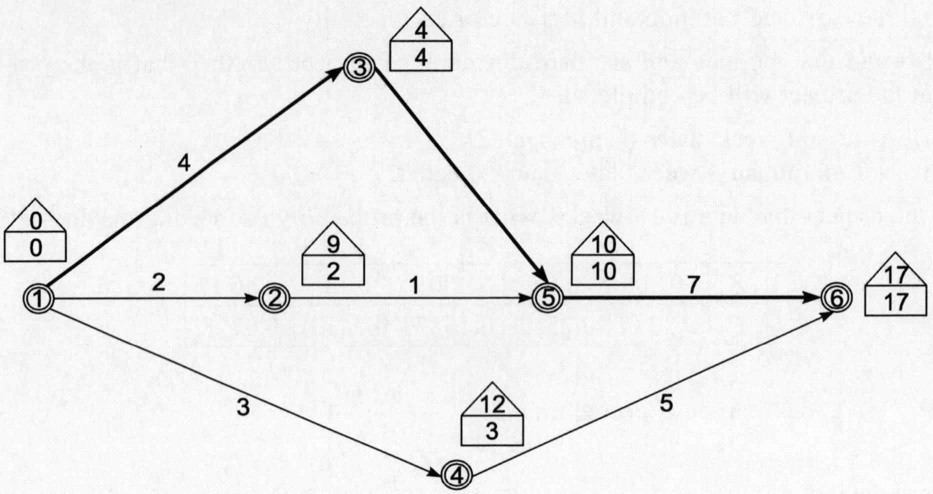

The critical path is 1→3→5→6.

(ii) The expected duration of the project is 17 days,
Variance of the project length is

$$V_{ij} = \sigma_t^2$$
$$= 1 + 4 + 4$$
$$= 9$$

Standard deviation along the project length is

$$\sigma_t = 3$$

We know that

$$Z = \frac{T_s - T_e}{\sigma_t}$$

(a) At least for week earlier, we have

$$Z = \frac{13 - 17}{3} = -\frac{4}{3}$$
$$= -1.33$$

The probability for $z = -1.33$ is 9.18% (using normal table)

(b) Not more than 4 weeks taken

$$Z = \frac{21 - 17}{3} = \frac{4}{3}$$
$$= 1.33$$

(iv) When the due date is 19 weeks.

$$Z = \frac{19 - 17}{3} = \frac{2}{3}$$
$$= 0.67.$$

Examples 14: For the network diagram below, the time estimates (in days) for each activity are indicated on the diagram. Determine: (i) The expected time standard deviation and variance for each activity, (ii) The probability of completing the project in 32 days, (iii) The project duration, (iv) Critical path.

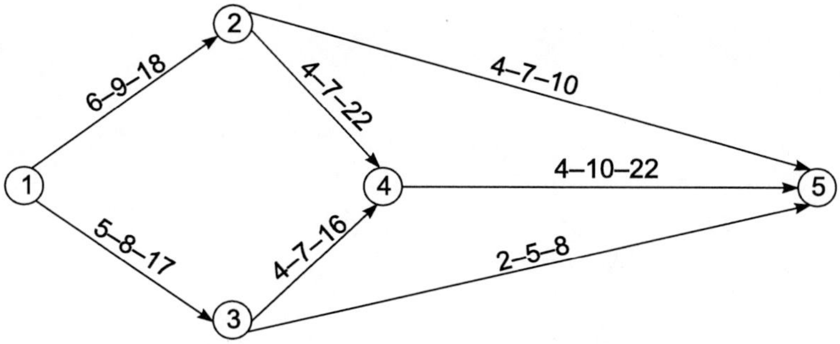

Solution: We know that the expected time $t_e = \frac{t_o - 4t_m + t_p}{6}$

Standard deviation $\sigma_t = \frac{t_p - t_o}{6}$

and variance $V_{ij} = \sigma_t^2 = \left(\frac{t_p - t_o}{6}\right)^2$

Activity	t_o	t_m	t_p	$t_e = \frac{t_o + 4t_m + t_p}{6}$	$\sigma_t = \frac{t_p - t_o}{6}$	$V_{ij} = \sigma_t^2 = \left(\frac{t_p - t_o}{6}\right)^2$
1-2	6	9	18	$\frac{6+4\times9+16}{6} = 10$	2	4
1-3	5	8	17	$\frac{5+4\times8+17}{6} = 9$	2	4
2-4	4	7	22	$\frac{4+4\times7+22}{6} = 9$	3	9
3-4	4	7	16	$\frac{4+4\times7+16}{6} = 8$	2	4
4-5	4	10	22	$\frac{4+4\times10+22}{6} = 11$	3	9
2-5	4	7	10	$\frac{4+4\times7+10}{6} = 7$	1	1
3-5	2	5	8	$\frac{2+4\times5+8}{6} = 5$	1	1

The network diagram for the project is

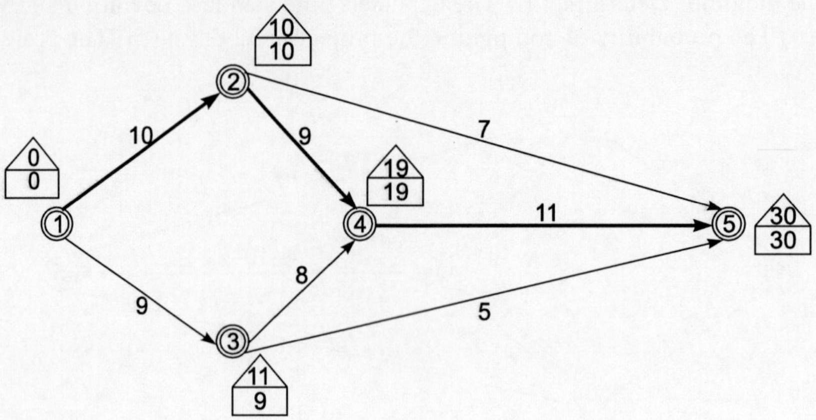

Activity	Finish E_i	Earliest time		Latest time	Float
		Start $E_i + t_{ij}$	Finish $L_i = L_j - t_{ij}$	Finish L_j	$L_j - (E_i + t_{ij})$
1-2	0	10	0	10	0
1-3	0	9	2	11	2
2-4	10	19	10	19	0
3-4	9	17	11	19	2
4-5	19	30	19	30	0
2-5	10	17	23	30	13
3-5	9	14	11	30	16

The critical path is $1 \to 2 \to 4 \to 5$. The project duration is 30 days.
Variance along the critical path

$$\sum V_{ij} = \sigma^2_{(1-2)} + \sigma^2_{(2-4)} + \sigma^2_{(4-5)}$$
$$= 4 + 9 + 9 = 22$$

Standard deviation along the critical path

$$\sigma_t = \sqrt{22}$$
$$= 4.69$$

Hence,
$$Z = \frac{T_s - T_e}{\sigma_t}$$
$$= \frac{32 - 30}{4.69}$$
$$= 0.42$$

Using normal distribution table, we get

$$P = 66\%$$

Example 15: The crash and normal duration and cost are given below:

Activity	Normal		Crash	
	Duration (days)	Cost	Duration (days)	Cost
1-2	4	140	1	230
1-3	3	140	1	160
1-4	3	200	1	240
2-5	5	100	2	200
3-6	2	50	1	80
4-6	10	150	9	180
5-6	7	200	5	250

Calculate the different minimum cost schedules by compressing the project duration by one unit at a time.

Solution: Using normal time duration, we have

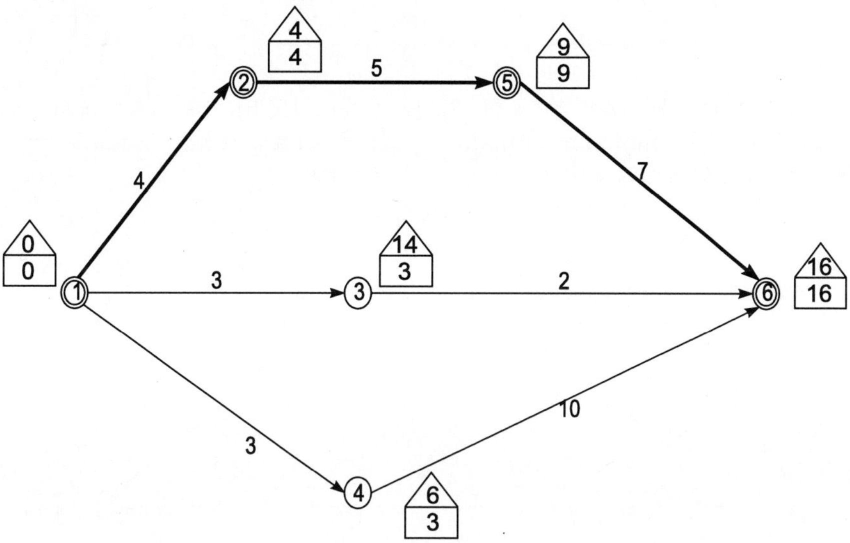

Here the critical path is $1 \to 2 \to 5 \to 6$ and normal project duration is 16 days.

$$\text{Total cost} = 980$$

The crash cost slope is

$$\text{Crash cost slope} = \frac{\text{Crash cost} - \text{Normal cost}}{\text{Normal time} - \text{Crash time}}$$

To find the optimum cost schedule, first we determine the cost slope of activities, as follows:

Activities	Crash cost slope = $\dfrac{\text{Crash cost} - \text{Normal cost}}{\text{Normal time} - \text{Crash time}}$
1-2	$\dfrac{230-140}{4-1} = 30$
1-3	$\dfrac{160-140}{3-1} = 10$
1-4	$\dfrac{200-100}{5-1} = 20$
2-5	$\dfrac{200-100}{5-1} = 33.33$
3-6	$\dfrac{80-50}{2-1} = 30$
4 – 6	$\dfrac{180-150}{10-9} = 30$
5 – 6	$\dfrac{250-200}{10-5} = 25$

To reduce the duration of the activities (1, 2), (2, 5) and (5, 6), first we choose an activity in critical activities which has minimum cost slope, i.e., (5, 6) and it has 25 minimum cost slope. So we compress the (5, 6) activity by one unit and, we get

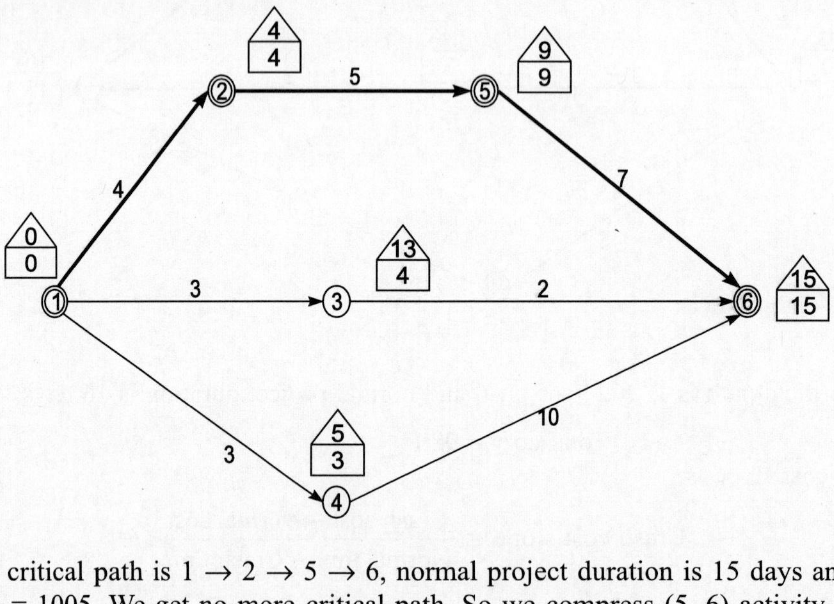

Here the critical path is $1 \to 2 \to 5 \to 6$, normal project duration is 15 days and total cost is 980 + 25 = 1005. We get no more critical path. So we compress (5, 6) activity by one unit and we have

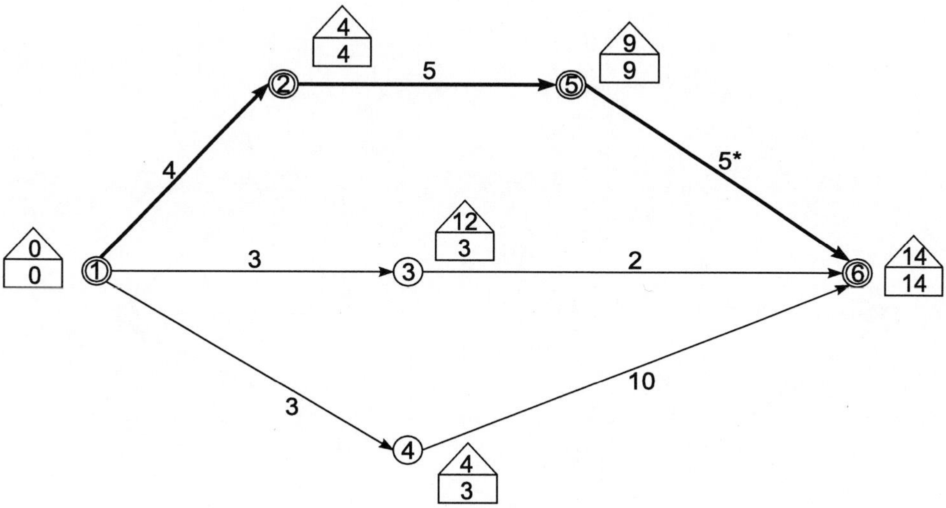

Here the critical path is $1 \to 2 \to 5 \to 6$, normal project duration is 14 days and total cost is 1005 + 25 = 1030. Again, we get no more critical path and (5, 6) activity reached to its crash value so it cannot be further reduced and put * over duration because we cannot compress an activity less than its crash duration. Now we see that in remaining critical activities, which has next highest minimum cost slope i.e. (1, 2) activities and its slope is 30. So compress the (1, 2) activity by one unit and we get

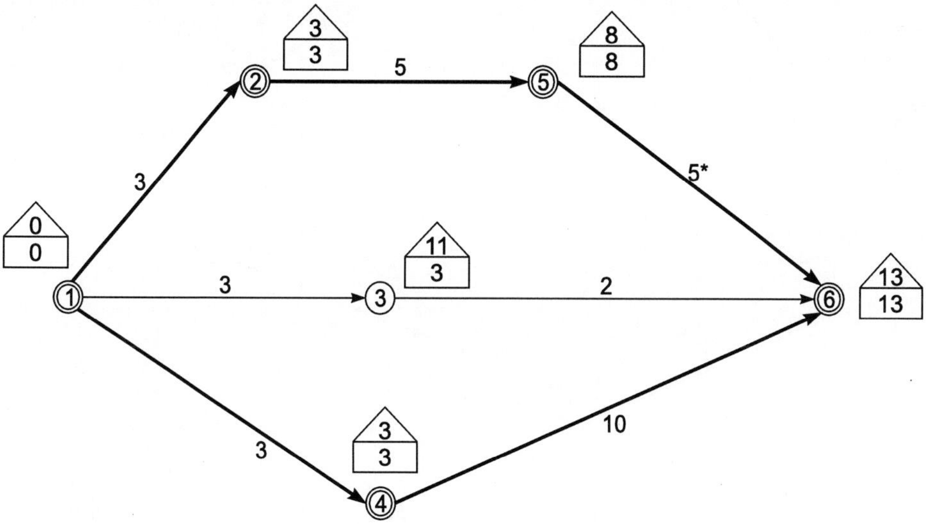

Here one more critical path has developed. The critical paths are $1 \to 2 \to 5 \to 6$ and $1 \to 4 \to 6$, Normal project duration is 13 days and total cost is 1030 + 30 = 1060. Now we compress the activities on both paths by same amount. So compress the (1, 2) activity of old path ($1 \to 2 \to 5 \to 6$) and (1,4) of new path ($1 \to 4 \to 6$) by one unit each and we get

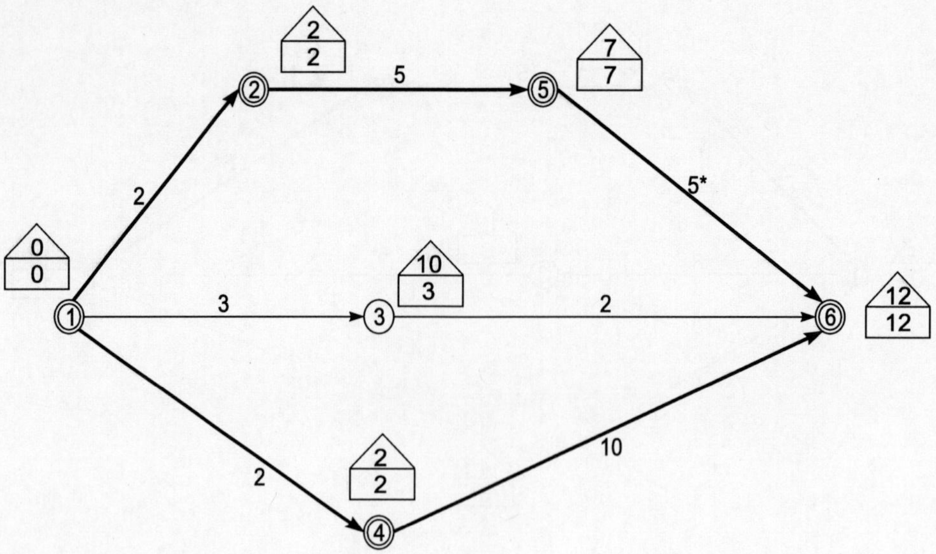

Here no more critical paths, normal project duration is 12 days and total cost is 1060 + 30 + 20 = 1110 so we compress (1, 2) and (1, 4) activities by one unit each, we get

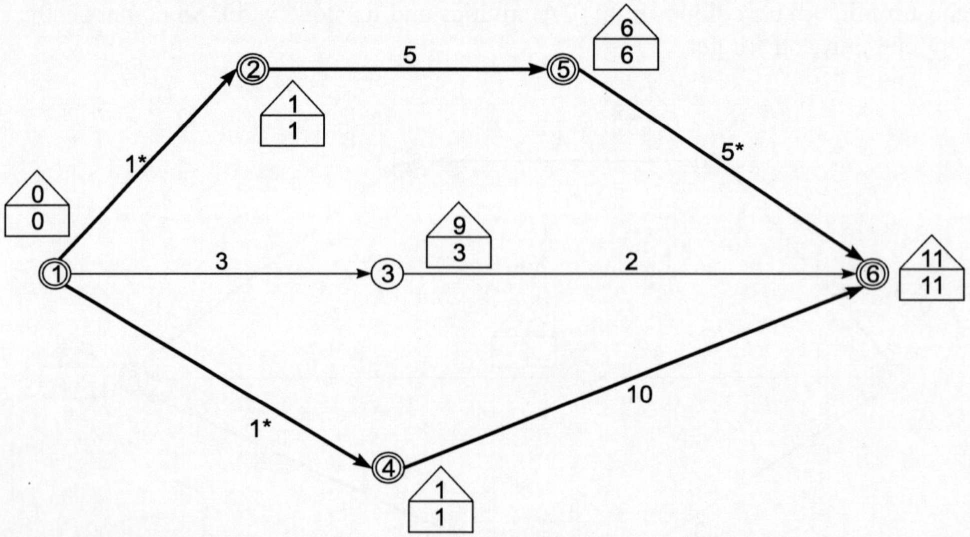

Again no more critical paths and both the activities (1, 2) and (1, 4) reached to own crash value, so we cannot compress these activities and put *over durations. The normal project duration is 11 days and total cost is 1110 + 30 + 20 = 1160. We have no choice, so we compress the (2, 5) and (4, 6) activities by one unit each, we get

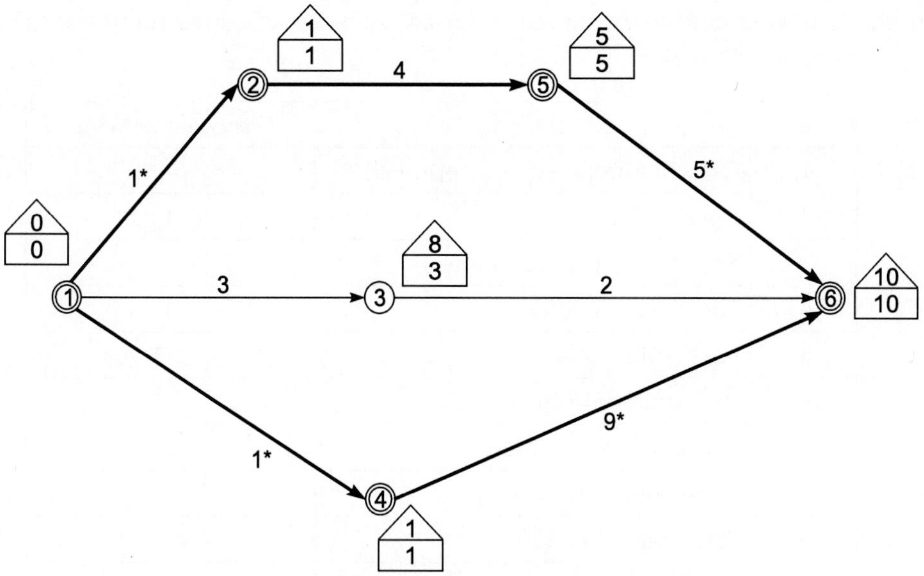

Again, no more critical paths and (4, 6) activity reached to its crash value. So all the activities of critical path (1 → 4 → 6) have reached to their crash value i.e., more compression is not possible. Hence, the normal project duration is 10 days and the total cost is 1160 + 33.33 + 30 = 1223.33.

EXERCISE 11.1

1. An assembly is to be made from two parts *X* and *Y*. Both parts must be turned on a lathe *Y* must be polished whereas *X* need not be polished. The sequence of activities, together with their predecessors, is given below:

Activity	*Description*	*Predecessor Activity*
A	Open work order	–
B	Get material for *X*	*A*
C	Get material for *Y*	*A*
D	Turn *X* on lathe	*B*
E	Turn *Y* on lathe	*B*, *C*
F	Polish *Y*	*E*
G	Assemble *X* and *Y*	*D*, *F*
H	Place	*G*

Draw a network diagram of activities for the project.

2. A new type of water pump is to be designed for an automobile. Prepare a network diagram from the following table showing the major activities for effective control of the project and the preceding activities.

No.	*Activity*	*Preceding activities*
1.	Drawing prepared and approved	–
2.	Cost analysis	1
3.	Tool feasibility (economics)	1
4.	Tool manufactured	3
5.	Favourable cost	2,3
6.	Raw-material procured	4,5
7.	Sub-assemblies ordered	5
8.	Sub-assemblies received	7
9.	Parts manufactured	4,6
10.	Final assembly	9,8
11.	Testing and shipment	10

3. Following are the activities which are to be performed for a building site preparation. Determine the precedence relationship and draw the network diagram:
   1. Clear the site
   2. Survey and layout
   3. Rough grade
   4. Excavate for sewer
   5. Excavate for electrical manholes
   6. Install sewer and bask fill
   7. Install electrical manholes
   8. Construct the boundary wall
4. A project consists of 6 jobs *A*, *B*, *C*, *D*, *E* and *F*. Number the events if

   (i) *B* and *C* depend only on *A*
   (ii) *D* depends on *B* but not on *C*
   (iii) Job *E* depends on *C* and *B*
   (iv) The project is completed when *D* and *E* are done

   Draw the network diagram.
5. A publisher has a contract with an author to publish a textbook. The simplified activities associated with the production of the textbook are given subsequently. Develop the associated network for the project:

Activity	*Description*	*Predecessor Activity*	*Duration (weeks)*
A	Manuscript proofreading by the editor	-	5
B	Sample pages prepared by typist	-	3
C	Book cover design	-	5
D	Preparation of diagram used in the book	-	4
E	Author's approval of edited manuscript and sample pages	*A, B*	3
F	Book typesetting	*E*	5
G	Author checks typed pages	*F*	3
H	Author checks art-work	*D*	2
I	Production of printing plates	*G, H*	3
J	Book production and binding	*C, I*	7

6. A task consists of activities *A*, *B*, *C*, *D*, *E* of durations 15, 15, 10, 8, 11 days, respectively. The details are as follows:

   (i) *A* and *B* can start simultaneously.
   (ii) *C* can start after 7 days and *D* after 5 days of starting *A*.
   (iii) *D* can start after 4 days of starting *C* and 7 days of starting *B*.
   (iv) *E* can start after *A* is 1/3 finished and *B* is completely finished.
   (vi) *A*, *C*, *D*, *E* are terminal activities.

   Construct the network diagram.

7. Draw the network diagram for the data given below and find:

   (i) Critical path
   (ii) Early start and late start times for each activity
   (i) Total float (slack) for each activity.

Activity	*Predecessor*	*Estimated time (weeps)*
A	-	3
B	-	5
C	-	4
D	*A*	2
E	*B*	3
F	C	9
G	*D, E*	8
H	*B*	7
I	*H, F*	9

8. Draw a network diagram showing the following relationships:

Activity	*Immediate Predecessor*
A	-
B	-
C	-
D	*A*
E	*B,C*
F	*A*
G	*C*
H	*D,E,F*
I	*D*
J	*G*
K	*G*
L	*H,J*
M	*K*
N	*I,L*

9. Tasks *A*, *B*, *C* …, *G*, *H*, *I* constitute a project. The notation $X < Y$ means that the task *X* must be finished before *Y* can begin with this notation

$$A < D, A < E, B < F, D < F, C < G, C < H, F < I, G < I$$

Draw a network diagram for the project of these tasks and find the minimum time of completion of the project, when the time (in days) of completion of each task is as follows:

Task	*A*	*B*	*C*	*D*	*E*	*F*	*G*	*H*	*I*
Time (days)	8	10	8	10	16	17	18	14	9

10. A project has the following time schedule:

Activity	1-2	1-3	1-4	2-5	3-6	3-7	4-6	5-8	6-9	7-8	8-9
Time (in weeks)	2	2	1	4	8	5	3	1	5	4	3

(i) Construct the network diagram
(ii) Compute total float for each activity
(iii) Critical path and its duration.

11. The following table gives the activities in a construction project and the time duration:

Activity	*Preceding activity*	*Normal time (days)*
(1,2)	-	20
(1,3)	-	25
(2,3)	(1,2)	10

Contd.

(2,4)	(1,2)	12
(3,4)	(1,3), (2,3)	5
(4,5)	(2,4), (3,4)	10

(i) Draw the network diagram of the project
(ii) Calculate the total float and free-float for each activity
(iii) Find the critical path and the project duration.

12. A small maintenance project consists of the following 10 jobs whose precedence relationships are identified by their node numbers:

Job	*Note numbers*	*Estimated duration (days)*
a	(1,2)	2
b	(2,3)	3
c	(2,4)	5
d	(3,5)	4
e	(3,6)	1
f	(4,6)	6
g	(4,7)	2
h	(5,8)	8
i	(6,8)	7
j	(7,8)	4

(i) Draw the network diagram for the project
(ii) Determine the early and late start and finish times for each job.
(iii) Which jobs are critical?
(iv) How much float (slack) does job (3, 5) have?

13. The following are the detail of estimated times of activities of a certain project.

Activity	*Immediate predecessors*	*Estimated time (weeks)*
A	-	2
B	*A*	3
C	*A*	4
D	*B,C*	6
E	-	2
F	*E*	8

(i) Find the critical path and the expected time for the project
(ii) Determine the earliest start time and earliest finish time for each activity
(iii) Calculate the total float (slack) for each activity.

14. What is PERT? Define optimistic time, pessimistic time and most likely time and explain how you will estimate the expected time to complete the activity in PERT technique.
15. A small project consists of seven activities, the details of which are given below:

Activity	*Immediate Predecessor*	*Duration (days)*		
		Optimistic t_o	*Most likely* t_m	*Pessimistic* t_p
A	-	1	3	7
B	*A*	2	6	14
C	*A*	3	3	3
D	*B*, *C*	4	10	22
E	B	3	7	15
F	*D*, *E*	2	5	14
G	*D*	4	4	4

(i) Draw the network diagram. Find the critical path, the expected project duration completion time and the next most critical path.
(ii) What project duration will have 95% confidence of completion?

16. Consider a project whose details are given below. Make the network. Find the critical path, variance of expected duration of project.

Activity	*Optimistic* (t_o)	*Most likely* (t_m)	*Pessimistic* (t_p)
(1,2)	6	8	11
(1,3)	19	20	23
(1,4)	27	33	41
(2,5)	17	18	21
(2,6)	16	20	26
(3,6)	7	9	13
(4,7)	8	10	13
(5,7)	8	8	10
(6,7)	4	4	6

17. Given the following formation of time estimates concerning various activities of a project

Activity	*1*	*2*	*3*	*4*	*5*	*6*
t_o	2	6	1	2	6	6
t_m	4	8	5	5	8	7
t_p	6	10	15	9	10	9

Find the expected duration, standard derivation and variance for each activity.

18. Draw a critical path and project duration for the following network.

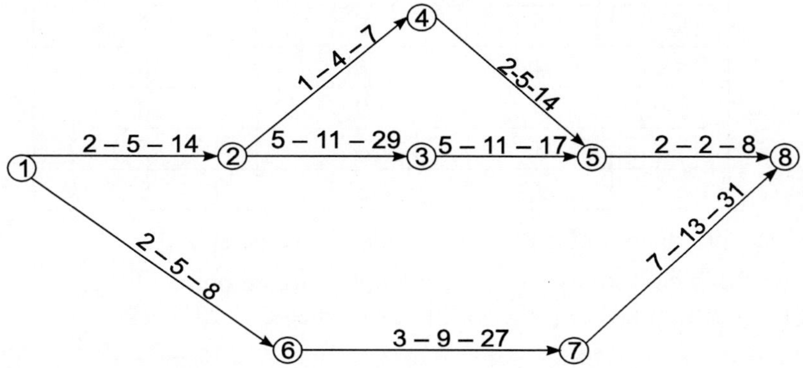

19. A project consists of activities *A*, *B*, *C*, *D*,…, *H*, *I*. The notation $X < Y$ means that the activity *X* must be completed before *Y* can start and $X, Y < W$ means that *W* will start only after completing of the activities *X* and *Y* with this notation:
Construct the network diagram for the following constraints:

$$A < D; \quad A < E; \quad B < F; \quad C < G; \quad D < H; \quad E, F < I.$$

The project has the following time schedules of the above activities:

Activity	*Optimistic* t_0	*Most likely time* t_m	*Pessimistic time* t_p
A	5	8	10
B	18	20	22
C	26	33	40
D	16	18	20
E	15	20	25
F	6	9	12
G	7	10	12
H	7	8	9
I	3	4	5

(i) Determine the expected task times and their variance, total float, earliest expected time, latest expected time and critical path.
(ii) The probability of an event occurring at the proposed completion data if the original contract time of completing the project is 41.5 weeks.
(iii) The duration of the project that will have 95% chance of being completed.

20. The following information is given:

Activity	*Optimistic*	*Most likely time*	*Pessimistic time*
1-2	3	3	3
2-3	3	6	9

Contd.

2-4	2	4	6
3-5	4	6	8
4-6	4	6	8
5-6	0	0	0
5-7	3	4	5
6-7	2	5	8

Draw the network diagram for the above table and calculate:

(i) Variance of each activity, critical path and expected project length.
(ii) The probability that the project will be completed in 23 weeks.

Given that

Z-value	1.90	1.91	1.92	1.93	1.44
Probability	0.9713	0.9719	0.9726	0.9732	0.9738

21. The following table gives data on normal time and cost and crash time and cost for a project:

Activity	*Normal*		*Crash*	
	Duration (days)	*Cost*	*Duration (days)*	*Cost*
1-2	6	60	4	100
1-3	4	60	2	200
2-4	5	50	3	150
2-5	3	45	1	65
3-4	6	90	4	200
4-6	8	80	4	300
5-6	4	40	2	100
6-7	3	45	2	80

The indirect cost per day is Rs. 10.

(i) Draw the network diagram for the project
(ii) Find the critical path
(iii) Determine the minimum total time and corresponding cost.

22. From the following data, find the optimum project duration and project cost:

Activity	*Normal*		*Crash*	
	Time (days)	*Cost (Rs.)*	*Time (days)*	*Cost (Rs.)*
A (1-2)	8	1000	6	2000
B (1-3)	4	1500	2	3500
C (2-4)	2	500	1	900
D (2-5)	10	1000	5	4000

Contd.

E(3-4)	5	1000	1	2000
F(4-5)	3	800	1	1000

It is given that indirect cost involved in the project is Rs. 700 per day.

ANSWERS

1.

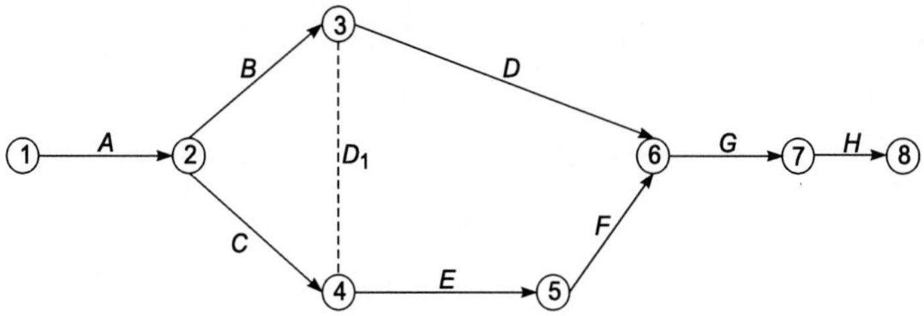

2.

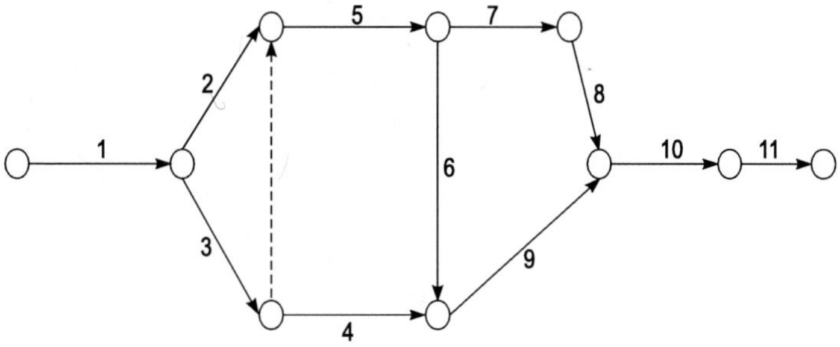

3.

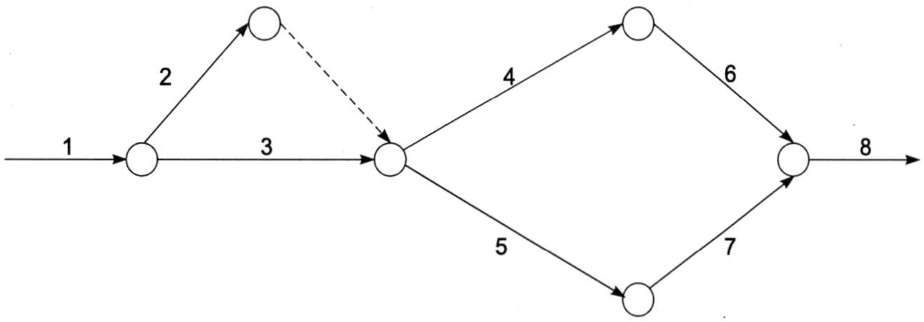

4.

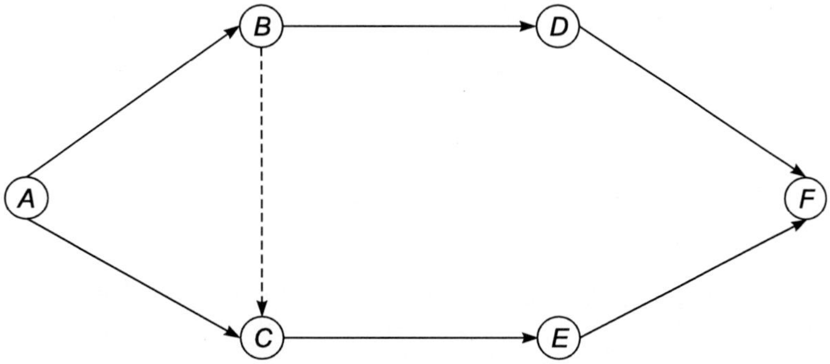

5.

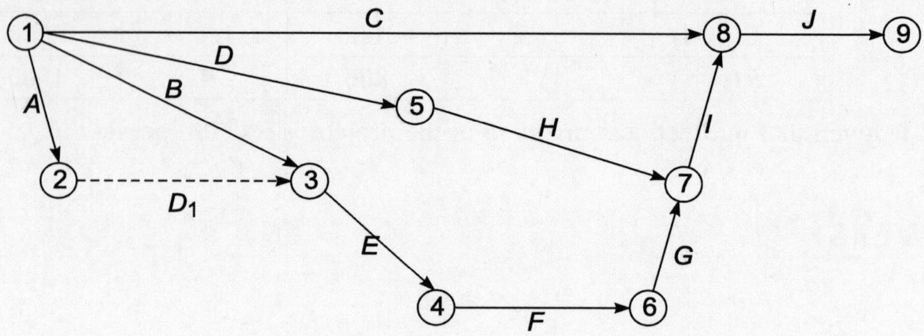

6.

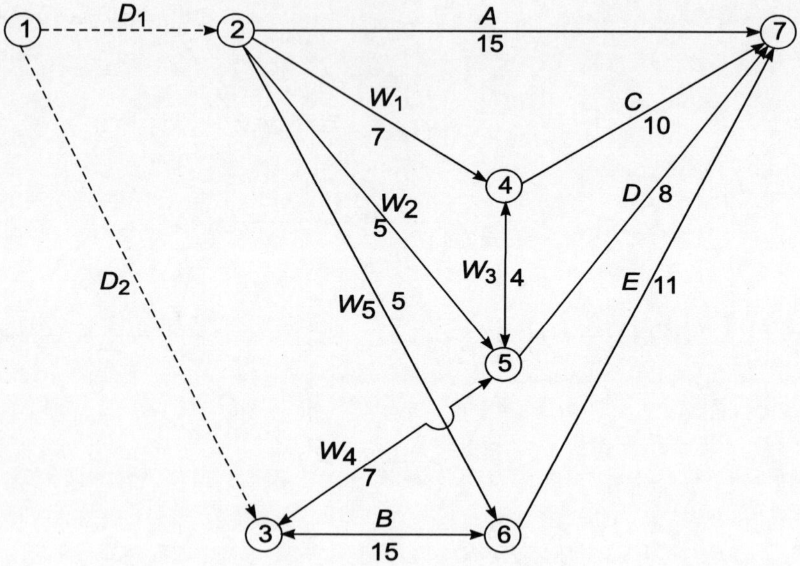

7.

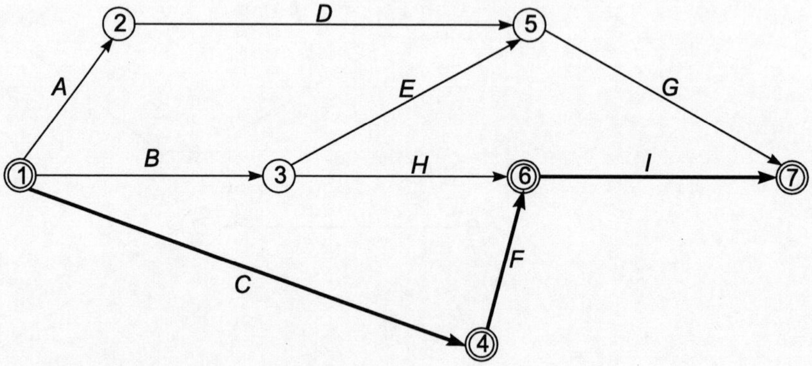

(i) $C \to F \to I$, project duration = 22 weeks.

(ii) and (iii)

Activity	*A*	*B*	*C*	*D*	*E*	*F*	*G*	*H*	*I*
Early start	0	0	0	3	5	4	8	5	13
Late start	9	1	0	12	11	4	14	6	13
Total float	9	1	0	9	6	0	6	1	0

8.

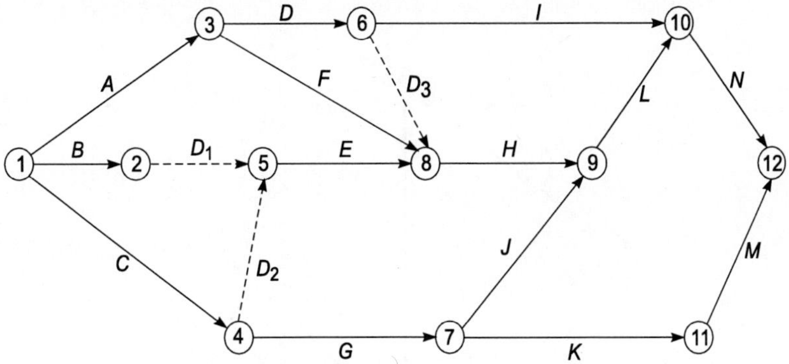

9.

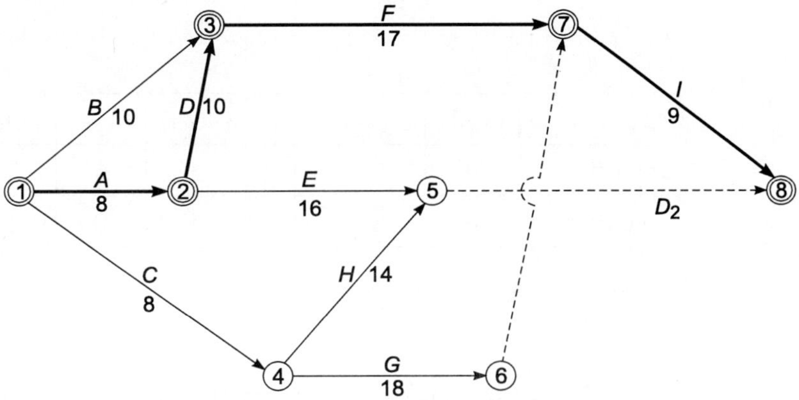

1 → 2 → 3 → 7 → 8, minimum time of completion of the project is 44 days.

10. (i)

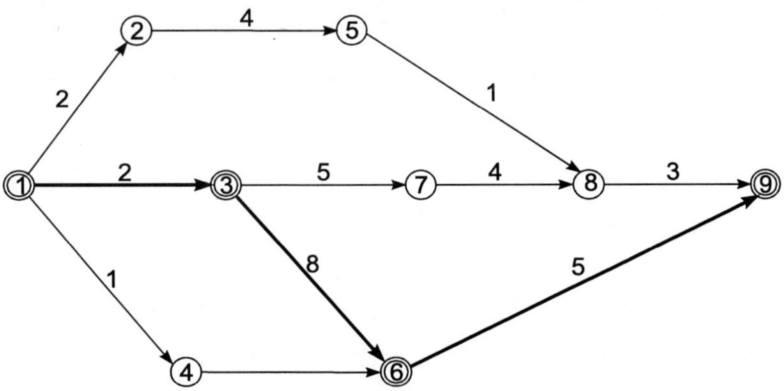

(ii)

Activity	1-2	1-3	1-4	2-5	3-6	3-7	4-6	5-8	6-9	7-8	8-9
Total float	5	0	6	5	0	1	6	5	0	1	1

(iii) $1 \to 3 \to 6 \to 9$, project duration = 15 weeks.

11. (i)

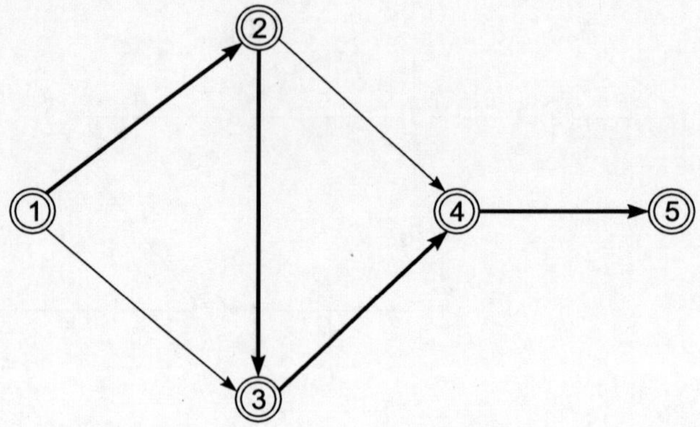

(ii)

Activity	(1, 2)	(1, 3)	(2, 3)	(2, 4)	(3, 4)	(4, 5)
Total float	0	5	0	3	0	0
Free float	0	5	0	3	0	0

(iii) $1 \to 2 \to 3 \to 4 \to 5$, project duration = 45 days.

12. (i)

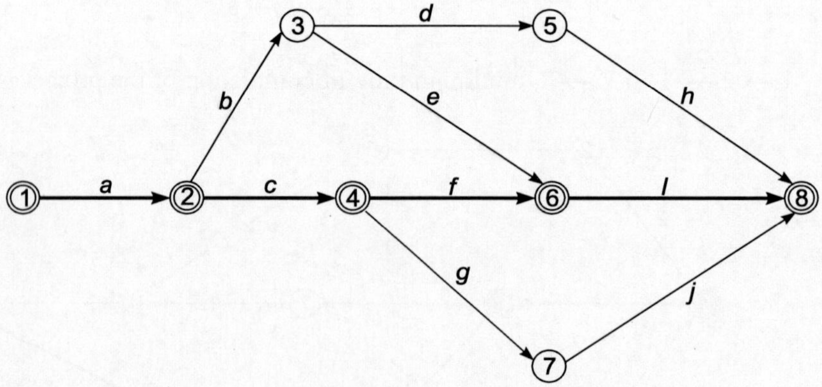

(ii)

Job	Earliest time		Latest time	
	Start	Finish	Start	Finish
1-2	0	2	0	2
2-3	2	5	5	8
3-4	2	7	2	7
3-5	5	9	8	12
3-6	5	6	12	13
4-6	7	13	7	13
4-7	7	9	14	16
5-8	9	17	12	20
6-8	13	20	13	20
7-8	9	13	16	20

(iii) Critical job are (1, 2), (2, 4), (4, 6) and (6, 8).
(iv) 3 days.

13. (i) $A \to C \to D \to F$, Project duration = 12 weeks
(ii) and (iii)

Activity	A	B	C	D	E	F
Earliest start time	0	2	2	6	0	2
Earliest finish time	2	5	6	12	2	10
Float (slack)	0	1	0	0	4	0

15. (i)

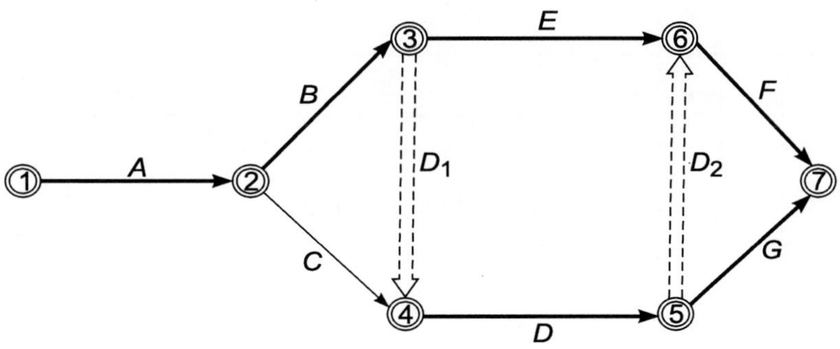

The critical path is $A \to B \to D \to F$; the expected project duration is 27 days. The next critical path is $A \to B \to D \to G$; the expected project duration is 25 days.

(ii) The project duration which has 95% probability of completion is

$$Z(1.64) = \frac{T_s - 27}{\sqrt{18}}$$

Or $$T_s = 34 \text{ days}$$

16.

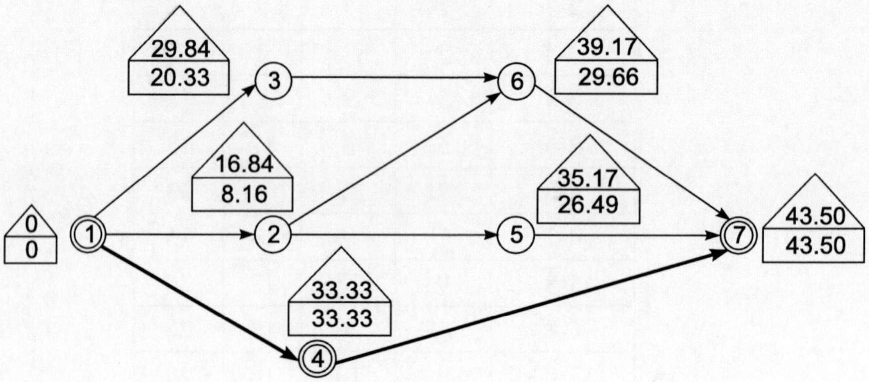

The critical path is $1 \to 4 \to 7$. The project duration is 43.50.
Variance of time schedule is 6.14.

17.

Activity	t_o	t_m	t_p	$t_e = \frac{t_o + 4t_m + t_p}{6}$	$-\sigma_t = \frac{t_p - t_o}{6}$	$V_{ij} = \sigma_t^2 = \left(\frac{t_p - t_o}{6}\right)^2$
1	2	4	6	$\frac{2+4\times4+6}{6} = 4$	$\frac{2}{3}$	$\frac{4}{9}$
2	6	8	10	$\frac{6+4\times8+10}{6} = 8$	$\frac{2}{3}$	$\frac{4}{9}$
3	1	5	15	$\frac{1+4\times5+15}{6} = 6$	$\frac{7}{3}$	$\frac{49}{9}$
4	2	5	9	$\frac{2+4\times5+9}{6} = 5.166$	$\frac{7}{3}$	$\frac{49}{9}$
5	6	8	10	$\frac{6+4\times8+10}{6} = 8$	$\frac{2}{3}$	$\frac{4}{9}$
6	6	7	9	$\frac{6+4\times7+9}{6} = 7.166$	$\frac{1}{2}$	$\frac{1}{4}$

18. $1 \to 2 \to 3 \to 5 \to 8$, project duration = 33 days.

19.

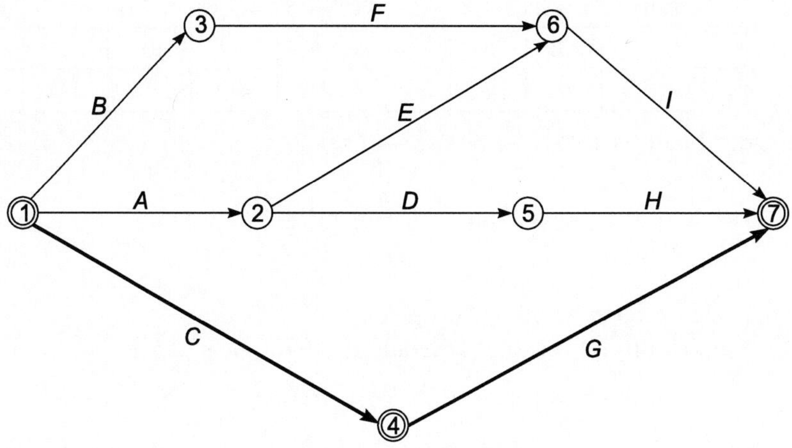

The critical path is $C \to G$ i.e., $1 \to 4 \to 7$.

Task	t_e	$V_{ij} = \sigma_t^2$	Earliest time		Latest time		Total float
			Start	Finish	Start	Finish	
A	7.8	0.69	0	7.8	9.0	16.8	9.0
B	20.0	0.44	0	20.0	9.0	29.8	9.8
C	33.0	5.34	0	33.0	0	33.0	0
D	18.0	0.44	7.8	25.8	16.8	34.8	9.0
E	20.0	2.78	0.8	27.8	1.8	38.8	11.0
F	9.0	1.00	20.0	29.0	29.8	38.8	9.8
G	9.8	0.69	33.0	42.8	33.0	42.8	0
H	8.0	0.11	25.8	33.8	34.8	42.8	9.0
I	4.0	0.11	29.0	33.8	38.8	42.8	9.8

Variance of time schedule is 6.03

(i) The probability of completing the project in 41.5 weeks is 0.30.
(ii) The duration of the project having 95% chance of completion is 46.85 weeks.

20.

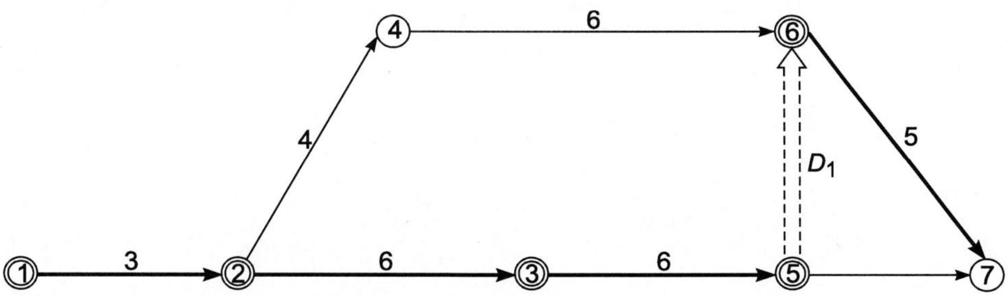

(i)

Activity	1 – 2	2 – 3	2 – 4	3 – 5	4 – 6	5 – 6	5 – 7	6 – 7
t_e	3	6	4	6	6	0	4	5
$V_{ij} = \sigma_t^2$	0	1	4/9	4/9	4/9	1	1/9	1

The critical path is $1 \to 2 \to 3 \to 5 \to 6 \to 7$.

The variance of the project length $\sigma_t^2 = 0 + 1 + \frac{4}{9} + 0 + 1$

$$= 2.444.$$

The expected project length = 20 weeks.

(ii) The probability of meeting schedule is 0.9726 or 97.26%.

21. (i)

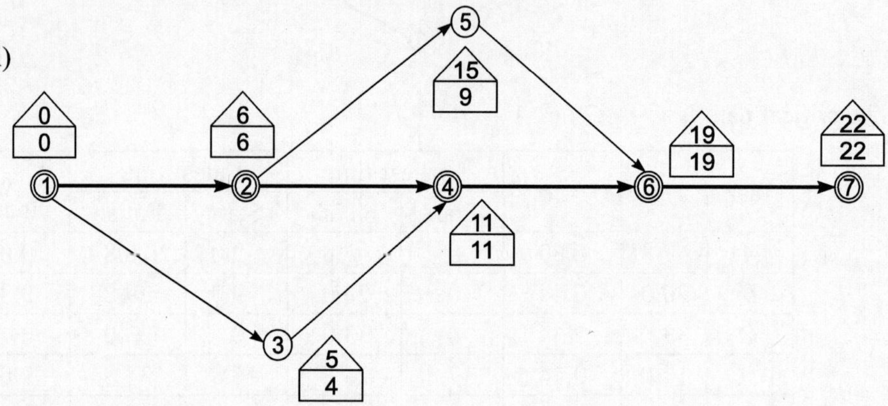

(ii) The critical path is $1 \to 2 \to 4 \to 6 \to 7$.

(iii) Minimum total time is 13 days and corresponding cost is 1145.

22. The optimum project duration is 12 days and project cost is 17800.

12 Chapter

Dynamic Programming

12.1 INTRODUCTION

Dynamic programming (DP) determines the optimum solution to n variable problem by decomposing it into n stages. Each stage is a single variable sub-problem. The computational advantage is that we will be optimizing a single variable problem instead of a n variable problem. The optimum solution at one stage will be utilized at the succeeding stage. Continuing the process n times we get the optimum solution of the given problem. The main contribution of DP is the principle of optimality a framework for decomposing the problem into n stages.

12.2 MULTI-DECISION PROCESS

In 1950, Richard Bellman developed a technique known as "Dynamic Programming" based on Bellman's principle for solving multistage decision problems. A multi-decision process, we divide the problem into different sub-problems. It is easier to solve these sub-problems being smaller in size and the optimum decision taken for finding the optimal solution of the next sub-problem. Proceeding this way we find the optimal solution of the problem as a whole. The various iterations of decision-making process are shown below:

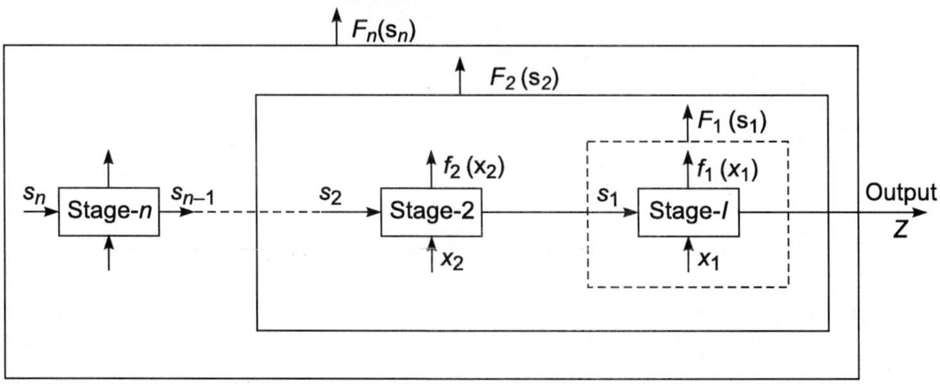

The procedure discussed above is called backward recursion, because the stage transformation function is of the type $s_i = T_i(s_{i+1}, x_{i+1})$. This is convenient when s_n is specified. When s_1 is specified, then it would be convenient to reverse the direction. This is called forward recursion.

12.3 APPLICATIONS OF DYNAMIC PROGRAMMING

12.3.1 Shortest Route Problem

Let $f_i(x_i)$ be the shortest distance to node i at the stage i. Define $d(x_{i-1}, x_i)$ = distance from node x_{i-1} to node x_i. Compute f_i from f_{i-1}. Using the following recursive equation

$$f_i(x_i) = \min [d(x_{i-1},x_i) + f_1(x_{i-1})] \quad i = 1, 2, 3, \ldots, n \ (n = \text{number of stages})$$

all feasible roots (x_{i-1}, x_i)

This is forward dynamic programming. The backward recursive equation is

$$f_i(x_i) = \min [d(x_1, x_{i+1}) + f_{i+1}(x_{i+1})] \quad i = 1, 2, 3, \ldots, n \ (n = \text{number of stages}).$$

all feasible roots (x_i, x_{i+1})

12.3.2 Travelling Salesman Problem

Example 1: A travelling salesman has to travel from city 1 to city 7. The distances between various cities are given in the following diagram:

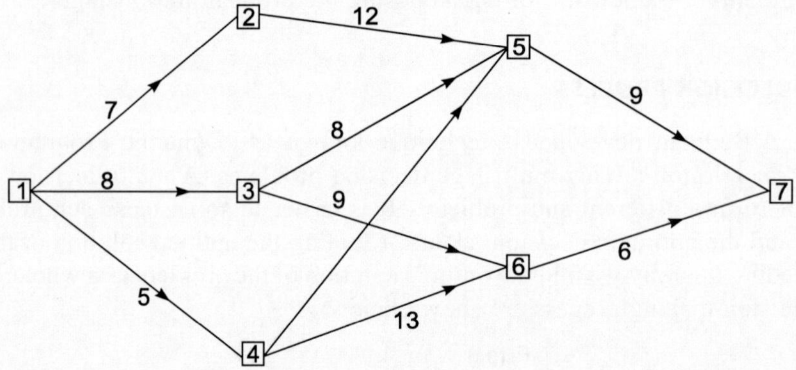

Using dynamic programming, find the minimum distance travelled by the salesman. Also find the optimal path.

Solution: We can solve this problem by exhaustively enumerating all the routes between nodes 1 to 7 (there are five such routes). However, in a large network, exhaustive enumeration is not efficient computationally. To solve this problem by dynamic programming, first we decompose it into stages.

Then we carry out the computation for each stage separately. The optimum result at each stage is used in finding the optimum at the succeeding stage. Here we are solving this problem by backward recursive equation which is given below:

$$f_i(x_i) = \min\, [d(x_i, x_{i+1}) + f_{i+1}\,(x_{i+1})] \quad i = 1, 2, 3, \ldots, n \; (n = \text{number of stages}).$$

all feasible roots (x_i, x_{i+1})

We have $f_4(x_4) = 0$, for $x_4 = 7$. The associated order of computation is $f_3 \rightarrow f_2 \rightarrow f_1$.

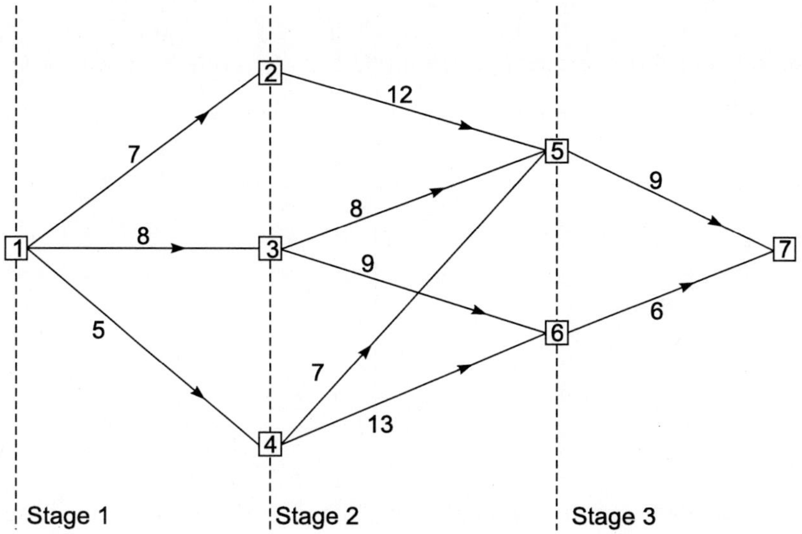

Stage 3: Node 7 ($x_4 = 7$) is connected to node 5 and 6 ($x_3 = 5$ or 6) with exactly one route each. Results can be summarized as

	$d(x_3, x_4)$	*Optimum Solution*	
x_3	$x_4 = 7$	$f_3(x_3)$	x_4^*
5	9	9	7
6	6	6	7

Stage 2: Route (2, 6) is not a feasible alternative as it does not exist. The optimum solution at stage 2 reads as follows:

	$d(x_2, x_2) + f_3(x_3)$		*Optimum Solution*	
x_2	$x_3 = 5$	$x_3 = 6$	$f_2(x_2)$	x_3^*
2	12 + 9 = 21	–	21	5
3	8 + 9 = 17	9 + 6 = 15	15	6
4	7 + 9 = 16	13 + 6 = 19	16	5

If you are in cities 2 or 4 the shortest route passes through city 5 and if you are in city 3, the shortest route passes through city 6.

Stage 1: From node 1, we have three alternative routes (1, 2), (1, 3) and (1, 4). Using $f_2(x_2)$ from stage 2, we can compute the following table:

	$d(x_1, x_2) + f_2(x_2)$			Optimum Solution	
x_1	$x_2 = 2$	$x_2 = 3$	$x_2 = 4$	$f_2(x_2)$	x_2^*
1	7 + 21 = 28	8 + 15 = 23	5 + 16 = 21	21	4

The optimum distance is 21 and optimum route is $1 \rightarrow 4 \rightarrow 5 \rightarrow 7$.

12.3.3 Solving a Linear Programming Problem Using Dynamic Programming

Let us consider a linear programming problem with two decision variables and m constraints as:

$$\text{Max } Z = c_1x_1 + c_2x_2$$

$$\text{s.t.} \quad \begin{array}{ccccc} \boxed{\begin{array}{c} a_{11}\ X_1 \\ a_{21}\ X_1 \\ \cdots \\ \cdots \\ \cdots \\ a_{m1}\ X_1 \end{array}} & \begin{array}{c} + \\ + \\ \cdots \\ \cdots \\ \cdots \\ + \end{array} & \boxed{\begin{array}{c} a_{12}\ X_2 \\ a_{22}\ X_2 \\ \cdots \\ \cdots \\ \cdots \\ a_{m2}\ X_2 \end{array}} & \begin{array}{c} \leq \\ \leq \\ \\ \\ \\ \leq \end{array} & \boxed{\begin{array}{c} b_1 \\ b_2 \\ \cdots \\ \cdots \\ \cdots \\ b_n \end{array}} \\ \text{Stage 1} & & \text{Stage 2} & & \text{Resources 3} \end{array}$$

Now we develop an algorithm for solving the problem by forward dynamic programming as:

Stage 1:

$$F_1(x_1) = \underset{\left\{0 \leq x_1 \leq \left(\frac{b_1 - a_{12}x_2}{a_{11}}, \frac{b_2 - a_{22}x_2}{a_{21}}, \ldots, \frac{b_m - a_{m2}x_2}{a_{m1}}\right)\right\}}{\text{Max } C_1 X_1}$$

$$0 \leq x_1 = \min = \underset{\left(\frac{b_1 \quad a_{12}x_2}{a_{11}}, \frac{b_2 - a_{22}x_2}{a_{21}}, \ldots, \frac{b_m - a_{m2}x_2}{a_{m1}}\right)}{\text{Max } C_1 X_1}$$

$$\text{Here } x_1^* = \text{Min}\left(\frac{b_1 - a_{12}x_2}{a_{11}}, \frac{b_2 - a_{22}x_2}{a_{21}}, \ldots, \frac{b_m - a_{m2}x_2}{a_{m1}}\right) \quad \ldots(1)$$

$$F_1(x_2) = \text{Max } c_1\left(\frac{b_1 - a_{12}x_2}{a_{11}}, \frac{b_2 - a_{22}x_2}{a_{21}}, \ldots, \frac{b_m - a_{m2}x_2}{a_{m1}}\right)$$

Stage 2: We have

$$F_2(x_2) = \text{Max}\left[c_1 \min\left(\frac{b_1 - a_{12}x_2}{a_{11}}, \frac{b_2 - a_{22}x_2}{a_{21}}, \ldots, \frac{b_m - a_{m2}x_2}{a_{m1}}\right) + c_2x_2\right] \quad \ldots(2)$$

$$\text{s.t.} \quad x_2 \leq \min\left\{\frac{b_1}{a_{12}}, \frac{b_2}{a_{22}}, \ldots, \frac{b_m}{a_{m2}}\right\} \quad \ldots(3)$$

Explain (3) gives upper bound on x_2. Now to determine min $\left(\frac{b_1 - a_{12}x_2}{a_{11}}, \frac{b_2 - a_{22}x_2}{a_{21}}, \ldots, \frac{b_m - a_{m2}x_2}{a_{m1}}\right)$. We proceed as follows:

(i) First take $\frac{b_1 - a_{12}x_2}{a_{11}}$ to be less than or equal to remaining $\frac{b_2 - a_{22}x_2}{a_{21}}, \frac{b_{22} - a_{32}x_2}{a_{31}}, \ldots,$

$\dfrac{b_{22} - a_{32}x_2}{a_{31}}$ one by one and find a condition on x_2. Thus, getting $(m - 1)$ conditions on x_2 and then take intersection of all these to get the maximum x_2 provided $a_{ij} \geq 0$ for i and j. Then we find x_1 from (1) and Max Z from (2). This will give one set of solution.

(ii) Now we take $\dfrac{b_2 - a_{22}x_2}{a_{21}}$ to be less than or equal to (i.e., minimum of remaining) $\dfrac{b_1 - a_{12}x_2}{a_{11}}$, $\dfrac{b_3 - a_{32}x_2}{a_{31}}, \ldots, \dfrac{b_m - a_{m2}x_2}{a_{m1}}$ and proceeding as in (i), we will get another set of solution.

Repeating this process until we get at most m such solutions. Among these solutions consider that one which gives the optimum solution of the problem.

Note: if $n \geq 3$ then the problem becomes complicated.

Example 2: Solve the following problem using dynamic programming:

$$\text{Max } Z = 10x_1 + 8x_2$$

s.t.

$$2x_1 + x_2 \leq 25$$
$$3x_1 + 2x_2 \leq 45$$
$$x_2 \leq 10$$

and

$$x_1, x_2 \geq 0.$$

Solution: The given problem can be written as:

$$\text{Max } Z = 10x_1 + 8x_2$$

s.t.

$$2x_1 + x_2 \leq 25$$
$$3x_1 + 2x_2 \leq 45$$
$$0.x_1 + x_2 \leq 10$$

and

$$x_1, x_2 \geq 0.$$

As there are two variables in the problem, we shall have two stages to solve the problem using dynamic programming. Let us solve it by using backward dynamic programming.

Stage 2:

$$F_2(x_2) = \underset{\left\{0 \leq x_2 \leq \left(25 - 2x_1, \frac{45 - 3x_1}{2}, 10\right)\right\}}{\text{Max } 8\,x_2}$$

$$0 \leq x_2 = \min \underset{\left(25 - 2x_1, \frac{45 - 3x_1}{2}, 10\right)}{\text{Max } 8\,x_2}$$

Here

$$x_2^* = \text{Min}\left(25 - 2x_1, \frac{45 - 3x_1}{2}, 10\right)$$

$$F_2(x_1) = \text{Max } 8\left(25 - 2x_1, \frac{45 - 3x_1}{2}, 10\right)$$

Stage 1: We have

$$F_1(x_1) = \text{Max}\left[10x_1 + 8\left\{\text{Min}\left(25-2x_1, \frac{45-3x_1}{2}, 10\right)\right\}\right] \quad \ldots(2)$$

s.t. $$x_1 \leq \text{Min}\left\{\frac{25}{2}, \frac{45}{3}\right\} = \frac{25}{2} \quad \ldots(3)$$

Now to calculate min $\left(25-2x_1, \frac{45-3x_1}{2}, 10\right)$. We proceed as follows:

Case I: Assuming 10 to be minimum, we have

$$10 \leq 25 - 2x_1 \Rightarrow x_1 \leq \frac{15}{2}$$

$$\text{Also } 10 \leq \frac{45-3x_1}{2} \Rightarrow x_1 \leq \frac{25}{3}, x_1 \geq 0.$$

Thus, we have $0 \leq x_1 \leq \frac{15}{2}$. Using equation (2), we get

$$\text{Max } Z = \text{Max } [10x_1 + 8 \times 10], \left(0 \leq x_1 \leq \frac{15}{2}\right)$$

$$= \text{Max}\left[10 \times \frac{15}{2} + 80\right]$$

$$= 155 \text{ when } x_1 = \frac{15}{2} \text{ and } x_2^* = 10. \quad \ldots(4)$$

Case II: Let min $\left(25-2x_1, \frac{45-3_{x1}}{2}, 10\right) = 25 - 2x_1$. Then we have

$$\text{Also} \quad 25 - 2x_1 \leq \frac{45-3_{x1}}{2} \Rightarrow x_1 \geq 5$$

$$25 - 2x_1 \leq 10 \Rightarrow x_1 \geq \frac{15}{2}, x_1 \geq 0.$$

Thus, we have $x_1 \geq \frac{15}{2}$. From equation (3), we have $x_1 \leq \frac{25}{2}$

Therefore, we have $\frac{15}{2} \leq x_1 \leq \frac{25}{2}$. Using equation (2), we get

$$\text{Max } Z = \text{Max}[10x_1 + 8(25 - 2x_1)], \left(\frac{15}{2} \leq x_1 \leq \frac{25}{2}\right)$$

$$= \text{Max}[200 - 6x_1]$$

$$= 155 \text{ when } x_1 = \frac{15}{2} \text{ and } x_2^* = 10. \quad \ldots(5)$$

Case III: Let min $\left(25-2x_1, \frac{45-3x_1}{2}, 10\right) = \frac{45-3x_1}{2}$. Then we have

$$\frac{45-3x_1}{2} \leq 25 - 2x_1 \Rightarrow x_1 \leq 5 \text{ (not applicable)}$$

Also $$\frac{45-3x_1}{2} \le 10 \Rightarrow x_1 \ge \frac{25}{3},\ x_1 \ge 0.$$

Thus, we have $x_1 \ge \frac{25}{3}$. From equation (3), we have $x_1 \le \frac{25}{2}$

Therefore, we have $\frac{25}{3} \le x_1 \le \frac{25}{2}$. Using equation (2), we get

$$\text{Max } Z = \text{Max}\left[10\,x_1 + 8\left(\frac{45-3x_1}{2}\right)\right],\ \left(\frac{25}{3} \le x_1 \le \frac{25}{2}\right)$$

$$= \text{Max}\,[10x_1 + 180 - 12x_1]$$
$$= \text{Max}\,[180 - 2x_1]$$

$$= \frac{490}{3} \text{ when } x_1 = \frac{25}{3} \text{ and } x_2^* = 10. \qquad \text{...(6)}$$

From (4), (5) and (6), the optimal solution is

$$\text{Max } Z = \text{Max}\left[155, 155, \frac{490}{3}\right]$$

$$= \frac{490}{3} \text{ which is for } x_1 = \frac{25}{3} \text{ and } x_2 = 10.$$

Example 3: Solve the following problem using dynamic programming:

$$\text{Max } Z = 500x_1 + 800x_2$$

s.t.
$$5x_1 + 6x_2 \le 60$$
$$x_1 + 2x_2 \le 16$$
$$x_1 \le 8$$
$$x_2 \le 6$$

and
$$x_1, x_2 \ge 0.$$

Solution: The given problem can be written as:

$$\text{Max } Z = 500x_1 + 800x_2$$

s.t.
$$5x_1 + 6x_2 \le 60$$
$$x_1 + 2x_2 \le 16$$
$$x_1 + 0.x_2 \le 8$$
$$0.x_1 + x_2 \le 6$$

and
$$x_1, x_2 \ge 0. \qquad \text{...(1)}$$

As there are two variables in the problem, we shall have two stages to solve the problem using dynamic programming. Let us solve it by using backward dynamic programming.

Stage 1: $$F_1(x_1) \quad \underset{\left\{0 \le x_1 \le \left(\frac{60-6x_2}{5},\, 16-2x_2,\, 8\right)\right\}}{\text{Max } 500\,x_1}$$

$$0 \le x_1 = \text{Min} \underset{\left(\frac{60-6x_2}{5},\, 16-2x_2,\, 8\right)}{\text{Max } 500\,x_1}$$

Here $$x_1^* = \text{Min}\left(\frac{60-6x_2}{5}, 16-2x_2, 8\right)$$

$$F_1(x_2) = \text{Max } 500\left(\frac{60-6x_2}{5}, 16-2x_2, 8\right)$$

Stage 2: We have

$$F_2(x_2) = \text{Max}\left\{500\ \text{Min}\left(\frac{60-6x_2}{5}, 16-2x_2, 8\right) + 800x_2\right\} \quad \ldots(2)$$

s.t. $$x_2 \le \text{Min}\left\{\frac{60}{6}, \frac{16}{2}, 6\right\} = 6 \quad \ldots(3)$$

Now to calculate min $\left(\frac{60-6x_2}{5}, 16-2x_2, 8\right)$. We proceed as follows:

Case I: Assuming 8 to be minimum, we have

$$8 \le \frac{60-6x_2}{5} \Rightarrow x_2 \le \frac{10}{3}$$

Also $$8 \le 16 - 2x_2 \Rightarrow x_2 \le 4,\ x_2 \ge 0.$$

Thus, we have $0 \le x_2 \le \frac{10}{3}$. Using equation (2), we get

$$\text{Max } Z = \text{Max}\,[500 \times 8 + 800\,x_2], \quad \left(0 \le x_2 \le \frac{10}{3}\right)$$

$$= \text{Max}\left[4000 + 800 \times \frac{10}{3}\right]$$

$$= \frac{20000}{3} \text{ when } x_2 = \frac{10}{3} \text{ and } x_1^* = 8. \quad \ldots(4)$$

Case II: Let min $\left(\frac{60-6x_2}{5}, 16-2x_2, 8\right) = \frac{60-6x_2}{5}$. Then we have

$$\frac{60-6x_2}{5} \le 8 \Rightarrow x_2 \ge \frac{10}{3}$$

Also $$\frac{60-6x_2}{5} \le 16 - 2x_2 \Rightarrow x_2 \le 5,\ x_2 \ge 0.$$

Thus, we have $\frac{10}{3} \le x_2 \le 5$. Using equation (2), we get

$$\text{Max } Z = \text{Max}\left[500\left(\frac{60-6x_2}{6}\right) + 800x_2\right], \quad \left(\frac{10}{3} \le x_2 \le 5\right)$$

$$= \text{Max}\,[6000 + 200x_2]$$

$$= 7000 \text{ when } x_2 = 5 \text{ and } x_1^* = 6. \quad \ldots(5)$$

Case III: Let min $\left(\frac{60-6x_2}{5}, 16-2x_2, 8\right) = 16 - 2x_2$. Then we have

$$16 - 2x_2 \leq 8 \Rightarrow x_2 \geq 4.$$

Also $$16 - 2x_2 \leq \frac{60 - 6x_2}{5} \Rightarrow x_2 \geq 5,\ x_2 \geq 0.$$

Thus, we have $x_2 \geq 5$.

From equation (3), we have $x_2 \leq 6$.

Therefore, $5 \leq x_2 \leq 6$. Now using equation (2), we have

Max
$$\begin{aligned} Z &= \text{Max}\,[500(16 - 2x_2) + 8x_2],\ (5 \leq x_2 \leq 6) \\ &= \text{Max}\,[8000 - 200x_2] \\ &= 7000 \text{ when } x_2 = 5 \text{ and } x_1^* = 6 \end{aligned} \quad \ldots(6)$$

From (4), (5) and (6), the optimal solution is

$$\begin{aligned} \text{Max } Z &= \text{Max}\left[\frac{20000}{3}, 7000, 7000\right] \\ &= 7000 \text{ which is for } x_1^* = 5 \text{ and } x_2 = 6. \end{aligned}$$

12.3.4 Calculus Method in Solving Dynamic Programming Problem

In this model, we shall discuss the problems which can be solved using dynamic programming techniques through calculus method. We illustrate this method by taking four cases:

Case I: When both the objective function and the constraint are in additive form.

Consider the problem

$$\text{Max (or Min) } Z = f_1(x_1) + f_2(x_2) + \ldots + f_n(x_n) = \sum_{i=1}^{n} f_i x_i$$

s.t. $$a_1x_1 + a_2x_2 + a_3x_3 + \ldots + a_nx_n \geq \text{ or } = \text{ or } \leq b$$

or $$\sum_{i=1}^{n} a_i\, x_i \geq \text{ or } = \text{ or } \leq b$$

and $$(a_1, a_2, a_3, \ldots, a_n) \geq 0,\ (x_1, x_2, x_3, \ldots, x_n) \geq 0 \text{ and } b > 0$$

or $$a_i \geq 0,\ x_i \geq 0 \text{ and } b > 0.$$

To solving the problem, first we define the state variables s_i's as

$$\begin{aligned} s_n &= a_1x_1 + a_2x_2 + a_3x_3 + \ldots + a_nx_n = \geq \text{ or } = \text{ or } \leq \text{b} \\ s_{n-1} &= a_1x_1 + a_2x_2 + a_3x_3 + \ldots + a_{n-1}\,x_{n-1} = s_n - a_nx_n = T_{n-1}(s_n, x_n) \\ s_{n-2} &= a_1x_1 + a_2x_2 + a_3x_3 + \ldots + a_{n-2}\,x_{n-2} = s_{n-1} - a_{n-1}\,x_{n-1} = T_{n-2}(s_{n-1}, x_{n-1}) \\ &\ldots\ \ldots\ \ldots\ \ldots\ \ldots\ \ldots\ \ldots \\ &\ldots\ \ldots\ \ldots\ \ldots\ \ldots\ \ldots\ \ldots \\ s_i &= a_1x_1 + a_2x_2 + a_3x_3 + \ldots + a_ix_i = s_{i+1} - a_{i+1}\,x_{i+1} = T_i(s_{i+1}, x_{i+1}) \\ &\ldots\ \ldots\ \ldots\ \ldots\ \ldots\ \ldots\ \ldots \\ &\ldots\ \ldots\ \ldots\ \ldots\ \ldots\ \ldots\ \ldots \\ s_2 &= a_1x_1 + a_2x_2 = s_3 - a_3x_3 = T_2(s_3, x_3) \\ s_1 &= a_1x_1 = s_2 - a_2x_2 = T_1(s_2, x_2) \end{aligned}$$

where $s_i = T_i(s_{i+1}, x_{i+1})$, $i = 1, 2, 3,\ldots, n-1$ are called state transformation functions connecting different states. The recursive relations are

$$F_i(s_i) = \text{Max (or Min) } [f_i(x_i) + F_{i-1}(s_{i-1})];\ i = 1, 2, 3,\ldots, n.$$

and $F_1(s_1) = f_1(x_1)$.

Now we can start with $F_1(s_1)$ and recursively optimize to get $F_2(s_2)$, $F_3(s_3)$, …, $F_n(s_n)$. This is forward dynamic programming. Also we can solve this type of problems by using backward dynamic programming technique.

Example 4: Solve the following problem using dynamic programming:

$$\text{Min } Z = x_1^2 + x_2^2 + x_3^2$$

s.t. $\quad x_1 + x_2 + x_3 \geq 15$

and $\quad x_1, x_2, x_3 \geq 0.$

Solution: We can solve it either using forward dynamic programming or backward dynamic programming. Let us first solve it by using forward dynamic programming.

Forward Dynamic Programming

As there are three variables in the problem, we shall have three stages to solve the problem using dynamic programming.

Now we define the state variables as:

$$s_1 = x_1 \quad \ldots(1)$$
$$s_2 = x_1 + x_2 \quad \ldots(2)$$
$$s_3 = x_1 + x_2 + x_3 \geq 15 \quad \ldots(3)$$

This gives

$$s_1 = x_1 = s_2 - x_2$$
$$s_2 = s_3 - x_3$$
$$s_3 = x_1 + x_2 + x_3 \geq 15$$

Recursive relations are

$$\begin{aligned} F_1(s_1) &= \min x_1^2 = \min (s_2 - x_2)^2 \\ F_2(s_2) &= \min [x_2^2 + F_1(s_1)] \\ &= \min (x_2^2 + x_1^2) \\ &= \min [x_2^2 + (s_2 - x_2)^2] \\ F_3(s_3) &= \min [x_3^2 + F_2(s_2)] \end{aligned}$$

For minimum of F_2, $\dfrac{dF_2}{dx_2} = 0$ gives $2x_2 - 2(s_2 - x_2) = 0 \Rightarrow x_2 = \dfrac{s_2}{2}$

Also $\dfrac{d^2F_2}{dx_2^2} = 2 + 4 = 4$ which is positive. Therefore, F_2 in minimum at $x_2 = \dfrac{s_2}{2}$

$$F_2(s_2) = \left[\left(\frac{s_2^2}{2}\right)^2 + \left(s_2 - \frac{s_2}{2}\right)^2\right] = \frac{s_2^2}{2}$$

Now $$F_3(s_3) = \min\left[x_3^2 + \frac{s_2^2}{2}\right] = \min\left[x_3^2 + \frac{(x_3 - x_3)}{2}\right]$$

For minimum of F_3, $\frac{dF_3}{dx_3} = 0$ gives $2x_3 - (s_3 - x_3) = 0 \Rightarrow x_3 = \frac{s_3}{3}$

Also $\frac{d^2F_3}{dx_3^2}$ 2 + 1 = 3 which is positive. Therefore, F_3 in minimum at $x_3 = \frac{s_3}{3}$

Thus, $$F_3(s_3) = \frac{s_3^2}{3}$$

Now $$\text{Min } Z = \min[x_1^2 + x_2^2 + x_3^2]$$
$$= \min F_3(s_3)$$
$$\frac{(15)^2}{3} = 75 \text{ as } s_3 = 15.$$

Min $Z = 75$, for $x_3 = 5$, $s_3 = 15$, $s_2 = s_3 - x_3 = 10$, $x_2 = \frac{s_2}{2} = 5$, $s_1 = s_2 - x_2 = 5$, $x_1 = s_1 = 5$.

Hence, $$\text{Min} Z = 75, \text{ for } x_1 = 5, x_2 = 5, x_3 = 5.$$

Backward Dynamic Programming

As there are three variables in the problem, we will have three stages to solve the problem using dynamic programming.

Now we define the state variables as:

$$s_1 = x_1 + x_2 + x_3 \geq 15 \quad \ldots(1)$$
$$s_2 = x_2 + x_3 = s_1 - x_1 \quad \ldots(2)$$
$$s_3 = x_3 = s_2 - x_2 \quad \ldots(3)$$

Recursive relations are

$$F_3(s_3) = \min x_3^2 = \min(s_2 - x_2)^2$$
$$F_2(s_2) = \min[x_2^2 + F_3(s_3)]$$
$$= \min(x_2^2 + x_3^2)$$
$$= \min[x_2^2 + (s_2 - x_2)^2]$$
$$F_1(s_1) = \min[x_1^2 + F_2(s_2)]$$

For minimum of F_2, $\frac{dF_2}{dx_2} = 0$ gives $2x_2 - 2(s_2 - x_2) = 0 \Rightarrow x_2 = \frac{s_2}{2}$

Also $\frac{d^2F_2}{dx_2^2} = 2 + 4 = 4$ which is positive. Therefore, F_2 in minimum at $x_2 = \frac{s_2}{2}$.

$$F_2(s_2) \left[\left(\frac{s_2^2}{2}\right)^2 + \left(s_2 - \frac{s_2}{2}\right)^2\right] = \frac{s_2^2}{2}$$

Now $$F_1(s_1) = \min\left[x_1^2 + \frac{s_2^2}{2}\right] = \min\left[x_1^2 + \frac{(s_1 - x_1)^2}{2}\right]$$

For minimum of F_1, $\frac{dF_1}{dx_1} = 0$ gives $2x_1 - (s_1 - x_1) = 0 \Rightarrow x_1 = \frac{s_1}{3}$

Also $\frac{d^2F_1}{dx_1^2}$ 2 + 1 = 3 which is positive. Therefore, F_1 is minimum at $x_1 = \frac{s_1}{3}$.

Thus, $$F_1(s_1) = \frac{s_1^2}{3}$$

Now
$$\begin{aligned} \min Z &= \min [x_1{}^2 + x_2{}^2 + x_3{}^2] \\ &= \min F_1(s_1) \\ &= \frac{(15)^2}{3} = 75 \text{ as } s_1 = 15. \end{aligned}$$

Min $Z = 75$, for $x_1 = 5$, $s_2 = s_1 - x_1 = 10$, $x_2 = \dfrac{s_2}{2} = 5$, $s_3 = s_2 - x_2 = 5$, $x_3 = s_3 = 5$.

Hence, $$\min Z = 75, \text{ for } x_1 = 5, x_2 = 5, x_3 = 5.$$

Case II: When the objective function is in multiplicative form and the constraint is in additive form. The case can be illustrated with the help of the following example:

Example 5: Solve the following problem using dynamic programming:

$$\text{Max } Z = x_1 x_2 x_3$$

s.t.
$$x_1 + x_2 + x_3 = 5$$
$$x_1, x_2, x_3 \geq 0.$$

Solution: We first define the state variables as:

$$s_1 = x_1 \quad \ldots(1)$$
$$s_2 = x_1 + x_2 \quad \ldots(2)$$
$$s_3 = x_1 + x_2 + x_3 = 5 \quad \ldots(3)$$

This gives

$$\begin{aligned} s_1 &= x_1 = s_2 - x_2 \\ s_2 &= x_1 + x_2 = s_3 - x_3 \\ s_3 &= x_1 + x_2 + x_3 = 5 \end{aligned}$$

Recursive relations are

$$\begin{aligned} F_1(s_1) &= f_1(s_1) = \max x_1 = s_1 \\ F_2(s_2) &= \max [x_2\, F_1(s_1)] = \max (x_2 .\, s_1) = \max [x_2 .\, (s_2 - x_2)] \\ F_3(s_3) &= \max [x_3\, F_2(s_2)] \end{aligned}$$

For maximum of F_2, $\dfrac{dF_2}{dx_2} = 0$ gives $s_2 - 2x_2 = 0 \Rightarrow x_2 = \dfrac{s_2}{2}$

Also $\dfrac{d^2F_2}{dx_2^2} = -2x_2$ which is negative. Therefore, F_2 is maximum at $x_2 = \dfrac{s_2}{2}$.

Thus, $$F_2(s_2) = \frac{s_2^2}{4}$$

Now $$F_3(s_3) = \max \left[x_3 \cdot \frac{s_2^2}{4} \right] = \max \left[x_3 \cdot \frac{(s_3 - x_3)^2}{4} \right]$$

For minimum of F_3, $\dfrac{dF_3}{dx_3}$ gives $(s_3 - x_3)^2 - 2x_3(s_3 - x_3) = 0$

$$\Rightarrow (s_3 - x_3)\,(s_3 - 3x_3) = 0$$

or $(s_3 - 3x_3) = 0$ [$s_3 - x_3 \neq 0$ because $x_2 = s_3 - x_3 > 0$, $x_2 \neq 0$ because if $x_2 = 0$, then max $Z = 0$.]

i.e., $$x_3 = \frac{s_3}{3}.$$

Also $\frac{d^2F_3}{dx_3^2} = 6x_3 - 4s_3 = -2s_3 < 0$ which is negative. Therefore F_3 in maximum at $x_3 = \frac{s_3}{3}$.

Thus, $$F_3(s_3) = \frac{s_3^3}{27} = \frac{(5)^3}{27} = \frac{125}{27} \text{ as } s_3 = 5.$$

Now $x_3 = \frac{s_3}{3} = \frac{3}{5}$, $s_2 = s_3 - x_3 = 5 - \frac{3}{5} = \frac{10}{3}$, $x_2 = \frac{s_2}{2} = \frac{10}{3}\cdot\frac{1}{2} = \frac{5}{3}$, $s_1 = s_2 - x_2 = \frac{10}{3} - \frac{5}{3} = \frac{5}{3}$,

$$x_1 = s_1 = \frac{5}{3}.$$

Hence, $\max Z = \frac{125}{27}$, for $x_1 = x_2 = x_3 = \frac{5}{3}$.

Case III: When the objective function is in additive form and the constraint is in multiplicative form. This case can be illustrated with the help of the following example:

Example 6: Solve the following problem using dynamic programming:

$$\text{Min } Z = x_1^2 + x_2^2 + x_3^2 + x_4^2$$

s.t. $$x_1 x_2 x_3 x_4 = 16$$

and $$x_1, x_2, x_3, x_4 \geq 0.$$

Solution: We define the state variables as:

$$s_1 = x_1 \qquad \ldots(1)$$
$$s_2 = x_1 x_2 \qquad \ldots(2)$$
$$s_3 = x_1 x_2 x_3 \qquad \ldots(3)$$
$$s_4 = x_1 x_2 x_3 x_4 \qquad \ldots(4)$$

This gives $s_1 = x_1 = \frac{s_2}{x_2}$, $s_2 = \frac{s_2}{x_2}$, $s_3 = \frac{s_4}{x_4}$

As there are four decision variables in the problem, we shall have four stages in the form of the following four recursive relations:

$$F_1(s_1) = f_1(s_1) = \min x_1^2 = s_1^2$$
$$F_2(s_2) = \min [x_2^2 + F_1(s_1)] = \left[x_2^2 + \frac{s_2^2}{x_2^2}\right]$$
$$F_3(s_3) = \min [x_3^2 + F_2(s_2)]$$
$$F_4(s_4) = \min [x_4^2 + F_3(s_3)]$$

For minimum of F_2, $\frac{dF_2}{dx_2} = 0$ gives $2x_2 - 2\frac{s_2^2}{x_2^3} = 0 \Rightarrow x_2 = \sqrt{s_2}$.

Also $\frac{d^2F_2}{dx_2^2} = 2 + 6\frac{s_2^2}{x_2^4}$ which is positive. Therefore, F_2 in minimum at $x_2 = \sqrt{s_2}$.

$$F_2(s_2) = \left[\left(\sqrt{s_2}\right)^2 + \frac{s_2^2}{\left(\sqrt{s_2}\right)^2} \right] 2s_2.$$

Now
$$F_3(s_3) = \min [x_3{}^2 + F_2(s_2)]$$
$$F_3(s_3) = \min [x_3^2 + 2s_2]$$

For minimum of F_3, $\frac{dF_3}{dx_3} = 0$ gives $2x_3 - 2\,\frac{s_3}{x_3^2} = 0 \Rightarrow x_3 = s_3{}^{1/3} \left(s_2 = \frac{s_3}{x_3} \right)$.

Also $\frac{d^2F_3}{dx_3^2}$ $2 + 4\,\frac{s_3}{x_3^3}$ which is positive.

Therefore, F_3 in minimum at $x_3 = s_3{}^{1/3}$.

Thus,
$$F_3(s_3) = 3s_3^{2/3}.$$

Now
$$F_4(s_4) = \min [x_4{}^2 + F_3(s_3)]$$
$$= \min [x_4^2 + 3s_3\, s_3^{2/3}] = \min \left[x_4^2 + 3\frac{s_4^{2/3}}{x_4^{2/3}} \right]$$

For minimum of F_4, $\frac{dF_4}{dx_4} = 0$ gives $2x_4 - 2\,\frac{s_4^{2/3}}{x_4^{5/3}} = 0 \Rightarrow x_4 = s_4{}^{1/4}$.

Also $\frac{d^2F_4}{dx_4^2} = 2 + \frac{10}{3}\frac{s_4^{2/3}}{x_4^{8/3}}$ which is positive. Therefore, F_4 in minimum at $x_4 = s_4{}^{1/4}$.

Thus,
$$F_4\,(s_4) = 4\,s_4{}^{1/2}.$$
$$x_4 = s_4{}^{1/4} = (16)^{1/4} = 2.$$

Now $s_3 = \frac{s_4}{x_4} = \frac{16}{2} = 8$, $x_3 = s_3{}^{1/3} = (8)^{1/3} = 2$, $s_2 = \frac{s_3}{x_3} = \frac{8}{2} = 4$,

$$x_2 = \sqrt{s_2} = \sqrt{4} = 2,\ s_1 = \frac{s_2}{x_2} = \frac{4}{2} = 2 = x_1.$$

Hence, Min $Z = 16$, for $x_1 = x_2 = x_3 = x_4 = 2$.

12.3.5 Computational Method in Solving Dynamic Programming Problem

Example 7: Solve the following problem using dynamic programming:

$$\text{Max } Z = x_1{}^3 + x_2{}^3 + x_3{}^3$$

s.t.
$$x_1\, x_2\, x_3 \le 5$$
$$x_1,\ x_2,\ x_3,\ \ge 0 \text{ and integer.}$$

Solution: The decision variables are discrete here, so we cannot use calculus method and solve the problem in tabular form. We define the state variables as:

$$s_1 = x_1 \qquad \ldots(1)$$
$$s_2 = x_1\, x_2 \qquad \ldots(2)$$
$$s_3 = x_1\, x_2\, x_3 \qquad \ldots(3)$$

This gives $s_1 = x_1 = \frac{s_2}{x_2}$ $s_2 = \frac{s_3}{x_3}$

The state transformations are shown in Tables 1 and 2. The above relation implies that $x_i = 1$, 2, 3, 4, 5. In Table 1, entries are $s_2 = s_3/x_3$. Observe that only integer values of s_2 are feasible. Similarly, Table 2, gives $s_1 = s_2/x_2$.

Table 1

x_3 / y_3	1	2	3	4	5
1	1	–	–	–	–
2	2	1	–	–	–
3	3	–	1	–	–
4	4	2	–	1	–
5	5	–	–	–	1

Table 2

x_3 / y_3	1	2	3	4	5
1	1	–	–	–	–
2	2	1	–	–	–
3	3	–	1	–	–
4	4	2	–	1	–
5	5	–	–	–	1

The recursive calculations are shown in Tables 3, 4 and 5.

Table 3

$x_1 = s_1$	1	2	3	4	5
$F_1(s_1) = f_1(x_1) = x_1^2$	1	8	27	64	125

Table 4

	$f_2(x_2) = x_2^2$					$F_1(s_1)$					$F_2(s_2)$
x_2 / s_2	1	2	3	4	5	1	2	3	4	5	Max
1	1	–	–	–	–	1	–	–	–	–	2
2	1	8	–	–	–	8	1	–	–	–	9
3	1	–	27	–	–	27	–	1	–	–	28
4	1	8	–	64	–	64	8	–	1	–	65
5	1	–	–	–	125	125	–	–	–	1	126

Table 5

	$f_3(x_3) = x_3^2$					$F_2(s_2)$					$F_3(s_3)$
x_2 / s_2	1	2	3	4	5	1	2	3	4	5	Max
1	1	–	–	–	–	2	–	–	–	–	3
2	1	8	–	–	–	9	2	–	–	–	10
3	1	–	27	–	–	28	---	2	–	–	29
4	1	8	–	64	–	65	9	–	2	–	66
5	1	–	–	–	125	126	–	–	–	2	127

The one optimal solution is $x_1 = 1$, $x_2 = 1$, $x_3 = 5$; Max $Z = 127$.

12.3.6 Capital Budgeting Problem

Example 8: A manufacturing company has to improve working of three plants P, Q and R, with available plants with costs and corresponding return are as follows (in Lakhs of Rs.)

Plan No.	Plant P		Plant Q		Plant R	
	Cost	Return	Cost	Return	Cost	Return
1	0	0	0	0	0	0
2	1	5	2	8	1	3
3	2	6	3	9	–	–
4	–	–	4	12	–	–

Find the optimal policy for budgeting the available capital.

Solution: Suppose x_1, x_2 and x_3 are the amounts allocated to plants P, Q and R respectively. The given problem can be formulated as:

$$\text{Max } Z = f_1(x_1) + f_2(x_2) + f_1(x_3)$$

s.t.

$$x_1 + x_2 + x_3 \leq 5$$

$$x_1, x_2, x_3 \geq 0 \text{ and integers.}$$

It is given that

Table 1 (a)

x_1	$f_1(x_1)$
0	0
1	5
2	6

Table 1 (b)

x_2	$f_2(x_2)$
0	0
2	8
3	9
4	12

Table 1 (c)

x_3	$f_3(x_3)$
0	0
1	3

The calculations are shown in Tables 2 (a) and (b).

Table 2 (a)

s_3 \ x_3	0	1
0	0	–
1	1	0
2	2	1
3	3	2
4	4	3
5	5	4

Entries are s_2

Table 2 (b)

s_2 \ x_2	0	2	3	4
0	0	–	–	–
1	1	–	–	–
2	2	0	–	–
3	–	1	0	–
4	–	2	1	0
5	–	–	2	1

Entries are s_1

Table 3

$x_1 = s_1$	0	1	2
$F_1(s_1) = f_1(x_1)$	0	5^*	6

Table 4

	$f_2(x_2)$				$F_1(s_1)$				$F_2(s_2)$
x_2 / s_2	0	2	3	4	0	2	3	4	Max
0	0	–	–	–	0	–	–	–	0
1	0	–	–	–	5	–	–	–	5
2	0	8	–	–	6	0	–	–	8
3	–	8	9	–	–	5	0	–	12
4	–	8	9	12	–	6	5*	0	14*
5	–	–	9	12	–		6	5	17

Table 5

	$f_3(x_3)$		$F_2(s_2)$		$F_3(s_3)$
x_3 / s_3	0	1	0	1	Max
0	0	–	0	–	0
1	0	3	5	0	5
2	0	3	8	5	8
3	0	3	12	8	12
4	0	3	14	12	15
5	0	3	17	14*	17*

Hence, one optimum solution is $x_3 = 1$, $x_2 = 3$, $x_1 = 1$ and maximum is 17.

12.3.7 Cargo Loading Problem

Example 9: We have to load a vessel with three items. The maximum allowable weight is 7 units. The weight per unit of different items and their values are given below. How many units of each item be loaded in the vessel so as to maximize the value of the loaded items.

Item	Weight/Unit	Value/Unit
1	1	20
2	3	90
3	2	70

Solution: Suppose x_1, x_2 and x_3 be the number of units of items 1, 2 and 3 loaded respectively. The given problem can be formulated as:

$$\text{Max } Z = 20x_1 + 90x_2 + 70x_3$$

$$\text{s.t.} \quad x_1 + 3x_2 + 2x_3 \leq 7$$

$$x_1, x_2, x_3 \geq 0 \text{ and integers.}$$

The decision variables are discrete here, so we cannot use calculus method and solve the problem in tabular form. Now we define the state variables as:

$$s_1 = x_1 \quad \ldots(1)$$

$$s_2 = x_1 + 3x_2 \quad \ldots(2)$$

$$s_3 = x_1 + 3x_2 + 2x_3 \leq 7 \quad \ldots(3)$$

This gives

$$s_1 = x_1 = s_2 - 3x_2$$

$$s_2 = s_3 - 2x_3$$

$$s_3 = x_1 + x_2 + x_3 \leq 7$$

Recursive relations are

$$F_1(s_1) = f_1(x_1)$$

$$F_2(s_2) = \text{Max}\,[f_2(x_2) + F_1(s_1)]$$

$$F_3(s_3) = \text{Max}\,[f_3(x_3) + F_2(s_2)]$$

From equation (3), we have

$$x_1 = 0, 1, 2, 3, 4, 5, 6, 7;$$

$$x_2 = 0, 1, 2;$$

$$x_3 = 0, 1, 2, 3.$$

The state transformations are shown in following Tables 1 and 2:

Table 1

s_3 \ x_3	0	1	2	3
0	0	–	–	–
1	1	–	–	–
2	2	0	–	–
3	3	1	–	–
4	4	2	0	–
5	5	3	1	–
6	6	4	2	0
7	7	5	3	1

Table 2

s_2 \ x_2	0	1	2
0	0	–	–
1	1	–	–
2	2	–	–
3	3	0	–
4	4	1	–
5	5	2	–
6	6	3	0
7	7	4	1

In Table 1, the entries are $s_2 = s_3 - 2x_3$. (Table 1 gives possible values of s_2). Then we prepare Table 2, in which entries are $s_1 = s_2 - 3x_2$ values (that is s_1 can assume for various combinations of s_2 and x_2 values). Now we prepare Tables 3, 4 and 5 for using recursive relations:

Table 3

$x_1 = s_1$	0	1	2	3	4	5	6	7
$F_1(s_1) = f_1(x_1) = 20\ x_1$	0**	20*	40	60	80	100	120	140

Table 4

	$f_2(x_2) = 90x_2$			$F_1(s_1)$			$f_2(x_2) = 90x_2$			$F_2(s_2)$
x_2 \ s_2	0	1	2	0	1	2	0	1	2	Max of rows
0	0	–	–	0	–	–	0	–	–	0
1	0	–	–	20^*	–	–	20^*	–	–	20^*
2	0	–	–	40	–	–	40	–	–	40
3	0	90	–	60	0^{**}	–	60	90^*	–	90^{**}
4	0	90	–	80	20	–	80	110	–	110
5	0	90	–	100	40	–	100	130	–	130
6	0	90	180	120	0	0	120	150	180	180
7	0	90	180	140	80	20	140	170	200	200

In Table 4:

(a) We enter values in possible position. The possible positions are dictated by Table 1.
(b) To get the $F_1(s_1)$ matrix, we read x_1 from Table 2 and then read $F_1(s_1)$ from Table 3.
(c) The last column gives the maximum with respect to x_2 for a fixed s_2.

The above is the solution of second sub-problem. Now we proceed to calculate $F_3(s_3)$, which will give the optimal value for this problem. To get $F_3(s_3)$, we construct Table 5.

Table 5

	$f_3(x_3) = 70x_3$				$F_2(x_2)$				$f_3(x_3) + F_2(x_2)$				$F_3(x_3)$
x_3 \ s_3	0	1	2	3	0	1	2	3	0	1	2	3	Max of Rows
0	0	–	–	–	0	–	–	–	0	–	–	–	0
1	0	–	–	–	20	–	–	–	20	–	–	–	20
2	0	70	–	–	40	0	–	–	40	70	–	–	70
3	0	70	–	–	90	20	–	–	90	90	–	–	90
4	0	70	140	–	110	40	–	–	110	110	140	–	140
5	0	70	140	–	130	90	20	–	130	180	180	–	160
6	0	70	140	210	180	110	40	0	180	180	180	210	210
7	0	70	140	210	200	130	90**	20*	200	200	230**	230*	230***

In Table 5, to get matrix $F_2(s_2)$,we read s_2 from Table 1 and then read corresponding $F_2(s_2)$ from Table 4. From the last column of Table 5, we see that largest $F_3(s_3) = 230$ is for s_3 and x_3. Now we trace back the entries which gave this largest value. These entries give the optimal solution. This is shown in table by marking the optimal entries by (*). From Table 4 and Table 3, we have $s_2 = 1$, $x_2 = 0$ and $s_1 = x_1 = 1$ respectively.

The optimal values are $x_1 = 1$, $x_2 = 0$, $x_3 = 3$ and maximum value is 230.

In this problem, Table 5 shows that there is an alternative optimal solution. This solution, we have shown by marking entries with (**). The alternate optimal solution is

$$x_3 = 2,\ x_2 = 1,\ x_1 = 0 \text{ and maximum value is 230.}$$

12.3.8 Reliability Problem

Example 10: An electric equipment has three components 1, 2, and 3 which are connected in series (i.e., if one component fails then the equipment does not work). To improve the reliability of the equipment each component is to be supplemented by one or two or three parallel units (in case the component fails then the supplementary unit works and consequently the system remains working) the following data gives the cost of parallel units and reliability at each component. The total money available for improvement of device is Rs. 10,000.

No. of parallel units	*Components 1*		*Components 2*		*Components 3*	
	Cost (in thousands of Rs.)	*R*	*Cost (in thousands of Rs.)*	*R*	*Cost (in thousands of Rs.)*	*R*
1	1	.4	2	.5	3	.7
2	3	.6	4	.6	4	.8
3	4	.8	5	.7	5	.9

In the above, *R* denotes the reliability of the equipment. Let x_i denote the money spent on attaching supplementary units to components *i*. Find x_i, such that the reliability of the equipment is maximized.

Solution: Suppose $f_i(x_1)$, $i = 1, 2, 3$ be the reliability of the ith component when x_i amount is spent on parallel units attached to the component. Here the components 1, 2 and 3 denote the stages 1, 2 and 3 respectively. The function $f_i(x_1)$ is the return function of stage *i*.

Using the principle of reliability, the reliability of the equipment is the product of reliabilities of its components. Thus, the problem is:

$$\text{Max } Z = f_1(x_1)\, f_2(x_2)\, f_1(x_3)$$

s.t. $$x_1 + x_2 + x_3 \leq 10$$

where x_1, x_2 and x_3 are specified in the above given table.

Using the given data, we have

Table 1 (a)

x_1	$f_1(x_1)$
1	.4
3	.6
4	.8

Table 1 (b)

x_2	$f_2(x_2)$
2	.5
4	.6
5	.7

Table 1 (c)

x_3	$f_3(x_3)$
3	.7
4	.8
5	.9

Now we define the state variables as:

$$s_1 = x_1 \quad \ldots(1)$$
$$s_2 = x_1 + x_2 \quad \ldots(2)$$
$$s_3 = x_1 + x_2 + x_3 \le 10 \quad \ldots(3)$$

This gives

$$s_1 = x_1 = s_2 - x_2$$
$$s_2 = s_3 - x_3$$
$$s_3 = x_1 + x_2 + x_3 \le 10$$

From Table 1, we see that the least value of $x_1 + x_2 + x_3$ is 6; therefore from the first relation above, we have $6 \le s_3 \le 10$. We observe that the least value of $x_1 + x_2$ is 3. The largest value of $s_3 - x_3$ is $10 - 3 = 7$. Thus $3 \le s_2 \le 7$. Finally, we know that $s_1 = x_1$. In view of above, we make Tables 2 and 3.

Table 2

s_3 \ x_3	3	4	5
6	3	–	–
7	4	3	–
8	5	4	3
9	6	5	4
10	7	6	5

Entries are $x_2 = s_3 - x_3$

Table 3

s_2 \ x_2	0	2	3
3	1	–	–
4	–	–	–
5	3	1	–
6	4	–	1
7	–	3	–

Entries are $x_1 = s_2 - x_2$

Calculations of recursive relation are shown in Table 4, 5 and 6:

Table 4

$x_1 = s_1$	1	3	4
$F_1(s_1) = f_1(x_1)$	.4	.6	.8*

Table 5

	$f_2(x_2)$			$F_1(s_1)$			$f_2(x_2)\ F_1(s_1)$			$F_2(s_2)$
s_2 \ x_2	2	4	5	2	4	5	2	4	5	Max
3	.5	–	–	.4	–	–	.2	–	–	.2
4	.5	–	–	–	–	–	–	–	–	–
5	.5	.6	–	.6	.4	–	.3	.24	–	.3
6	.5	.6	.7	.8*	–	.4	.4*	–	.28	.4*
7	.5	.6	.7	–	.6	–	–	.36	–	.36

Table 6

	$f_3(x_3)$			$F_2(s_2)$			$f_3(x_3)\ F_2(s_2)$			$F_3(s_3)$
x_3 / s_3	3	4	5	3	4	5	3	4	5	Max
6	.7	–	–	.2	–	–	.14	–	–	.14
7	.7	.8	–	–	.2	–	–	.16	–	.16
8	.7	.8	.9	.3	–	.2	.21	–	.18	.21
9	.7	.8	.9	.4	.3	–	.28	.24	–	.2
10	.7	.8	.9	.36	.4*	.3	.252	.32	.27	.32*

Hence, maximum reliability is .32.

EXERCISE 12.1

1. What is multistage decision problem?
2. Bellman's principle of optimality.
3. Define the following terms in dynamic programming:
   (i) Stage (ii) State
   (iii) Optimal policy (iv) Optimal decision rule
4. Outline the solution procedure of a linear programming problem with two decision variables and m constraints using the dynamic programming problem.
5. A travelling salesman has to travel from city 1 to city 10. The distances between various cities are given in the following diagram:

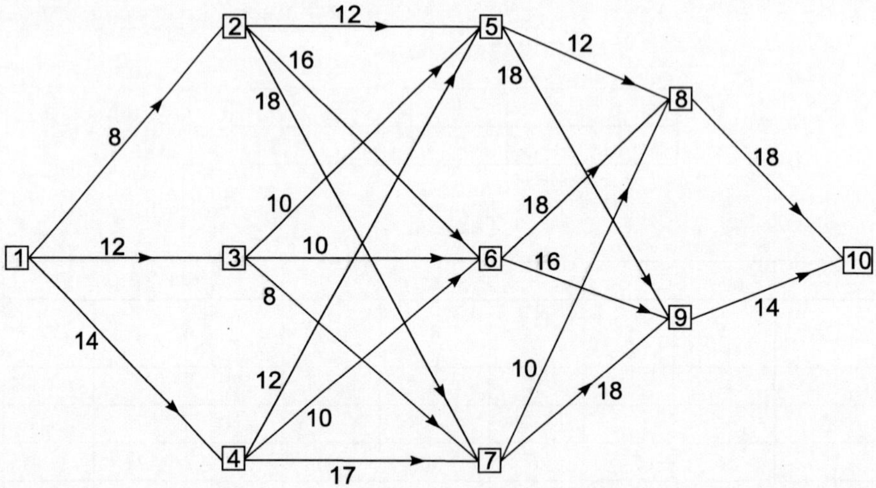

Using dynamic programming, find the minimum distance travelled by the salesman. Also find the optimal path.

6. A salesman has to travel from city 1 to city 12. The cost of the travel and available routes are shown in the diagram given below. Determine the minimum cost and the corresponding route.

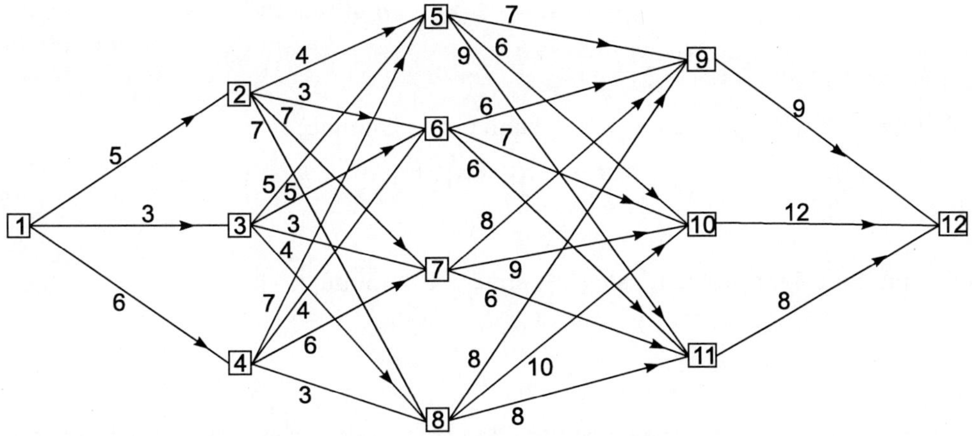

7. Solve the following travelling salesman problem by the dynamic programming technique:

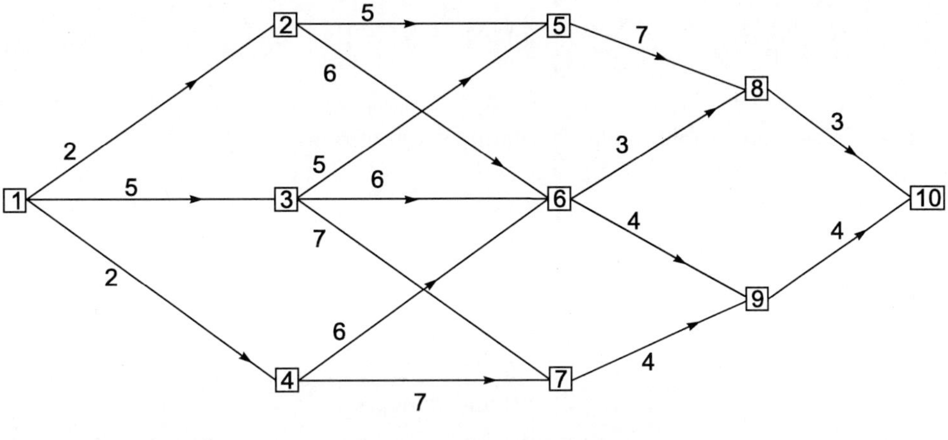

8. Solve the following problem using dynamic programming:

$$\text{Min } Z = x_1^2 + x_2^2 + x_3^2$$

s.t $\quad x_1 + x_2 + x_3 \geq 10$

and $\quad x_1, x_2, x_3 \geq 0.$

9. Solve the following problem using dynamic programming:

$$\text{Min } Z = x_1^2 + x_2^2 + 2x_3^2$$

s.t. $\quad x_1 + x_2 + 2x_3 = 12$

and $\quad x_1, x_2, x_3 \geq 0.$

10. Solve the following problem using dynamic programming:

$$\text{Max } Z = x_1^2 + 2x_2^2 + 4x_3$$

s.t. $x_1 + x_2 + x_3 = 8$

and $x_1, x_2, x_3 \geq 0.$

11. Solve the following problem using dynamic programming:

$$\text{Min } Z = x_1^2 + 2x_2^2 + 4x_3$$

s.t. $x_1 + x_2 + x_3 = 6$

and $x_1, x_2, x_3 \geq 0.$

12. Solve the following problem using dynamic programming:

$$\text{Min } Z = x_1^2 + x_2^2 + x_3^2$$

s.t. $x_1 x_2 x_3 = 27$

and $x_1, x_2, x_3 \geq 0.$

13. Solve the following problem using dynamic programming:

$$\text{Max } Z = x_1 x_2 x_3$$

s.t. $x_1 + x_2 + x_3 = 12$

and $x_1, x_2, x_3 \geq 0.$

14. This problem divides the given quantity into a given number of parts. Let c be the quantity to a be divided in n number of parts $(x_1, x_2, \ldots, x_n)$. Then the problem can be expressed as:

$$\text{Max } Z = x_1 x_2 \ldots x_n$$

s.t. $x_1 + x_2 + \ldots + x_n = c$

and $x_1, x_2, \ldots\ldots, x_n \geq 0.$

15. Solve the following problem using dynamic programming:

$$\text{Max } Z = x_1 x_2 x_3 x_4$$

s.t. $x_1 + x_2 + x_3 + x_4 = 12$

and $x_1, x_2, x_3, x_4 \geq 0.$

16. Solve the following problem using dynamic programming:

$$\text{Min } Z = x_1^2 + x_2^2 + x_3^2$$

s.t. $x_1 x_2 x_3 = 6$

$x_1, x_2, x_3 \geq 0$ and integers.

17. Examination is to begin on next Monday. A boy on the night of Thursday has three days (Friday, Saturday and Sunday) at his disposal to prepare for the examination. He has to appear in three subjects i.e., Mathematics, Physics and Chemistry. He decides to devote a day to one subject only, i.e.,

	Subject		
Study days	Mathematics	Physics	Chemistry
0	2	2	4
1	4	6	6
2	6	6	6
3	10	8	8

dicides not to read two or more subjects in one day. Thus, he devotes zero, one, two, three days to a subject. His assessment of grades ($A = 10$, $B = 8$, $C = 6$, $D = 4$, $E = 2$) on the basis of his sessionals is as shown in previous table. How should he plan his study to maximize total grade point?

18. A student has to take examination in three courses, A, B and C. He has three days available for study. He feels it would be best to devote one whole day to the study of the same course, so that he may study a course for one day, two days, or three days or not at all. His estimates of the improvement in grades he may get by the study are as follows:

Study days \ Course	A	B	C
0	0	1	0
1	1	1	1
2	1	3	3
3	3	4	3

How should he study so that he maximizes the sum of improvements of his grades?

19. (Capital Budgeting Problem) A businessman has four houses in a hill station. He intends to spend Rs. 8 lakhs in converting these houses into hotels. The money x_i required to convert i^{th} house ($i = 1, 2, 3, 4$) into a H_3, H_2 or H_1 type hotel and the corresponding returns are given in the following tables. Maximize the total return. Note that house number 2 cannot be converted into a H_1 class hotel. All data is in lakhs of Rs.

Hotel Type	*House No. 1*		*House No. 1*		*House No. 1*		*House No. 1*	
	x_1	Return	x_2	Return	x_3	Return	x_4	Return
	0	0	0	0	0	0	0	0
H_3	2	3	1	2	1	2	3	6
H_2	3	5	4	6	2	3	4	7
H_1	5	7	–	–	4	6	5	8

20. (Cargo Loading Problem) A 4-ton vessel is to be loaded with one or more of three items. The following table gives the unit weight, w_i, in tons and the unit revenue, r_i, in thousand rupees for items i. How should the vessel be loaded to maximize the total return?

Item i	w_i	r_i
1	2	31
2	3	47
3	1	14

21. (Reliability Problem) An electric equipment has three components 1, 2, 3 and 4 which are connected in series (i.e., if one component fails then the equipment does not work). To improve the reliability of the equipment each component is to be supplemented by one or two or three or four parallel units (in case the component fails then the supplementary unit works and consequently the system remains working). The following data gives the

cost of parallel units and reliability of each component. The total money available for improvement of device is Rs. 15 lakh.

No. of parallel units	Component 1		Component 2		Component 3		Component 4	
	Cost (in thousands of Rs.)	R	Cost (in thousands of Rs.)	R	Cost (in thousands of Rs.)	R	Cost (in thousands of Rs.)	R
1	2	.6	1	.4	1	.3	2	.4
2	4	.7	3	.5	2	.4	3	.5
3	6	.8	5	.7	3	.5	4	.6
4	8	.9	7	.8	4	.6	6	.8

In the above, R denotes the reliability of the equipment. Let x_i denote the money spent on attaching supplementary units to components i. Find x_i, such that the reliability of the equipment is maximized.

ANSWERS

5. 1→3→7→8→10; 48 units.
6. 1→3→7→11→12; Rs. 20
7. 1→2→6→8→10 or 1→4→6→8→10; 14 units.
8. $x_1 = x_2 = x_3 = 10/3$. Min. $Z = 100/3$.
9. $x_1 = x_2 = x_3 = 3$. Min. $Z = 36$.
10. $x_1 = 8, x_2 = x_3 = 0$. Max. $Z = 64$.
11. $x_1 = x_2 = x_3 = 2$. Min. $Z = 20$.
12. $x_1 = x_2 = x_3 = 3$. Min. $Z = 27$.
13. $x_1 = x_2 = x_3 = 4$. Max. $Z = 64$.
14. $x_1 = x_2 = \ldots = x_n = (c/n)$. Max. $Z = (c/n)^n$.
15. $x_1 = x_2 = x_3 = x_4 = 3$. Max. $Z = 81$.
16. $x_1 = 1, x_2 = 3, x_3 = 2$. Min. $Z = 14$.
17. $x_1 = x_2 = x_3 = 1$ or $x_1 = 2, x_2 = 1, x_3 = 0$ or $x_1 = 3, x_2 = 0, x_3 = 0$. Maximum total grade point is = 16.
18. $x_1 = 1, x_2 = 0, x_3 = 2$. Maximum value is 5.
19. First, second, third and fourth houses into H_2, H_3, H_3 and H_3 type hotels respectively. Maximum return is Rs. 15 lakh.
20. $x_1 = 2, x_2 = 0, x_3 = 0$. Maximum value is 62,000.
21. $x_1 = 2, x_2 = 5, x_3 = 4, x_4 = 4$. Maximum reliability is 1512.

13 Chapter

Non-Linear Programming Problem (Unconstrained Optimization Techniques)

13.1 INTRODUCTION

There are a number of non-linear optimization problems which have analytical solution. However, a lot many non-linear programming programmes are also there which do not fall in analytically solvable category and thus defy any general analytical approach. The only way to tackle such problems is to choose a point to start with, find the value of the function to be optimized at that point, take a tentative short step from the point in some direction and find the function value at the new point. If the new value is better (less in minimum problem and large in maximum problem) than the previous value, then discard the first point in favour of the second point, otherwise stay at the first point and move in some other direction. Go on doing this till a point is found from where no further change in its value is possible in any direction and the function value is the optimum within some prescribed tolerable limit.

It is not intended here to go into the theoretical discussion about the derivations and rates of convergence of various methods. We shall present here only the outlines of a few of the procedures.

13.2 UNCONSTRAINED NON-LINEAR OPTIMIZATION PROBLEMS

To find the solution of unconstrained problem, we divide the solution method in two categories:

1. Direct Search Methods
2. Indirect Search Methods

13.3 DIRECT SEARCH METHODS

13.3.1 Univariate Method

In the univariate method, we start from a given fixed point and move in one direction at a time through optimum step size to get the minimum of function $f(X) = f(x_1, x_2, \ldots, x_n)$. If the starting point is not given, we can choose it arbitrarily say origin and complete the movement in all directions getting the improved value of f at each movement and a new base point to start with for the next movement. The process will be over when no further improvement in the value of the given function to be minimized is possible in any one of the direction of a cycle. Movement covering all the directions moving in one direction at a time, completes one cycle. We shall consider functions in two variables x_1 and x_2 only, although the procedural method described below can be extended for functions of more than two variables also.

Procedure

(i) Choose a fixed point say $X_0(x_0, y_0)$ and a search direction $u_1 = (1, 0)^T$. Take a step of size λ_1 in chosen direction and find optimum $\lambda_1 = \lambda_1^*$ so that $f(X_0 + \lambda_1 u_1)$ is minimum. This will be given a new point $X_2 = X_0 + \lambda_1^* u_1$.

(ii) Repeat (i) by taking optimum step length λ_2^* from X_2 in the direction $u_2 = (0, 1)^T$ to arrive at a further new point $X_3 = X_2 + \lambda_2^* u_2$.

(iii) Step (i) and (ii) will complete one cycle of iterations and repeat them for further cycle of iterations until the variation in the value of f is very small (or up to the desired accuracy).

Example 1: Minimize $f(X) = f(x_1, x_2) = x_1^2 + 3x_2^2 - x_1x_2$ using the univariate method taking

$X_1 = \begin{pmatrix} 1 \\ 1 \end{pmatrix}$ as the starting point (base point).

Solution: Cycle 1

Step 1: We consider unit vectors $u_1 = (1, 0)^T$ and $u_2 = (0, 1)^T$ in x_1 and x_2 directions respectively.

Start from $X_1 = \begin{pmatrix} 1 \\ 1 \end{pmatrix}$, where $f(X_1) = 3$. We take a step size λ_1 in x_1 direction to arrive

$$X_2 = X_1 + \lambda_1 \begin{pmatrix} 1 \\ 0 \end{pmatrix} = \begin{pmatrix} 1 \\ 1 \end{pmatrix} + \begin{pmatrix} \lambda_1 \\ 0 \end{pmatrix} = \begin{pmatrix} 1+\lambda_1 \\ 1 \end{pmatrix}$$

To find optimum step size λ_1, we Min $f\begin{pmatrix} 1+\lambda_1 \\ 1 \end{pmatrix} = (1 + \lambda_1)^2 + 3 - (1 + \lambda_1)$

For minimum of f $\dfrac{df}{d\lambda_1} = 2(1 + \lambda_1) - 1 = 0 \Rightarrow \lambda_1 = -\dfrac{1}{2}$

and $\dfrac{d^2f}{d\lambda_1^2} = 2 > 0 \Rightarrow$ f is minimum for $\lambda_1 = -\dfrac{1}{2}$.

Therefore, new point $X_2 = \begin{pmatrix} 1+\lambda_1 \\ 1 \end{pmatrix} = \begin{pmatrix} 1/2 \\ 1 \end{pmatrix}$, $f(X_2) = \dfrac{1}{4} + 3 - \dfrac{1}{2} = \dfrac{11}{4} = 2.75.$

Step 2: Now to find optimum step size λ_2 in x_2 direction.

Minimize $f[X_2 + \lambda_2 (0, 1)^T] = f\left(\begin{pmatrix} 1/2 \\ 1 \end{pmatrix} + \begin{pmatrix} 0 \\ \lambda_2 \end{pmatrix}\right) = f\begin{pmatrix} 1/2 \\ 1+\lambda_2 \end{pmatrix} = \dfrac{1}{4}\ 3(1 + \lambda_2)^2 = \dfrac{1}{2}(1 + \lambda_2)$

For minimum of f $\dfrac{df}{d\lambda_2} = 6(1 + \lambda_2) - \dfrac{1}{2} = 0 \Rightarrow \lambda_2 = \dfrac{11}{12}$

and $\dfrac{d^2 f}{d\lambda_2^2} = 6 > 0 \Rightarrow f$ is minimum for $\lambda_2 = -\dfrac{11}{12}$

New point $X_3 = X_2 + \lambda_2(0, 1)^T = \begin{pmatrix} 1/2 \\ 1+\lambda_2 \end{pmatrix} = \begin{pmatrix} 1/2 \\ 1/12 \end{pmatrix}, f(X_3) = \dfrac{1}{4} + \dfrac{3}{144} - \dfrac{1}{48} = 0.25.$

Thus, one cycle of iterations is completed.

Cycle 2

Step 3: Now considering X_3 as the base point, to find optimum step size λ_3 in x_1-direction.

Minimize $f[X_3 + \lambda_3(1, 0)^T]$

$$= f\left(\begin{pmatrix} 1/2 \\ 1/12 \end{pmatrix} + \begin{pmatrix} \lambda_3 \\ 0 \end{pmatrix}\right) = f\begin{pmatrix} 1/2+\lambda_3 \\ 1/12 \end{pmatrix} = \left(\frac{1}{2} + \lambda_3\right)^2 + 3\left(\frac{1}{12}\right)^2 - \frac{1}{12}\left(\frac{1}{2} + \lambda_3\right)$$

For minimum of f, $\dfrac{df}{d\lambda_3} = 2\left(\dfrac{1}{2} + \lambda_3\right) = \dfrac{1}{12} 0 \Rightarrow \lambda_3 = -\dfrac{11}{24}$

$$\frac{d^2 f}{d\lambda_3^2} = 2 > 0 \Rightarrow f \text{ is minimum for } \lambda_3 \frac{11}{24}.$$

$\therefore$ New point $X_4 = \begin{pmatrix} 1/2 + (-11/24) \\ 1/12 \end{pmatrix} = \begin{pmatrix} 1/24 \\ 1/12 \end{pmatrix}$

$$f(X_4) = \left(\frac{1}{24}\right)^2 + 3\left(\frac{1}{12}\right)^2 - \left(\frac{1}{24}\right)\left(\frac{1}{12}\right) = \frac{11}{576} = 0.01909722.$$

Step 4: Now considering X_4 as the base point, to find optimum step size λ_4 in x_2-direction.
Minimize $f[X_4 + \lambda_4(0, 1)^T]$

$$= f\left(\begin{pmatrix} 1/24 \\ 1/12 \end{pmatrix} + \begin{pmatrix} 0 \\ \lambda_4 \end{pmatrix}\right) = f\begin{pmatrix} 1/24 \\ 1/12+\lambda_4 \end{pmatrix} = \left(\frac{1}{24}\right)^2 3\left(\frac{1}{12} + \lambda_4\right)^2 - \frac{1}{24}\left(\frac{1}{12} + \lambda_4\right)$$

For minimum of f, $\dfrac{df}{d\lambda_4} = 6\left(\dfrac{1}{12} + \lambda_4\right) - \dfrac{1}{24} = 0 \Rightarrow \lambda_4 = -\dfrac{11}{144}.$

$\dfrac{d^2 f}{d\lambda_4^2} = 6 > 0 \Rightarrow f$ is minimum for $\lambda_4 = -\dfrac{11}{144}.$

$$\therefore \text{ New point } X_5 = \begin{pmatrix} 1/24 \\ (1/12)-(11/24) \end{pmatrix} = \begin{pmatrix} 1/24 \\ 1/144 \end{pmatrix},$$

$$f(X_5) = \left(\frac{1}{24}\right)^2 + 3\left(\frac{1}{144}\right)^2 - \left(\frac{1}{24}\right)\left(\frac{1}{144}\right) = \frac{33}{(144)^2} = 0.00159$$

Thus, minimum $f(X) = 0.00159$, when $x_1 = 1/24$ and $x_2 = 1/144$.

Note: From the above steps we see that as the base point moves from (1, 1) towards $\left(\frac{1}{24}, \frac{1}{144}\right)$ $\approx$ (0, 0), $f(X)$ is becoming smaller and smaller and is tending to zero.

13.3.2 Pattern Search Methods

In univariate method starting from a fixed point (the base point) search is made for the minimum of the given function by moving in the axial directions (the directions of axes) only. It can be observed that the rate of convergence in univariate method to arrive at the optimal point is very slow. To accelerate the rate of convergence a number of algorithms have been derived. Here we describe two of them:

(i) Hooke and Jeeve's method
(ii) Powell's method

13.3.2.1 Hooke and Jeeve's Method

In this method, we start from a fixed point say X_1. Step size in all the directions is taken to be constant (i.e., $\Delta x_1 = \Delta x_2 = \ldots = \Delta x_n$ = (constant) and search is made in each direction u_i, u_i is the unit vector in the direction of x_i – axis whose i^{th} component is 1 and all other components are 0. Search is made first in positive and then if necessary in the negative direction of each axis to arrive at a temporary base point. A temporary base point Y_{kj} obtained from X_k by perturbing the jth component of $X_k (Y_{k0} = X_k)$, defined as

$$Y_{ki} = \begin{cases} Y_{k,i-1} + \Delta x_i\, u_i \text{ if } f^+ = f\left(Y_{k,i-1} + \Delta x_i u_i\right) < f = f\left(Y_{k,i-1}\right) \\ Y_{k,i-1} + \Delta x_i\, u_i \text{ if } f^- = f\left(Y_{k,i-1} - \Delta x_i u_i\right) < f = f\left(Y_{k,i-1}\right) \\ Y_{k,i-1} \qquad\qquad \text{if } f\left(Y_{k,i-1}\right) = f \le \min\left(f^+, f^-\right) \end{cases}$$

This process of finding new base point is continued for $i = 1, 2, 3, \ldots, n$ until all directions are exhausted (covered).

After one cycle of covering all the axial directions reaching at the final temporary base point of the cycle say Y_{1n}, n being the number of design variables in the problem. If Y_{1n} is different from $Y_{10} = X_1$, then a new base point is taken as $X_2 = Y_{1n}$ and a single step of optimum size λ^* is taken in the direction $S = X_2 - X_1$. This means the pattern of the preceding set of axial steps is repeated for further distance. After this, search along the axes is again resumed and a second new pattern is generated to go further ahead. This is why the method is called pattern search method. These steps are repeated till a desired accuracy is achieved or the change in the value of the function satisfies the given condition in the problem. We shall confine our discussion by considering functions in two variables ($n = 2$), which can be extended for more than two variables easily.

Example 2: Minimize $f(X) = 2x_1^2 + x_2^2 + 2x_1x_2 + x_1 - x_2$ by Hooke and Jeeve's method, taking starting base point as $X_1 = (0, 0)$ and $\Delta x_1 = \Delta x_2 = 0.8$.

Solution: **Step 1:** Set $Y_{10} = X_1 = (0, 0), f(X_1) = 0$. Moving in direction u_1, from base point X_1,

$$f^+(X_1 + \Delta x_1 u_1) = f^+[(0, 0) + 0.8(1, 0)] = f^+(0.8, 0) = 2.08 \not< f(X_1)$$
$$f^-(X_1 - \Delta x_1 u_1) = f^-(-0.8, 0) = 0.48 \not< f(X_1)$$

As f^+, f^- are not less than $f(X_1)$, so the movement in u_1 direction is not beneficial and hence is discarded and $Y_{11} = X_1$ and $f(Y_{11}) = f(X_1) = 0$.

Step 2: Resuming movement in u_2-direction,

$$f^+(X_1 + \Delta x_2 u_2) = f^+[(0, 0) + 0.8(0, 1)] = f^+(0, 0.8) = -0.16 < f(X_1)$$

As $Y_{12} = X_1 + \Delta x_2 u_2 = (0, 0.8)$ is different from X_1, the new base point is taken as $Y_{12} = X_2 = (0, 0.8)$ and $f(X_2) = -0.16$.

Step 3: As movement has been made in both the axial directions u_1 and u_2, the third movement is to be made in the first pattern direction $S_{p1} = X_2 - X_1 = (0, 0.8) - (0, 0) = (0, 0.8)$ through step length λ from X_2 so that $f(X_2 + \lambda S_{p1})$ is minimum.

For minimum of $f[X_2 + \lambda(X_2 - X_1)] = f\{0, 0.8(1 + \lambda)\} = [0.8(1 + \lambda)]^2 - 0.8(1 + \lambda)$,

$$\frac{df}{d\lambda} \; 2 \times 0.8(1 + \lambda) - 8.0 = 0 \Rightarrow \lambda = -\frac{3}{8} \text{ and}$$

$$\frac{d^2f}{d\lambda^2} = 1.6 > 0 \Rightarrow f \text{ is minimum for } \lambda^* = -\frac{3}{8}.$$

Thus, we get the new base point $Y_{20} = X_3 = X_2 + \lambda^*(X_2 - X_1) = (0, 0.8) - (3/8)(0, 0.8) = (0, 0.5)$ and $f(X_3) = f(0, 0.5) = -0.25$. This completes the first cycle of iterations along both the axial directions u_1, u_2 and first pattern direction S_{p1}.

Second cycle of iterations: Step 4: First we move in u_1 direction starting with base point $X_3 = (0, 0.5)$.

$$f^+(X_3 + \Delta x_1 u_1) = f^+[(0, 0.5) + (0.8, 0)] = f^+(0.8, 0.5) = 2.63 \not< f(X_3)$$
$$f^-(X_3 - \Delta x_1 u_1) = f^-(-0.8, 0.5) = -0.57 < f(X_3) = -0.25$$

Thus, new temporary base point is $Y_{21} = X_3 - \Delta x_1 u_1 = (-0.8, 0.5), f(Y_{21}) = -0.57$

Step 5: Moving in u_2– direction from base point $Y_{21} = (-0.8, 0.5)$ to Y_{22}, we have

$$f^+(Y_{21} + \Delta x_2 u_2) = f^+(-0.8, 1.3) = -1.21 < f(Y_{21})$$

Thus, new base point $Y_{22} = X_4 = Y_{21} + \Delta x_2 u_2 = (-0.8, 1.3)$ and $f(X_4) = -1.21$.

Step 6: Now after exhausting both the axial directions in second cycle of iterations we move along the second pattern search direction $S_{p2} = X_4 - X_3$ starting from X_4 through optimal step length λ so that $f(X_4 + \lambda S_{p2}) = f[(-0.8, 1.3) + \lambda(-0.8, 0.8)] = f[-0.8(1 + \lambda), 1.3 + \lambda(0.8)]$ is minimum.

To find minimum of $f(\lambda)$, $\frac{df}{d\lambda} = 1.28\lambda - 0.32 = 0 \Rightarrow \lambda = 0.25$

and $\frac{d^2 f}{d\lambda^2} = 1.28 \succ 0 \Rightarrow f$ is minimum for $\lambda = 0.25$

Thus, we get the point $X_5 = Y_{30} = X_4 + \lambda S_{p2} = (-1, 1.5)$ and $f(X_5) = -1.25$.

Third cycle of iterations: Step 7: Moving in u_1 direction from $X_5 = (-1, 1.5)$.

$$f^+(X_5 + \Delta x_1 u_1) = f^+[(-1, 1.5) + 0.8(1, 0)] = f^+(-0.2, 1.5) = 0.03 \not< f(X_5)$$
$$f^-(X_5 - \Delta x_1 u_1) = f^-[(-1, 1.5) - 0.8(1, 0)] = f^-(-1.8, 1.5) = 0.03 \not< f(X_5), \text{ So } Y_{31} = X_5.$$

Step 8: Making movement in u_2 direction from $Y_{31} = X_5$,

$$f^+(X_5 + \Delta x_2 u_2) = f^+[(-1, 1.5) + 0.8(0, 1)] = f^+(-1, 2.3) = -0.61 \not< f(X_5)$$
$$f^-(X_5 - \Delta x_2 u_2) = f^-[(-1, 1.5) - 0.8(0, 1)] = f^-(-1, 0.7) = -0.61 \not< f(X_5)$$

This shows that the movement from X_5 in any of the axial directions produces no change in the value of the function f at X_5. Thus, X_5 is the optimal solution. Therefore, Min $f = -1.25$, when $x_1 = -1$ and $x_2 = 1.5$.

13.3.2.2 Powell's Method

Let f be the function of n variables which is to be minimized and $u_1, u_2, \ldots, u_n$ be the axial directions. In Powell's method starting from a fixed point search is made sequentially in the directions u_n; $u_1, u_2, \ldots, u_n$ in the first cycle and along $S_{p1}, u_2, u_3, \ldots u_{n-1}, u_n, S_{p1}$ in the second cycle and $S_{p2}, u_3, u_4, \ldots u_{n-1}, u_n, S_{p1}, S_{p2}$ in the third cycle and so on until the minimum point is reached. Where S_{pi}'s are the pattern search directions defined as $S_{pi} = X_i - X_{i-n}$, n is the number of design (decision) variables. In particular taking $n = 2$, Search is made in the directions u_2, u_1, u_2 and S_{p1}, u_2, S_{p1} and so on. Thus, in the second cycle one of the axial directions is replaced by the pattern search direction. We illustrate the method through examples.

Note: In Powell's method any of u_1 or u_2 can be taken as the first search direction and hence the cycle of iterations can also be in the order u_1; u_2, u_1 and S_{p1}; u_1, S_{p1}. Here u_2 is discarded in second cycle of iteration. Powell's method being the method of conjugate directions, it converges in at most two cycles of iterations.

Example 3: Minimize $f(x_1, x_2) = 2x_1^2 + x_2^2 + 2x_1x_2 + x_1 - x_2$ using Powell's method taking $X_1 = (0, 0)$ as the starting point.

Solution: Given $f = 2x_1^2 + x_2^2 + 2x_1x_2 + x_1 - x_2$ and base point $X_1 = (0, 0)$.

Cycle 1: In the first cycle of iteration, search will be made in the directions u_2, u_1, u_2

Taking step of size λ_1 in the direction S_2 from X_1 to reach at X_2 such that

$$f(X_2) = f(X_1 + \lambda_1 u_2) = \lambda_1^2 - \lambda_1 \text{ is minimum.}$$

For minimum $\quad \dfrac{df}{d\lambda_1}\ 2\lambda_1 - 1 = 0 \ \Rightarrow \lambda_1 = \dfrac{1}{2}$

$$\frac{d^2 f}{d\lambda_1^2} = 2 > 0 \Rightarrow \text{Minimum.}$$

So $X_2 = X_1 + \lambda_1 u_2 = (0, 0) + (1/2)(0, 1) = (0, 1/2), f(X_2) = f(\,0, 1/2\,) = -0.25 < f(X_1)$.

Next moving from X_2 in the direction of $u_1 = (\,1, 0)$ through step length λ_2 arriving at point $X_3 = X_2 + \lambda_2 u_1$. We find λ_2 so that $f(X_3) = f(\,\lambda_2\,,\,1/2\,) = 2\lambda_2{}^2 + 2\lambda_2 - 0.25$ is minimum.

For f minimum, $\quad \dfrac{df}{d\lambda_2} = 4\lambda_2 + 2 = 0 \Rightarrow \lambda_3 = -\dfrac{1}{2}$

$$\frac{d^2 f}{d\lambda_2^2} = 4 >= 0 \Rightarrow \text{Minimum.}$$

So $X_3 = X_2 + \lambda_2 u_1 = (\,0, ½\,) - ½(\,1, 0) = (\,-½\,,\,½\,)$ and $f(X_3\,) = -0.75 < f(X_2) = -0.25$.
Now from X_3, move to X_4 taking step length of size λ_3 in the direction of $u_2 = (\,0, 1)$.

$$X_4 = X_3 + \lambda_3\,(\,0, 1) = (\,-½, ½\,) + (\,0\,,\,\lambda_3\,) = (\,-½\,,\,½ + \lambda_3\,).$$

Now we determine λ_3 such that $f(X_4) = f(\,-½\,,\,½ + \lambda_3\,) = \lambda_3{}^2 - \lambda_3 - 0.75$ is minimum.

For minimum $\quad f, \dfrac{df}{d\lambda_3} = 2\lambda_3 - 1 = 0 \Rightarrow \lambda_3 = \dfrac{1}{2}$

$$\frac{d^2 f}{d\lambda_3^2} = 2 > 0 \Rightarrow \text{Minimum.}$$

Thus, $\quad X_4 = (-½, 1)$ and $f(X_4\,) = -1 < f(X_3) = -0.75$.

This completes one cycle of iterations.

Cycle 2: We generate the pattern search direction S_{p1} as

$$S_{p1} = X_4 - X_2 = (-½, 1) - (0, ½) = (-½, ½)$$

Now from X_4, we move to X_5 by taking step length λ_4 in the direction of $S_{p1} = (-½, ½)$.

$$X_5 = X_4 + \lambda_4 S_{p1} = (-½\,,\,1) + \lambda_4\,(-½\,,\,½) = [-½\,(\,1 + \lambda_4\,),\ 1 + ½\lambda_4]$$

We now determine λ_4 such that $f(X_5) = 0.25\lambda_4{}^2 - 0.5\,\lambda_4 - 1$ is minimum.

For minimum $\quad f, \dfrac{df}{d\lambda_4} = 0.5\lambda_4 - 0.5 = 0 \Rightarrow \lambda_4 = 1$

$$\frac{d^2 f}{d\lambda_4^2} = 0.5 > 0 \Rightarrow f \textit{ is } \text{minimum, where } \lambda_4 = 1.$$

$\therefore \quad X_5 = X_4 + \lambda_4\,{}_{Sp1} = (\,-½, 1\,) + (-1/2)(½, ½) = (-1, 3/2)$ and $f(X_5\,) = -1.25 < f(X_4)$.

Now from X_5 we move to X_6 by taking step length λ_5 in the direction of u_2 (0, 1).

$$X_6 = X_5 + \lambda_5 u_2 = (-1, 3/2) + \lambda_5 (0, 1) = (-1\ , 3/2 + \lambda_5).$$

We determine λ_5 such that $f(X_6) = f(-1, 3/2 + \lambda_5) = \lambda_5^2 - (5/4)$ is minimum.

For minimum $\quad f, \dfrac{df}{d\lambda_5} = 2\lambda_5 = 0 \Rightarrow \lambda_5 = 0.$

$\therefore \quad X_6 = (-1, 3/2) = X_5.$

Which ($\lambda_5 = 0$) shows that f cannot be minimized in the direction of u_2 and there is no other direction to move. Thus, min $f = -1.25$, when $x_1 = -1$, $x_2 = 3/2$.

13.4 INDIRECT SEARCH METHODS

Indirect search methods use the derivatives along with finding the value of function at the search point. These methods, therefore, are also called "Gradient Methods".

13.4.1 Steepest Descent (Cauchy's) Method

This method is also called "Cauchy's method". In this method like other methods, we start from an initial point called the base point and movement is made in the direction of steepest direction.

Procedure Method

Step 1: Let X_1 be the starting point. Let $f(x_1, x_2)$ be the function to be minimized. A function decreases most rapidly in the negative direction of gradient. Calculate $f(X_1)$ and $S_1 = -\nabla f(X_1)$.

Step 2: Find the optimum step length λ_1 in this direction to arrive at the point $X_2 = X_1 + S_1 \lambda_1$ so that $f(X_2) < f(X_1)$ proceed this way till one of the following conditions is satisfied and terminate the process

(i) $\left|\dfrac{\partial f}{\partial x_i}\right| < \epsilon$, ϵ is a small positive number.

(ii) $|X_{i+1} - X_i| < \epsilon$, i.e., change in design vectors in the consecutive iterations is small.

(iii) Relative change in the value of f at two consecutive steps is small i.e.,

$$\left|\frac{f(X_{i+1}) - f(X_i)}{f(X_i)}\right| < \epsilon.$$

Example 4: Minimize $f(x_1, x_2) = 2x_1^2 + x_2^2 + 2x_1x_2 + x_1 - x_2$, starting from the point $X_1 = (0,0)$ and using Steepest descent method, taking $\epsilon = 0.01$.

Solution: $f(x_1, x_2) = 2x_1^2 + x_2^2 + 2x_1x_2 + x_1 - x_2$

$$\therefore \quad \nabla f(X) = \left(\frac{\partial f}{\partial x_1}, \frac{\partial f}{\partial x_2}\right) = (4x_1 + 2x_2 + 1,\ 2x_2 + 2x_1 - 1)$$

Initial step: Given $X_1 = (0, 0)$ and hence $-\nabla f(X_1) = (-1, 1)$.

Iteration I: We find λ_1 which minimizes

$$\begin{aligned} f[X_1 - \lambda_1 \nabla f(X_1)] &= f[(0, 0) + \lambda_1(-1, 1)] = f(-\lambda_1, \lambda_1) \\ &= 2\lambda_1^2 + \lambda_1^2 - 2\lambda_1^2 - \lambda_1 - \lambda_1 \\ &= \lambda_1^2 - 2\lambda_1 \end{aligned}$$

For minimum of f, $\dfrac{\partial f}{\partial \lambda_1} = 2\lambda_1 - 2 = 0 \Rightarrow \lambda_1 = 1$ and $\dfrac{\partial^2 f}{\partial \lambda_1^2} = 2 > 0$ (minimum)

Thus new point $\quad X_2 = X_1 - \lambda_1 \nabla f(X_1) = (-1, 1).$

$$\nabla f(X_2) = (-1, -1).$$

$$|\nabla f(X_2)| \not< \in \Rightarrow X_2 \text{ is not optimal point.}$$

Iteration II: Now we find λ_2 which minimizes

$$f[X_3] = f[X_2 - \lambda_2 \nabla f(X_2)] = f[(-1, 1) - \lambda_2(-1, -1)] = f(-1 + \lambda_2, 1 + \lambda_2)$$

or $\quad f(X_3) = 5\lambda_2^2 - 2\lambda_2 - 1$

Now we find λ_2 which minimizes $f(X_3)$ as a function of λ_2.

For minimum, $\quad \dfrac{\partial f}{\partial \lambda_2} = 10\lambda_2 - 2 = 0 \Rightarrow \lambda_2 = 1/5$ and $\dfrac{\partial^2 f}{\partial \lambda_2^2} = 10 > 0$ (minimum)

$$X_3 = X_2 - \lambda_2 \nabla f(X_2) = (-1 + (1/5), 1 + (1/5)) = (-0.8, 1.2).$$

$$\nabla f(X_3) = (0.2, -0.2)$$

and $\quad |\nabla f(X_3)| \not< 0.01 \Rightarrow X_3$ is not optimal point.

Iteration III: Next we find λ_3 which minimizes

$$\begin{aligned} f[X_4] &= f[X_3 - \lambda_3 \nabla f(X_3)] = f[(-0.8, 1.2) - \lambda_3(0.2, -0.2)] \\ &= f(-0.8 - 0.2\lambda_3, 1.2 + 0.2\lambda_3) \end{aligned}$$

or $\quad f(X_4) = 0.04\lambda_3^2 - 0.08\lambda_3 - 1.2$

Now we find λ_3 which minimizes $f(X_3)$ as a function of λ_3.

For minimum, $\dfrac{\partial f}{\partial \lambda_3} = 0.08\lambda_3 - 0.08 = 0 \Rightarrow \lambda_3 = 1$ and $\dfrac{\partial^2 f}{\partial \lambda_3^2} = 0.08 > 0$ (minimum)

$$X_4 = X_3 - \lambda_3 \nabla f(X_3) = (-1, 1.4) \text{ and } f(X_4) = -1.24$$

$$\nabla f(X_4) = (-0.2, -0.2)$$

and $\quad |\nabla f(X_4)| \not< 0.01 \Rightarrow X_4$ is not an optimal point.

Iteration IV: Further we find λ_4 such that $f(X_5) = f[(X_4 - \lambda_4 \nabla f(X_4)]$ is minimum.

For $f(X_5) = f[X_4 - \lambda_4 \nabla f(X_4)] = 0.2\lambda_4^2 - .08\lambda_4 - 1.24$ minimum,

For minimum, $\quad \dfrac{\partial f}{\partial \lambda_4} = 0.4\lambda_4 - 0.08 = 0 \Rightarrow \lambda_4 = 0.2$ and $\dfrac{\partial^2 f}{\partial \lambda_4^2} = 0.4 > 0$ (minimum)

$\therefore X_5 = X_4 - \lambda_4 \nabla f(X_4) = (-1, 1.4) - 0.2(-0.2, -0.2) = (-0.96, 1.44)$ and $\nabla f(X_4) = (0.04, -0.04)$

and $\quad |\nabla f(X_5)| \not< 0.01 \Rightarrow X_5$ is not an optimal point.

Iteration V: Next to minimize $f(X_6) = f(X_5 - \lambda_5 \nabla f(X_5)) = .0016\,\lambda_5^2 - 0.0032\lambda_5 - 1.248$

$$\frac{\partial f}{\partial \lambda_5} = 0.0032\lambda_5 - 0.0032 = 0 \Rightarrow \lambda_5 = 1 \text{ and } \frac{\partial^2 f}{\partial \lambda_5^2} = 0.0032 > 0 \text{ and hence minima}$$

$$X_6 = X_5 - \lambda_5 \nabla f(X_5) = (-1, 1.48) \text{ and } \nabla f(X_6) = (-0.04, 0.04)$$

$|\nabla f(X_6)| \not< 0.01 \Rightarrow X_6$ is not an optimal point.

Iteration VI: Next to minimize $f(X_7) = f(X_6) - \lambda_6\, \nabla f(X_6) = .0016\lambda_6^2 - 1.2496$

$$\frac{\partial f}{\partial \lambda_6} = 0.0032\lambda_6 = 0 \Rightarrow \lambda_6 = 0 \text{ and } \frac{\partial^2 f}{\partial \lambda_6^2} = 0.0032 > 0 \Rightarrow f(X_7) \text{ is minimum.}$$

With $X_7 = f(X_6) - \lambda_6 \nabla f(X_6) = (-1, 1.48)$

and $\lambda_6 = 0 \Rightarrow$ further improvement in f is not possible alternatively $X_6 = X_7 = (-1, 1.48)$ $\Rightarrow |X_7 - X_6| = (0, 0) < \in = (.01, .01)$.

Thus $\min f = f(X_6) = f(X_7) = -1.2496 \approx -1.25$, when $x_1 = -1$ and $x_2 = 1.48$.

13.5 CONSTRAINED OPTIMIZATION PROBLEMS

To find the solution of constrained problem, we divide the solution method in two categories:

1. Direct Methods
2. Indirect Methods.

13.6 DIRECT METHODS

13.6.1 Complex Method

Complex method deals with the constrained optimization problems of the type

Minimize $Z = f(X)$ ...(1)

Subject to $g_i(X) \le 0, j = 1, 2, 3, \ldots, m$ ...(2)

$X = (x_1, x_2, \ldots, x_n)^T$ and $x_i^{(l)} \le x_i \le x_i^{(u)}, I = 1, 2, \ldots, n$ ...(3)

$x_i^{(l)}$ = lower bound on x_i and $x_i^{(u)}$ = upper bound on x_i. Conditions in (3) are called side constraints. For a minimizations problem in n variables if we consider k points where $k \ge n + 1$ then the figure formed on joining them is known as complex. We shall consider the minimization problem in two variables and take $k = 2n = 4$. These four points will form the vertices of the complex.

Working procedure for the case of two variables x_1 and x_2

Step 1: In complex method one point X_1 is given and remaining $(2n - 1) = 3$ ($n = 2$ here) points, X_2, X_3 and X_4 are obtained one at a time by using random members $r_{i,j}$, $0 < r_{i,j} < 1$. Calculate

$$x_{i,j} = x_i^l + r_{i,j}\,[x_i^{(u)} - x_i^{(l)}],\ I = 1, 2 \text{ and } j = 2, 3, 4$$

$x_{i,j} = {}_i^{\text{th}}$ component of the point X_j. It is worth noticing that point X_2, X_3 and X_4 thus generated satisfy side constraint (3) but may not satisfy all constraints in (2). In case a point say X_4 is not satisfying all the constraints in (2), then a new point denoted as $X_4^{(1)}$ is obtained by moving X_4

half way towards (in the direction of) the centroid $X_0 = (X_1 + X_2 + X_3)/3$ of the remaining points X_1, X_2, X_3 i.e., $X_4^{(1)} = (X_0 + X_j)/2$. We check if new point $X_4^{(1)}$ satisfies all constraints in (2) or not. If not then we further get a new point $X_4^{(2)}$ by moving $X_4^{(1)}$ half way towards the centroid X_0 given above. We proceed this way until a feasible point X_4 satisfying (2) is obtained. Thus we get four feasible points X_1, X_2, X_3 and X_4 all satisfying conditions (2) and (3) and they form the vertices of starting complex.

Step 2: Calculate $f(X_1), f(X_2), f(X_3)$ and $f(X_4)$ and mark the largest and the least value obtained. Let $f(X)$ be largest at $X_n = X_4$ say and smallest at X_l then the process of reflection is used to find a new point X_r as $X_r = (1 + \alpha) X_0 - \alpha X_h$ where $\alpha \geq R$ and X_0 is the centroid of all vertices except X_h that is $X_0 = (X_1 + X_2 + X_3)/3$. Now check the feasibility of X_r.

(i) If X_r is feasible and $f(X_r) < f(X_h)$, then X_h is replaced by X_r and move on step 2.

(ii) If $f(X_r) \geq f(X_h)$, a new trial point X_r is found by taking new α = old $\alpha/2$ and is tested for further satisfaction of the condition $f(X_r) < f(X_h)$. We continue the process until the condition $f(X_r) < f(X_h)$ is satisfied and this way value of α will become smaller and smaller.

(iii) If X_r is not obtained in any manner such that $f(X_r) < f(X_h)$ then we neglect the whole reflection process and new reflection process is started by taking X_h which gives the second largest value of the function. Now for the convergence, the procedure ends when the distance between any two vertices among X_1, X_2, X_3, X_4 becomes smaller than the prescribed value of $\in$.

Example 5: Minimize $f(X) = f(x_1, x_2) = (x_1 - 1)^2 + (x_2 - 1.5)^2 - 0.25$,

Subject to $x_1 + x_2 \leq 4$, $0 \leq x_1 \leq 2$ and $1 \leq x_2 \leq 3$ by complex method with $X_1 = \begin{pmatrix} 0.7 \\ 1.1 \end{pmatrix}$.

Solution: Here constraint is $g(X) = x_1 + x_2 - 4 \leq 0$ and side constraints are $0 \leq x_1 \leq 2$ & $1 \leq x_2 \leq 3$.

Consider 4 vertices X_1, X_2, X_3 and X_4 of the complex with $X_1 = \begin{pmatrix} 0.7 \\ 1.1 \end{pmatrix}$

Let us choose the random numbers

$$r_{1,2} = 0.4,\ r_{1,3} = 0.6,\ r_{1,4} = 0.8$$
$$r_{2,2} = 0.5,\ r_{2,3} = 0.7,\ r_{2,4} = 0.9$$

to find the remaining three vertices X_2, X_3 and X_4 using the formula

$$x_{i,j} = x_i^{(l)} + r_{ij}[x_i^{(u)} - x_i^{(l)}],\ i = 1, 2;\ j = 1, 2, 3, 4.$$

where $x_{i,j}$ is the ith component of vertex X_j.
Here $x_1^{(l)} = 0$, $x_1^{(u)} = 2$ and $x_2^{(l)} = 1$, $x_2^{(u)} = 3$.
Therefore,

$$x_{1,2} = x_1^{(l)} + r_{1,2}[x_1^{(u)} - x_1^{(l)}] = 0.8$$
$$x_{1,3} = x_1^{(l)} + r_{1,3}[x_1^{(u)} - x_1^{(l)}] = 1.2$$
$$x_{1,4} = x_1^{(l)} + r_{1,4}[x_1^{(u)} - x_1^{(l)}] = 1.6$$
$$x_{2,2} = x_2^{(l)} + r_{2,2}[x_2^{(u)} - x_2^{(l)}] = 2$$
$$x_{2,3} = x_2^{(l)} + r_{2,3}[x_2^{(u)} - x_2^{(l)}] = 2.4$$
$$x_{2,4} = x_2^{(l)} + r_{2,4}[x_2^{(u)} - x_2^{(l)}] = 2.8$$

Thus, the first simplex consists of the vertices

$$X_1 = \begin{pmatrix} 0.7 \\ 1.1 \end{pmatrix}, X_2 = \begin{pmatrix} 0.8 \\ 2 \end{pmatrix}, X_3 = \begin{pmatrix} 1.2 \\ 2.4 \end{pmatrix} \text{ and } X_4 = \begin{pmatrix} 1.6 \\ 2.8 \end{pmatrix},$$

$\therefore$ $g(X_1) = x_1 + x_2 = 0.7 + 1.1 = 1.8 \le 4$

Similarly, $g(X_2) = 2.8 \le 4$, $g(X_3) = 3.6 \le 4$ and $g(X_4) = 4.8 > 4$.

$\therefore$ $g(x)$ is not satisfied at vertex X_4 and hence it is replaced by some point in the feasible region. X_0, the centroid of satisfying vertices is

$$X_0 = \frac{X_1 + X_2 + X_3}{3} = \begin{pmatrix} 0.9 \\ 1.83 \end{pmatrix}$$

Thus, new X_4 vertex is $X_4 = X_4^{(1)} = \dfrac{X_0 + X_4}{2} = \begin{pmatrix} 1.25 \\ 2.315 \end{pmatrix}$ and $g(X_4^{(1)}) = 3.565 \le 4$.

Thus, $X_4^{(1)}$ lies in the feasible region and the initial complex has the vertices X_1, X_2, X_3 and $X_4 = X_4^{(1)}$ with

$$X_1 = \begin{pmatrix} 0.7 \\ 1.1 \end{pmatrix}, X_2 = \begin{pmatrix} 0.8 \\ 2 \end{pmatrix}, X_3 = \begin{pmatrix} 1.2 \\ 2.4 \end{pmatrix} \text{ and } X_4 = X_4^{(1)} = \begin{pmatrix} 1.25 \\ 2.315 \end{pmatrix},$$

and $f(X_1) = 0, f(X_2) = 0.04, f(X_3) = 0.60$ and $f(X_4^{(1)}) = 0.4767$.

$f(X_3) = 0.60$ gives the maximum value and $f(X_1) = 0$ is the minimum value
So take $X_3 = X_h$ with $f(X_h) = 0.6$ and $X_1 = X_l, f(X_l) = 0$
The new centroid X_0 is obtained as

$$X_0 = \frac{X_1 + X_2 + X_4^{(1)}}{3} = \begin{pmatrix} 0.917 \\ 1.805 \end{pmatrix} \text{ and } f(X_0) = -0.15.$$

$f(X_0) < f(X_h) \Rightarrow f(X)$ is decreasing from $X_h (= X_3)$ toward X_0. So let us find X_r using reflection process as

$$X_r = (1 + \alpha) X_0 - \alpha X_h$$

Taking $\alpha = 1$, $X_r = 2X_0 - X_h \begin{pmatrix} 0.634 \\ 1.21 \end{pmatrix}$ and $f(X_r) = -0.034944$.

As X_r feasible and $f(X_r) < f(X_h)$. So we proceed further by considering the different values of α as

(i) Taking $\alpha = 0.1$ then we have

$$X_r^{(1)} = 1.1\, X_0 - 0.1\, X_h = (1.1) = \begin{pmatrix} 0.917 \\ 1.805 \end{pmatrix} - 0.1 \begin{pmatrix} 1.2 \\ 2.4 \end{pmatrix} = \begin{pmatrix} 0.8887 \\ 1.7455 \end{pmatrix}$$

and $f(X_r^{(1)}) = -0.1773$.

(ii) Taking $\alpha = 0.2$ then we have

$$X_r^{(2)} = 1.2\, X_0 - 0.2\, X_h = (1.2) \begin{pmatrix} 0.917 \\ 1.805 \end{pmatrix} - 0.2 \begin{pmatrix} 1.2 \\ 2.4 \end{pmatrix} = \begin{pmatrix} 0.8604 \\ 1.686 \end{pmatrix}$$

and $f(X_r^{(2)}) = -0.1959$.

(iii) Taking $\alpha = 0.3$ then we have

$$X_r^{(3)} = 1.3\,X_0 - 0.3\,X_h = (1.3)\begin{pmatrix}0.917\\1.805\end{pmatrix} - 0.3\begin{pmatrix}1.2\\2.4\end{pmatrix} = \begin{pmatrix}0.8321\\1.6265\end{pmatrix}$$

and $f(X_r^{(3)}) = -0.2058$.

(iv) Taking $\alpha = 0.4$ then we have

$$X_r^{(4)} = 1.4\,X_0 - 0.4\,X_h = (1.4)\begin{pmatrix}0.917\\1.805\end{pmatrix} - 0.4\begin{pmatrix}1.2\\2.4\end{pmatrix} = \begin{pmatrix}0.8038\\1.567\end{pmatrix}$$

and $f(X_r^{(4)}) = -0.2070$.

(v) Taking $\alpha = 0.5$ then we have

$$X_r^{(5)} = 1.5\,X_0 - 0.5\,X_h = (1.5)\begin{pmatrix}0.917\\1.805\end{pmatrix} - 0.5\begin{pmatrix}1.2\\2.4\end{pmatrix} = \begin{pmatrix}0.7755\\1.5075\end{pmatrix}$$

and $f(X_r^{(5)}) = -0.1995$.

Since the decreasing of the values of f continues up to $X_r^{(4)}$, so let us replace the vertex $X_h = X_3$ with the highest values by $X_r^{(4)}$ to get the new complex with vertices

$$X_1 = \begin{pmatrix}0.7\\1.1\end{pmatrix} \text{ and } f(X_1) = 0$$

$$X_2 = \begin{pmatrix}0.8\\1.1\end{pmatrix} \text{ and } f(X_2) = 0.04$$

$$X_3 = X_r^{(4)} = \begin{pmatrix}0.8038\\1.567\end{pmatrix} \text{ with } f(X_3) = f(X_r^{(4)}) = -0.207016$$

and
$$X_4 = \begin{pmatrix}1.25\\2.315\end{pmatrix} \text{ with } f(X_4) = f(X_4^{(1)}) = -0.476725$$

Thus, $f(X_4) = f(X_4^{(1)})$ gives the maximum values and $f(X_3) = f(X_r^{(4)})$ gives the minimum value.

So
$$X_h = X_4 \begin{pmatrix}1.25\\2.315\end{pmatrix} \text{ with } f(X_4) = -0.476725$$

and
$$X_l = X_3 = X_r^{(4)} = \begin{pmatrix}0.8038\\1.567\end{pmatrix} \text{ with } f(X_3) = f(X_l) = -0.207016$$

The centroid
$$X_0 = (X_1 + X_2 + X_3)/3 = \begin{pmatrix}0.7679\\1.5557\end{pmatrix} \text{ with } f(X_0) = -0.19303$$

$$|f(X_l) - f(X_0)| = 0.014.$$

Thus, if the desired accuracy is $\in = 0.01$, then the above solution is accepted and therefore Min $f(X) = -0.2070$ at X_3, i.e., when $x_1 = 0.8038$ and $x_2 = 1.5557$.

13.6.2 Zoutendijk Method (Method of Feasible Direction)

Consider the problem

$$\text{Minimize } f(X) \qquad \ldots(1)$$

Subject to $g_j(X) \le 0, j = 1, 2, \ldots, m$ ...(2)

$$X = (x_1, x_2, \ldots, x_m)^T$$

In the method of feasible direction, we choose the starting point X_i ($i = 1$ for starting point) which satisfies all the constraints in (2) and move to a better point X_{i+1} (the point where value of the function in (1) is lesser than that it has at X_i) as per the iterative formula given as

$$X_{i+1} = X_i + \lambda_i S_i$$

where X_i = Starting point for i^{th} iteration.

S_i is the direction in which to move.

λ_i = length of the step to be taken in the direction of S_i.

X_{i+1} is the point obtained at the end of ith iteration

Care should be taken to choose λ_i such that new point X_{i+1} is in feasible region. Further the search direction is found such that:

(i) even a very small movement in that direction violets none of the constraints in (2) and
(ii) the value of the function $f(X)$ in (1) decreases.

Feasible Direction: A direction S which satisfies condition (i) is called the feasible direction.

One way to find the feasible direction is: If at X_i, $\left\{\frac{d}{d\lambda}\left[g_j(X_i + \lambda_i S_i)\right]\right\}_{\lambda_1 = 0} = S_i^T \nabla g_j(X_i) \le 0$

then direction S_i is feasible direction.

Usable Feasible Direction: A direction S_i which satisfies both the conditions given in (i) and (ii) is called usable feasible direction. Further a direction S_i will be a usable feasible direction if

(a) $$\left\{\frac{d}{d\lambda}\left[f(X_i + \lambda_i S_i)\right]\right\}_{\lambda_i = 0} = S_i^T \nabla f(X_i) > 0 \qquad \ldots(3)$$

(b) $$\left\{\frac{d}{d\lambda}\left[g_j(X_i + \lambda_i S_i)\right]\right\}_{\lambda_i = C} = S_i^T \nabla g_j(X_i) \le 0 \qquad \ldots(4)$$

We can reduce the value of f by taking step lengths in such a direction S_i as described above. The process terminates at X_i if $\nabla f(X_i) = 0$. Zoutendijk method is a method of feasible directions.

Example 6: Using the method of feasible directions due to Zoutendijk method

Minimize $f(X) = x_1^2 + x_2^2 - 2x_1 - 3x_2 + 3$

Subject to $g_1(X) = x_1 + x_2 \le 4$ taking the starting point $X_1 = \begin{pmatrix} 0 \\ 0 \end{pmatrix}$

Solution: Given $X_1 = \begin{pmatrix} 0 \\ 0 \end{pmatrix}$, $f(X) = x_1^2 + x_2^2 - 2x_1 - 3x_2 + 3$, $x_1 + x_2 \le 4$

Thus, $g_1(X) = x_1 + x_2 - 4.$

Then $f(X_1) = 3$ and $g(X_1) = -4 < 0.$

Step 1: Here $g(X_1) < 0$ so search direction S_1 is $S_1 = -\nabla f(X_1) = -\begin{Bmatrix} \partial f/\partial x_1 \\ \partial f/\partial x_2 \end{Bmatrix}_{X_1=(0,0)} = \begin{pmatrix} 2 \\ 3 \end{pmatrix}$

After normalization, we have $S_1 = \begin{pmatrix} 2/3 \\ 1 \end{pmatrix}$

Step 2: To obtain a new point X_2 we take a step of length λ_1 in the direction of $-\nabla f(X_1)$ to arrive at $X_2 = X_1 + \lambda_1 S_1 = [(2/3)\lambda_1, \lambda_1]$

$$f(X_2) = f[(2/3)\lambda_1, \lambda_1] = (13/9)\lambda_1^2 - (13/3)\lambda_1 + 3$$

To find λ_1 for minimum f,

$\dfrac{\partial f}{\partial \lambda_1} = \dfrac{26}{9}\lambda_1 - \dfrac{13}{3} = 0 \Rightarrow \lambda_1 = 1.5$ and $\dfrac{\partial^2 f}{\partial \lambda_1^2} = \dfrac{26}{9} > 0$. Thus, f is minimum when $\lambda_1 = 1.5$ giving $X_2 = (0, 0) + 1.5\,(2/3, 1) = (1, 1.5)$ and $g(X_2) = -1.5 < 0.$

So the new search direction S_2 is $S_2 = -\nabla f(X_2) = \begin{Bmatrix} \partial f/\partial x_1 \\ \partial f/\partial x_2 \end{Bmatrix}_{X_1=(1,1.5)} = \begin{pmatrix} 0 \\ 0 \end{pmatrix}$, which implies that there is no search direction available to get minimum f. Hence, minimum $f = -0.25$

At $X = X_2 = (1, 1.5)$, i.e., $x_1 = 1$ and $x_2 = 1.5$.

13.7 INDIRECT METHODS

13.7.1 Transform Techniques

In case constraints $g_i(X)$ in a non-linear optimization problem have certain simple forms in independent variables, then it may be possible to make use of transformation of these variables in such a way that constraints are automatically satisfied. Thus, it is possible to convert a constrained optimization problem into unconstrained one using transformation techniques.

Some of typical transformations of independent variables are:

(i) If lower and upper bounds on x_i are given as:

$l_i \le x_i \le u_i,$ ...(1)

then these can be satisfied by transforming the variable x_i as

$x_i = l_i + (u_i - l_i)\sin^2 y_i$ ...(2)

where y_i is the new variable which can take any value.

(ii) If $x_i \in (0, 1)$, then we can use one of the following transformations

(a) $x_i = \sin^2 y_i$ (b) $x_i = \cos^2 y_i$ (c) $x_i = \dfrac{e^{y_i}}{e^{y_i} + e^{-y_i}}$ (d) $x_i = \dfrac{y_i^2}{1 + y_i^2}$ ...(3)

(iii) If the design variable is restricted to assume only positive values then one of the following transformations can be used:

(a) $x_i = |y_i|$ (b) $x_i = y_i^2$ or (c) $x_i = e^{yi}$ (4)

(iv) If $x_i \in (-1,1)$, then we can use

$$\text{(a) } x_i = \sin y_i \quad \text{(b) } x_i = \cos y_i \quad \text{or} \quad \text{(c) } x_i = \frac{2y_i}{1+y_i^2} \tag{5}$$

Note that, to use above transformations, the constrained function $g_i(X)$ should be simple. In case it is not possible to eliminate all the constraints by change of variables, then it is better not to use the transformation method.

Example 7: A courier service does not accept rectangular packets of more than 42 cm in length. If {length + 2(width + height)} is at most of 72 cm, then compute the maximum volume of the rectangular packet.

Solution: Let x_1, x_2 and x_3 be the length, width and height of the rectangular packet then the formulation of the problem is

$$\text{Max } f(X) = x_1x_2x_3 \quad \ldots(1)$$

$$\text{Subject to} \quad x_1 + 2x_2 + 2x_3 \le 72, \quad \ldots(2)$$

$$x_1 \le 42, \quad \ldots(3)$$

$$x_1, x_2, x_3 \ge 0 \quad \ldots(4)$$

Let us transform x_i's into y_i's by taking $y_1 = x_1$, $y_2 = x_2$ and $y_3 = x_1 + 2x_2 + 2x_3$

These imply that $x_3 = (1/2)(y_3 - y_1 - 2y_2)$ ...(5)

Thus, constraints (2), (3) and (4) can be written as:

$$0 \le y_1 \le 42,$$

$$0 \le y_2 \le 36,$$

and

$$0 \le y_3 \le 72, \quad \ldots(6)$$

The upper bound for y_i's in (6) can easily be obtained (say for y_2, taking $x_1 = x_3 = 0$ in (2), we get $2\,x_2 \le 72$. Thus $x_2 = y_2 \le 36$).

The constrained in (6) are automatically satisfied if we define z_1, z_2 and z_3 as:

$$y_i = l_i + (u_i - l_i)\sin^2 z_i \quad \ldots(7)$$

Using (6) and (7), we have

$$y_1 = l_1 + (u_1 - l_1)\sin^2 z_1 = 42\sin^2 z_1 \quad \ldots(8)$$

$$y_2 = l_2 + (u_2 - l_2)\sin^2 z_2 = 36\sin^2 z_2 \quad \ldots(9)$$

$$y_3 = l_3 + (u_3 - l_3)\sin^2 z_3 = 72\sin^2 z_3 \quad \ldots(10)$$

So the original problem reduces to

$$\text{Max} \quad f = (1/2)\, y_1y_2(y_3 - y_1 - 2y_2)$$

$$= (1/2)\,(42\sin^2 z_1)\,(36\sin^2 z_2)(72\sin^2 z_3 - 42\sin^2 z_1 - 72\sin^2 z_2)$$

$$\text{Subject to} \quad 0 \le \sin^2 z_i \le 1,\ i = 1, 2, 3 \quad \ldots(11)$$

For minimum f,

$$\frac{\partial f}{\partial z_1} = 0 \Rightarrow 1512\sin z_1 \cos z_1 \sin^2 z_2 \left(\sin^2 Z_3 - \frac{7}{6}\sin^2 z_1 - 72\sin^2 z_2\right) = 0 \quad \ldots(12)$$

$$\frac{\partial f}{\partial z_2} = 0 \Rightarrow 108864\sin^2 z_1 \sin z_2 \cos z_2 \left(\sin^2 z_3 - \frac{7}{12}\sin^2 z_1 - 2\sin^2 z_2\right) = 0 \quad \ldots(13)$$

$$\frac{\partial f}{\partial z_3} = 0 \Rightarrow 108864 \sin^2 z_1 \sin^2 z_2 \sin^2 z_3 \cos z_3 = 0 \qquad ...(14)$$

From (14), $\sin^2 z_1 = 0 \Rightarrow z_1 = 0 \Rightarrow y_1 = 0 \Rightarrow f = 0$ which is not acceptable.
$\sin^2 z_2 = 0 \Rightarrow z_2 = 0 \Rightarrow y_2 = 0 \Rightarrow f = 0$ which is not acceptable.
If $\sin z_3 = 0 \Rightarrow z_3 = 0 \Rightarrow y_3 = 0 \Rightarrow f$ is negative which is not acceptable.
If $\cos z_3 = 0 \Rightarrow z_3 = \pi/2$ ...(15)

From (12), $\sin^2 z_3 - \frac{7}{6}\sin^2 z_1 - 72 \sin^2 z_2 = 0 \Rightarrow 7 \sin^2 z_1 + 6 \sin^2 z_2 = 6$ ...(16)

From (13), $\sin^2 z_3 - \frac{7}{12}\sin^2 z_1 - 2 \sin^2 z_2 = 0 \Rightarrow 7 \sin^2 z_1 + 24 \sin^2 z_2 = 12$ ...(17)

(17) – (16) $\Rightarrow 18 \sin^2 z_2 = 6 \Rightarrow \sin^2 z_2 = 1/3$.
From (16), $7 \sin^2 z_1 + 6(1/3) = 6 \Rightarrow \sin^2 z_1 = 4/7$.
Hence, $y_1 = 42(4/7) = 24$, $y_2 = 36(1/3) = 12$, $y_3 = 72$.
$\therefore \quad x_1 = 24, x_2 = 12, x_3 = 12$ and Max $f = 3456$ cm^3.

13.7.2 Penalty Function Methods

Consider Minimize $f(X)$ ...(1)
Subject to $g_j(X) \le 0$, $i = 1, 2, \ldots, m$

$$h_i(X) = 0,\ i = 1, 2, \ldots, p,\ X \in E_n$$

Penalty method transformations problem (1) into a sequence of problems each with no constraint. To solve problem (1), penalty method creates the effect of constrains by bringing in the modification in the objective function just as done in Big-M method of artificial variable technique in the simplex method. Thus, in order to solve constrained optimization problem given in (1), we introduce an auxiliary unconstrained function

$$F(X, r_k) = f(X) + P(X, r_k) \qquad ...(2)$$

where $P(X, r_k)$ is a function of constraints $g_j(X)$ and $h_i(X)$ and r_k is a positive parameter such that

$$\lim_{r_k \to 0} \min F(X, rk) = \min f(X).$$

Formulation of the penalty function

On the basis of the formulation of penalty function, the penalty function method is divided into two categories:

1. The interior penalty method
2. The exterior penalty method

In the interior penalty method, the form of penalty function $P(X, r_k)$, which is a function of constraint functions $g_j(X)$ and $h_i(X)$ and a positive parameter r_k is

$$P(X, r_k) = -r_k \sum_{j=1}^{m} \frac{1}{g_j(X)} + \frac{1}{\sqrt{r_k}} \sum_{i=1}^{p} h_i^2(X) \qquad ...(3)$$

In the interior penalty method the minimum of the auxiliary function $F(X, r_k)$ as defined above in (2) is approaching to the minimum of the objective function $f(X)$ from points inside the feasible region as r_k tends to zero.

In exterior penalty method we take the form of penalty function $P(X, r_k)$ as

$$P(X, r_k) = \frac{1}{r_k}\sum_{j=1}^{m}[\max(0, g_j(X))]^2 + \frac{1}{r_k}\sum_{i=1}^{p} h_i^2(X) \qquad \ldots(4)$$

and minimize the auxiliary function $F(X, r_k)$ for a sequence of decreasing values of r_k, i.e., minimum of $F(X, r_k)$ approaches to minimum of $f(X)$ from the points of infeasible region as r_k tends to zero.

We illustrate the method through the following examples.

Example 8: Minimize $f(X) = f(x_1, x_2) = (x_1 + 2)^3 + 3x_2 + 1$

Subject to $x_1 \geq 2,\ x_2 \geq 0$.

Solution: The given problem can be rewritten as

$$f(X) = (x_1 + 2)^3 + 3x_2 + 1$$

Subject to $2 - x_1 \leq 0,\ -x_2 \leq 0$.

Hence, the auxiliary unconstrained problem becomes

$$\text{Minimize } F(X, r_k) = (x_1 + 2)^3 + 3x_2 + 1 - r_k\left(\frac{1}{2 - x_1} - \frac{1}{x_2}\right)$$

For minimization of F,

$$\frac{\partial F}{\partial x_1} = 3\,(x_1 + 2)^2 - \frac{r_k}{(2 - x_1)}\,2 = 0 \Rightarrow (x_1^2 - 4)^2 = \frac{r_k}{3} \text{ or } x_1 = \left(4 + \sqrt{\frac{r_k}{3}}\right)^{1/2}$$

$$\frac{\partial F}{\partial x_2} = 3 - \sqrt{\frac{r_k}{x_2^2}} \Rightarrow x_2 = \sqrt{\frac{r_k}{3}}$$

Thus, $x_1 = \left(4 + \sqrt{\frac{r_k}{3}}\right)^{1/2}$ and $x_2 = \sqrt{\frac{r_k}{3}}$ are the possible feasible values of x_1 and x_2. Now as $r_k \to 0$, $x_1 = 2$ and $x_2 = 0$.

$$\lim_{r_k \to 0} \min F(X, r_k) \min f(X) = 65, \text{ when } x_1 = 2 \text{ and } x_2 = 0.$$

The solution of the given problem tending to $x_1 = 2$ and $x_2 = 0$ can be seen from the following table as $r_k \to 0$ through four values 10, 0.1, 0.001 and 0.0001 given to r_k for $k = 1, 2, 3$ and 4.

k	r_k	x_1	x_2	$F(X, r_k)$	$f(X)$
1	10	2.414	10.826	117.556	92.478
2	0.1	2.045	0.183	70.508	67.739
3	0.001	2.005	0.018	65.529	65.274
4	0.0001	2.000	0.000	65.000	65.000

The above table shows, how the values of auxiliary function $F(X, r_k)$ are approaching to the values of given function $f(X)$ as r_k is tending to zero through positive values.

Example 9: Minimize $f(X) = 2x$
Subject to $x \geq 3$ using interior penalty method.

Solution: The given problem can be written as

$$\text{Min } f(X) = 2x$$

Subject to $g_1(x) = 3 - x \leq 0.$

The auxiliary unconstrained problem can be written as

Minimize $F(X, r_k) = 2x - r_k\left(\dfrac{1}{3-x}\right)$ such that $\lim\limits_{r_k \to 0} \min F(X, r_k) = \min f(X)$

For minimization of F,

$$\frac{\partial F}{\partial x} = 2 - \frac{r_k}{(3-x)^2} = 0 \Rightarrow x = 3 + \sqrt{\frac{r_k}{2}}$$

$$\frac{\partial^2 F}{\partial x^2} = \frac{-2r_k}{(3-x)^3} \succ 0 \text{ at } x = 3\sqrt{\frac{r_k}{2}}.$$

Now as $r_k \to 0$ at $x = 3$.

$\lim\limits_{r_k \to 0} \min F(X, r_k) \min f(X) = 6$, when $x = 3$.

Note that this is the problem in one variable x.

Example 10: Minimize $f(X) = x_1^2 + 2x_2^2$
Subject to $2x_1 + 5x_2 \leq 10$ using exterior penalty method.

Solution: The given problem can be written as

$$f(X) = x_1^2 + 2x_2^2$$

Subject to $g_1(x) = 2x_1 + 5x_2 - 10 \leq 0$

The auxiliary function $F(X, r_k)$ in the exterior penalty method is given as

Minimize $F(X, r_k) = x_1^2 + 2x_2^2 + \dfrac{1}{r_k}\,[\text{Max. } (0, 2x_1 + 5x_2 - 10)]^2$

For minimization of F, $(r = 1/r_k)$

$$\frac{\partial F}{\partial x} = 2x_1 + \frac{4}{r_k}(2x_1 + 5x_2 - 10) = 0 \Rightarrow (2 + 8r)x_1 + 20rx_2 - 40r = 0 \quad \ldots(1)$$

$$\frac{\partial F}{\partial x_2}\; 4x_2 + 10r(2x_1 + 5x_2 - 10) = 0 \Rightarrow 20rx_1 + (4 + 50r)x_2 - 100r = 0 \quad \ldots(2)$$

From (1) and (2), we have

$$\frac{x_1}{160r} = \frac{x_2}{200r} = \frac{1}{8+132r} \Rightarrow x_1 = \frac{40r}{2+33r} = \frac{40}{2r_k+33}$$

and $$x_2 = \frac{50r}{2 + 33r} = \frac{50}{2r_k + 33}$$

Now as $r_k \to 0,\ x_1 = \dfrac{40}{33}$ and $x_2 = \dfrac{50}{33}$

and minimum $f(X) = (40/33)^2 + 2(50/33)^2 = 200/33.$

EXERCISE 13.1

1. Write a short note on the following:
   (i) Powell's Method
   (ii) Hooke and Jeeve's Method
   (iii) Complex Method
   (iv) Penalty Method
2. What do you mean by feasible directions? Describe in detail the procedure used in the method of feasible direction.
3. Minimize $f(X) = f(x_1, x_2) = x_1^2 + x_2^2 - 2x_1 - 4x_2$ using univariate method taking $X_1 = \begin{pmatrix} 0 \\ 0 \end{pmatrix}$ as the starting point (base point).
4. Minimize $f(X) = f(x_1, x_2) = x_1^2 + 3x_2^2 + x_1 - 17x_2 - x_1x_2$ using univariate method taking $X_1 = \begin{pmatrix} 0 \\ 0 \end{pmatrix}$ as the starting point (base point).
5. Minimize $f(X) = f(x_1, x_2) = 2x_1^2 + x_2^2 + x_1 - x_2 + 2x_1x_2$ using univariate method taking $X_1 = \begin{pmatrix} 0 \\ 0 \end{pmatrix}$ as the starting point (base point).
6. Minimize $f(X) = 3x_1^2 + x_2^2 - 2x_1x_2 - 4x_1 - 3x_2$ by Hooke and Jeeve's method, taking starting base point as $X_1 = (0, 0)$ and $\Delta x_1 = \Delta x_2 = 1$.
7. Minimize $f(x_1, x_2) = 4x_1^2 + 3x_2^2 - 5x_1x_2 - 8x_1$ using Powell's method taking $X_1 = (0, 0)$ as the starting point.
8. Minimize $f(x_1, x_2) = 2x_1^2 + x_2^2$ starting from the point $X_1 = (1, 2)$ and using steepest descent method solve up to two iterations.
9. Minimize $f(X) = f(x_1, x_2) = (x_1 - 1)^2 + (x_2 - 2)^2$,
   Subject to $x_1 + x_2 \le 4$, $x_1 - x_2 \le 2$, $x_1 \ge 0$ and $x_2 \ge 0$ by complex method with $X_1 = \begin{pmatrix} 0.5 \\ 1.5 \end{pmatrix}$
10. Using the method of feasible directions due to Zoutendijk method
    Minimize $f(X) = x_1^2 + x_2^2 - 4x_1 - 4x_2 + 8$

    Subject to $g_1(X) = x_1 + 2x_2 - 4 \le 0$ taking the starting point $X_1 = \begin{pmatrix} 0 \\ 0 \end{pmatrix}$

11. Find the dimensions of a rectangular prism type box that has the largest volume when the sum of its length, width and height is limited to a maximum value of 60 cm and its length is restricted to a maximum value of 36 cm.
12. Minimize $f(X) = x_1^2 + x_2^2 - 6x_1 - 8x_2 + 10$
    Subject to $4x_1^2 + x_2^2 \leq 16$, $3x_1 + 5x_2 \leq 15$
    Using (i) Interior penalty method (ii) Exterior penalty method.

ANSWERS

3. $x_1 = 1$, $x_2 = 2$.
4. $x_1 = 1$, $x_2 = 3$, $f(X) = -25$.
5. $x_1 = -1$, $x_2 = 1.5$, $f(X) = -1.25$.
6. $x_1 = 1.75$, $x_2 = 3.2.5$, $f(X) = -8.375$.
7. $x_1 = 48/23$, $x_2 = 40/23$.
8. $f(X) = 1.0336$.
9. $x_1 = 1$, $x_2 = 2$.
10. $x_1 = 1.6$, $x_2 = 1.2$, $f(X) = 0.8$.
11. $x_1 = 20$, $x_2 = 20$, $x_3 = 20$ and Max $f = 8000$ cm^3.
12. $x_1 = 3$, $x_2 = 4$.

INDEX